Unraveling the Mechanisms of Zn Efficiency in Crop Plants: From Lab to Field Applications

Unraveling the Mechanisms of Zn Efficiency in Crop Plants: From Lab to Field Applications

Editor

Gokhan Hacisalihoglu

MDPI • Basel • Beijing • Wuhan • Barcelona • Belgrade • Manchester • Tokyo • Cluj • Tianjin

Editor
Gokhan Hacisalihoglu
Florida A&M University
USA

Editorial Office
MDPI
St. Alban-Anlage 66
4052 Basel, Switzerland

This is a reprint of articles from the Special Issue published online in the open access journal *Plants* (ISSN 2223-7747) (available at: https://www.mdpi.com/journal/plants/special_issues/Zn_Efficiency).

For citation purposes, cite each article independently as indicated on the article page online and as indicated below:

LastName, A.A.; LastName, B.B.; LastName, C.C. Article Title. *Journal Name* **Year**, *Volume Number*, Page Range.

ISBN 978-3-0365-3427-5 (Hbk)
ISBN 978-3-0365-3428-2 (PDF)

Cover image courtesy of Gokhan Hacisalihoglu

Contents

About the Editor

Gokhan Hacisalihoglu is a Professor in the Biological Sciences Department at Florida A&M University, USA. He is also a Courtesy Professor at Florida State University, USA. He earned his Ph.D. in plant biology from Cornell University, USA. He has several years of experience teaching a variety of university courses from first year General Biology to Plant Anatomy & Development, and Scientist Life Skills. He has won many teaching and research awards, including the FAMU Teaching Innovation Award, FAMU Advanced Teacher of the Year Award, FAMU Teacher of the Year Award, FAMU Research Excellent Award, and Best Professor of the Arts & Sciences. He is a member of the American Society of Plant Biologists. He was a visiting fellow at the US Department of Agriculture, Michigan State University, Florida State University, Cornell University, and the University of Florida. He has received a Fulbright Specialist Award from Japan as an expert consultant in plant biology. Dr. G. Hacisalihoglu has authored 2 other textbooks, From Growing to Biology: Plants 1e and Plant Biology Concepts and Laboratory Manual. Subscribe to his YouTube Channel at: https://www.youtube.com/c/DrhaciExplains. Learn more at Twitter @GHAgriFoodbio.

Editorial

Unraveling the Mechanisms of Zinc Efficiency in Crop Plants: From Lab to Field Applications

Gokhan Hacisalihoglu

Department of Biological Sciences, Florida A&M University, Tallahassee, FL 32307, USA; gokhan.h@famu.edu

Citation: Hacisalihoglu, G. Unraveling the Mechanisms of Zinc Efficiency in Crop Plants: From Lab to Field Applications. *Plants* **2022**, *11*, 177. https://doi.org/10.3390/plants11020177

Received: 4 January 2022
Accepted: 7 January 2022
Published: 11 January 2022

Publisher's Note: MDPI stays neutral with regard to jurisdictional claims in published maps and institutional affiliations.

Global food security and sustainability in the time of pandemics (COVID-19) and a growing world population are important challenges that will require optimized crop productivity under the anticipated effects of climate change. Agricultural sustainability in the time of a growing world population will be one of the major challenges in the next 50 years. Zinc (Zn) is one of the most important essential mineral nutrients required for metabolic processes, so a shortage of Zn constrains crop yield and quality worldwide. Zinc efficiency and higher growth and yield when there is low Zn supply make it a promising sustainable solution for developing cultivars that are zinc efficient.

Future crop plants need to be more Zn efficient with sustainable food yields under sub-optimal Zn conditions. Therefore, there is a substantial value in biological research aimed at understanding how plants uptake and utilize Zn.

A total of 11 articles are included in this Special Issue of *"Plants"* that provide an overview of current developments and trends in the times of high-throughput genomics and phenomics data analysis. Furthermore, this Special Issue presents research findings in various experimental models and areas ranging from maize to *Medicago* (alfalfa), flax, and sorghum.

Hacisalihoglu [1] outlines the variety of advances that took place in plant Zn efficiency research. Furthermore, it addresses why we need to study Zn in plants and the current understanding of Zn transport, uptake, storage, Zn efficiency under sub-optimal Zn regimes, and biofortification breeding efforts especially in food crop plants

Mallikarjuna et al. [2] describe a comparative transcriptome analysis of Fe and Zn deficiency in maize. In their contribution, they reported low-Zn mediated changes in transcriptome and differentially expressed candidate genes that could be further used in maize breeding programs.

Anisimov et al. [3] have used barley plants to describe root Zn uptake and distribution patterns together with assessing soil buffer capabilities.

Cardini et al. [4] have used alfalfa plants to investigate molecular mechanisms of Zn transport. They highlight 12 putative Zn transporter genes as well as differential expressions of MsHMA4, MsNAS1, MsZIP2, and MsHMA4 in roots and shoots.

Desta et al. [5] focused on how plant available Zn is affected by landscape position as well as its association with soil pH and other soil factors in Ethiopia.

Lozano-Gonzales et al. [6] have used cucumber plants to investigate potential effects of silicon application under differential Zn conditions. They discuss further results related to plant stress recovery.

Grujcic et al. [7] have used maize plants to investigate the nitrogen (N) effect on Zn, Fe, and Se biofortification efforts. They discuss further results related to N fertilization on improving micronutrient status in maize.

Reynolds-Marzal et al. [8] have used wheat plants to focus on the efficiencies of Zn and Se in a two-year field experiment in Spain. They discuss further results related to Zn and Se biofortification in wheat.

Petschinger et al. [9] have used moss plants to focus on metal tolerance. They discuss further results related to cell shape and cell wall thickness in mosses.

Hacisalihoglu and Armstrong [10] have used flax and sorghum plants to determine genetic variability and diversity for seed nutritional traits. In their contribution, they discuss further results related to daily value (% DV) and identified the top 12 flax and sorghum varieties for climate change stress and future food security.

Overall, the contributions to this Special Issue topic spans the full spectrum of Zn in plants and soils, cellular mechanisms, gene expressions, and biofortification. This Special Issue is an excellent summary of current progress with future outlooks that illustrates our increased knowledge on Zn and provides the foundation for further future research on the improvement of Zn nutrition in plants.

Finally, we encourage the readers to visit the articles published in this Special Issue of "Unraveling the Mechanisms of Zn Efficiency in Crop Plants: From Lab to Field Applications".

Funding: This research received no external funding.

Institutional Review Board Statement: Not applicable.

Informed Consent Statement: Not applicable.

Data Availability Statement: Not applicable.

Conflicts of Interest: The authors declare no conflict of interest.

References

1. Hacisalihoglu, G. Zinc (Zn): The Last Nutrient in the Alphabet and Shedding Light on Zn Efficiency for Future of Crop Production under Suboptimal Zn. *Plants* **2020**, *9*, 1471. [CrossRef]
2. Mallikarjuna, M.G.; Thirunavukkarasu, N.; Sharma, R.; Shiriga, K.; Hossain, F.; Bhat, J.S.; Mithra, A.C.R.; Marla, S.S.; Manjaiah, K.M.; Rao, A.R.; et al. Comparative Transcriptome Analysis of Iron and Zinc Deficiency in Maize (*Zea mays* L.). *Plants* **2020**, *9*, 1812. [CrossRef] [PubMed]
3. Anisimov, V.S.; Anisimova, L.N.; Sanzharov, A.I. Zinc Plant Uptake as Result of Edaphic Factors Acting. *Plants* **2021**, *10*, 2496. [CrossRef] [PubMed]
4. Cardini, A.; Pellegrino, E.; White, P.J.; Mazzolai, B.; Marscherpa, M.C.; Ercoli, L. Transcriptional Regulation of Genes Involved in Zinc Uptake, Sequestration and Redistribution Following Foliar Zinc Application to *Medicago sativa*. *Plants* **2021**, *10*, 476. [CrossRef] [PubMed]
5. Desta, M.K.; Broadley, M.R.; McGrath, S.P.; Hernandez-Allica, J.; Hassall, K.L.; Gameda, S.; Amede, T.; Haefele, S.M. Plant Available Zinc Is Influenced by Landscape Position in the Amhara Region, Ethiopia. *Plants* **2021**, *10*, 254. [CrossRef] [PubMed]
6. Lozano-Gonzales, J.M.; Valverde, C.; Hernandez, C.D.; Martin-Esquinas, A.; Hernandez-Apaolaza, L.H. Beneficial Effect of Root or Foliar Silicon Applied to Cucumber Plants under Different Zinc Nutritional Statuses. *Plants* **2021**, *10*, 2602. [CrossRef] [PubMed]
7. Grujcic, D.; Yazici, A.M.; Tutus, Y.; Cakmak, I.; Sing, B.R. Biofortification of Silage Maize with Zinc, Iron and Selenium as Affected by Nitrogen Fertilization. *Plants* **2021**, *10*, 391. [CrossRef] [PubMed]
8. Reynolds-Marzal, D.; Rivera-Martin, A.; Santamaria, O.; Poblaciones, M.J. Combined Selenium and Zinc Biofortification of Bread-Making Wheat under Mediterranean Conditions. *Plants* **2021**, *10*, 1209. [CrossRef] [PubMed]
9. Petschinger, K.; Adlassnig, W.; Sabovljevic, M.S.; Lang, I. Lamina Cell Shape and Cell Wall Thickness Are Useful Indicators for Metal Tolerance—An Example in Bryophytes. *Plants* **2021**, *10*, 274. [CrossRef] [PubMed]
10. Hacisalihoglu, G.; Armstrong, P.R. Flax and Sorghum: Multielement Content and Nutritional Value within 220 Varieties and Potential Selection for Future Climates to Sustain Food Security. *Plants* **2022**, *11*, 111.

Review

Zinc (Zn): The Last Nutrient in the Alphabet and Shedding Light on Zn Efficiency for the Future of Crop Production under Suboptimal Zn

Gokhan Hacisalihoglu

Department of Biological Sciences, Florida A&M University, Tallahassee, FL 32307, USA; gokhan.h@famu.edu

Received: 20 September 2020; Accepted: 29 October 2020; Published: 31 October 2020

Abstract: At a global scale, about three billion people have inadequate zinc (Zn) and iron (Fe) nutrition and 500,000 children lose their lives due to this. In recent years, the interest in adopting healthy diets drew increased attention to mineral nutrients, including Zn. Zn is an essential micronutrient for plant growth and development that is involved in several processes, like acting as a cofactor for hundreds of enzymes, chlorophyll biosynthesis, gene expression, signal transduction, and plant defense systems. Many agricultural soils are unable to supply the Zn needs of crop plants, making Zn deficiency a widespread nutritional disorder, particularly in calcareous (pH > 7) soils worldwide. Plant Zn efficiency involves Zn uptake, transport, and utilization; plants with high Zn efficiency display high yield and significant growth under low Zn supply and offer a promising and sustainable solution for the production of many crops, such as rice, beans, wheat, soybeans, and maize. The goal of this review is to report the current knowledge on key Zn efficiency traits including root system uptake, Zn transporters, and shoot Zn utilization. These mechanisms will be valuable for increasing the Zn efficiency of crops and food Zn contents to meet global needs for food production and nutrition in the 21st century. Furthermore, future research will address the target genes underlying Zn efficiency and the optimization of Zn efficiency phenotyping for the development of Zn-efficient crop varieties for more sustainable crop production under suboptimal Zn regimes, as well food security of the future.

Keywords: zinc; sustainability; food security; seed quality; zinc efficiency; staple foods; crops

1. Introduction

It is estimated that global food crop production must double in order to feed the increased world population of 10 billion by the year 2050 [1]. Zinc (Zn) deficiency, together with vitamin A and iron (Fe) deficiencies, are the most common nutritional disorders, especially in developing countries [2]. Research shows that 17.3% of people worldwide are at Zn deficiency risk [3]. Zn is one of the 17 essential mineral nutrients and plays an important role in plant growth, function, gene expression, structures of enzymes, photosynthesis, pollen development, sugar transformation, protein synthesis, membrane permeability, signal transduction, and auxin metabolism [4–6]. Plants take up the Zn from the soil and soil Zn deficiency has become a critically important abiotic stress factor, affecting over 49% of arable lands worldwide (Figure 1a) [5–8]. Zn deficiency negatively affects plant growth, causing stunting short internodes, small leaves, and interveinal leaf chlorosis, as well as delayed maturity and necrotic tissue death in severe cases [4]; therefore, adequate Zn is essential for crop yield and quality. Moreover, the use of synthetic fertilizers is often insufficient to alleviate soil Zn deficiency.

In order to reduce Zn deficiency throughout the susceptible regions, research has been conducted in various countries that are low in Zn, such as Turkey, Australia, Brazil, India, and China [8–11]. Plants with high Zn efficiency exhibit high yield and significant growth under low Zn supply [9]. Identifying, developing, and growing Zn-efficient crop varieties could provide approaches for managing low-Zn stress in soils to minimize yield and quality losses [4]. Moreover, elucidating the mechanisms of Zn efficiency will provide important information for improving crop nutrition, as well as sustainable global food systems [4,8,11].

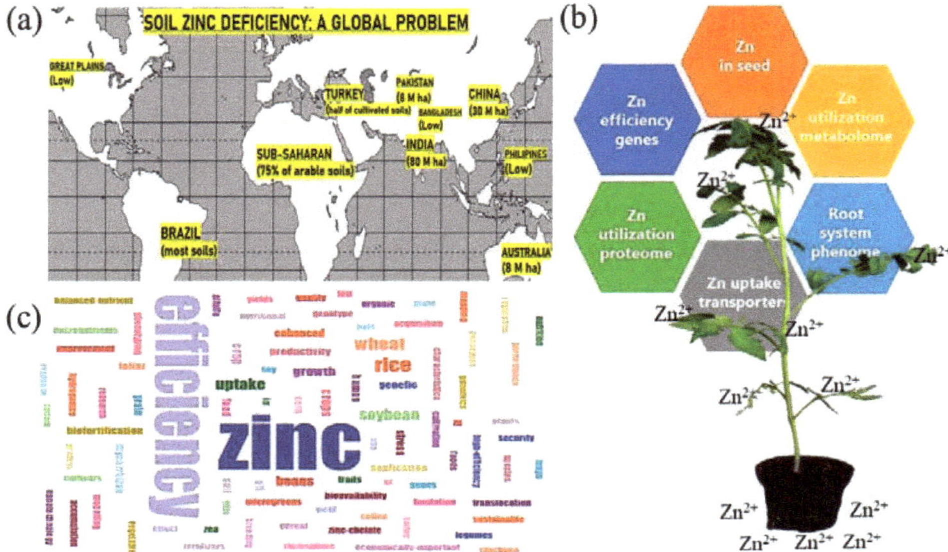

Figure 1. Overview of Zn deficiency and plant Zn efficiency. (**a**) Zn deficiency world map showing major countries and regions with low-Zn soils; (**b**) potential plant Zn efficiency approaches; and (**c**) word cloud of plant Zn efficiency keywords from the literature.

Zn is also essential for human nutrition and development, therefore highlighting the importance of improved Zn contents in staple food crops such as rice, wheat, maize, beans, and others [12]. Understanding the mechanisms of Zn transport and distribution within crops could inform efforts to improve the Zn content of key foods. For example, an effective approach in recent years is biofortification (biological fortification), enriching crops using transgenic techniques, agronomic practices, or conventional crop breeding, which offers sufficient levels of Zn via cereals, vegetables, beans, and fruits to the targeted regions worldwide [13].

This review will focus on advances in the strategies of how crop plants respond to low Zn availability to cope with low-Zn stress conditions, as well as current knowledge of Zn efficiency and future research directions.

2. Soil Zn Deficiency

Zn, a divalent cation, was established as an essential micronutrient for higher green plants in 1926 by Sommer and Lipman [14]. The type of soils affected by Zn deficiency include all soils with low Zn availability, such as high pH calcareous soils, intensively cropped soils, sandy soils, and high P soils [15]. About half of soils are naturally low in Zn worldwide [5]. When it comes to low-Zn soils, there are many countries with soils extensively deficient in Zn [4,5]. For example, Zn is mostly deficient in the majority of soils in Bangladesh, Brazil, Pakistan, the Philippines, and Sudan. Furthermore, Zn is deficient in approximately 75% of the arable soils in sub-Saharan Africa, 50% of the cultivated soils in India, 50% of the cultivated soils in Turkey, 45% of soils in western Australia, and 33% of the soils in China (Figure 1) [4,5]. It has been reported that there is Zn deficiency in the Great Plains and western regions of the United States [16] and sandy soils in Florida [17]. It appears that the use of synthetic fertilizers is not necessarily sufficient for alleviating sub-soil Zn deficiency. Therefore, identifying, developing, and growing Zn-efficient crop varieties are preferred ways to manage low-Zn stress in soils to minimize yield and quality losses [4,9]. Hundreds of genotypes of wheat (*Triticum aestivum*) [8], beans (*Phaseolus vulgaris*) [9], chickpeas (*Cicer arietinum*) [18,19], and rice (*Oryza sativa*) [20,21] were screened for Zn efficiency to accomplish this goal. Plant Zn efficiency screening refers to both visual symptom rating systems as well as biomass and yield under low and sufficient Zn conditions [9,10]. More recent high-throughput phenotyping systems will be beneficial for improving plant Zn efficiency assessment and prediction (Figure 1b,c). The development of cereal or vegetable cultivars with higher Zn efficiencies suitable for low-Zn soils is important for sustainable agricultural production and reduced fertilizer input, as well as population growth. Furthermore, the availability of Zn-efficient cultivars will increase the cultivation of them worldwide.

3. Evidence of Natural Genetic Variation for Plant Zn Efficiency: A Large Untapped Resource for Overcoming Low-Zn Stress

Soil Zn deficiency can cause negative impacts on yield and therefore economic losses [7]. One key approach for crop improvement is identifying beneficial natural alleles and using association studies to reveal the mechanisms underlying natural variation in Zn efficiency. Therefore, exploring natural variation can be beneficial for crop breeding and selection. Indeed, many crop species and varieties show considerable variation in Zn efficiency. While plant species such as alfalfa, carrots, oats, peas, rye, and sunflower are considered Zn efficient, apples, beans, citrus, cotton, flax, grapes, lettuce, onions, pecans, rice, soybeans, spinach, and sweet corn are considered Zn inefficient. Moreover, plant species such as barley, canola, potatoes, sorghum, sugar beet, tomato, and wheat display medium-level Zn efficiency (Figure 2a,b) [3,5].

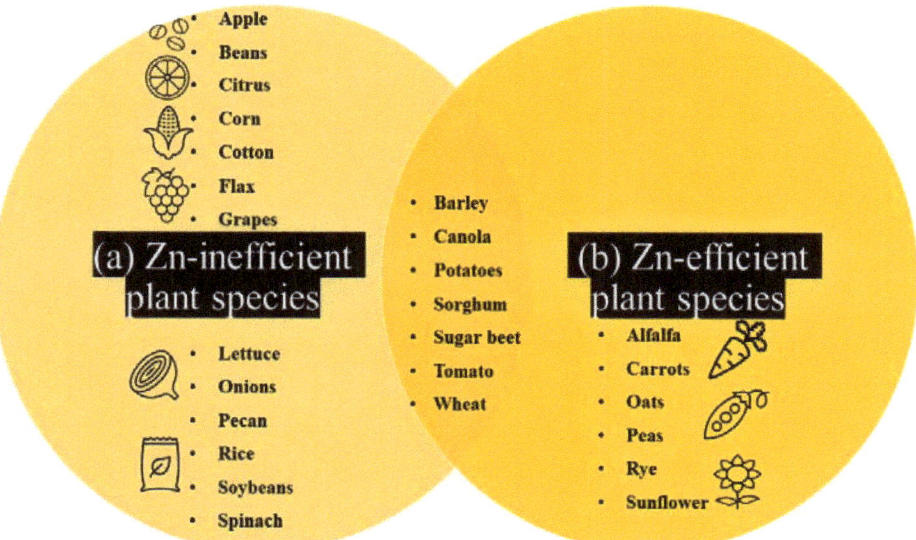

Figure 2. Venn diagram showing the Zn-inefficient plant species (**a**) and Zn-efficient plant species (**b**); overlap area: mildly Zn-efficient plant species.

It is well known that if researchers can identify crop traits that improve Zn efficiency, growers could have improved yields in Zn-poor soils worldwide. Significant genotypic differences for Zn efficiency have been observed in many crop species, such as rice (*Oryza* sativa) [20.21], wheat (*Triticum aestivum*) [9], common beans (*Phaseolus vulgaris*) [10], maize (*Zea mays*) [22], sorghum (*Sorghum bicolor*) [23], soybeans (*Glycine max*), tomatoes (*Solanum lycopersicum*) [24], chickpeas (*Cicer arietinum*) [18,19], barley (*Hordeum vulgare*) [25], and pigeon peas (*Cajanus cajan*) [26]. There is increasing importance for Zn-efficient cultivars that could adapt to and cope with low-Zn soils. Moreover, while the above list is not exhaustive, there are certain staple crop species with a broad screening of a large number of genotypes in low-Zn soil [27,28]. For the past two decades, there has been substantial research into the Zn efficiency of wheat, beans, and rice. Taking wheat (*Triticum aestivum*) as an example, several studies have shown that wheat genotypes differ widely in their Zn efficiency when grown in low-Zn alkali soils of Central Turkey [29], southern Australia [27], China [30], and Brazil [31]. As a result, there are few Zn-efficient genotypes identified based on extensive field studies [29,32].

Common bean (*Phaseolus vulgaris*) is a prevalent, protein-rich legume crop that is extremely sensitive to low-Zn stress in soil (Figure 2a). A large number of screenings of common bean genotypes in Zn-deficient soil experiments have identified the most Zn-efficient genotypes [10]. Blair et al. [33] investigated Zn accumulation in common beans, utilizing low-mineral (DOR364) and high-mineral (G19833) genotypes and identified the linkage group B11 as an important locus for the Zn efficiency trait.

Rice (*Oryza sativa*) is one of the most important staple food crops for humans and feeds over half of the world population. In the U.S. alone, rice is an economically important commodity with a yearly economic value of USD 3 billion [34]. Zn deficiency in rice was first reported in the 1960s in the U.S. [35]. Furthermore, rice is mainly cultivated in soils with low Zn availability (Figures 1a and 2a). Recent studies showed have revealed that there is a wide genetic variation in Zn efficiency in rice, and Zn in seeds is negatively correlated with yield [21]. Recently developed high-throughput phenotyping systems will improve the assessment and prediction of Zn efficiency.

Maize (*Zea mays*) is the third most important cereal crop globally and the first crop with reported Zn deficiency symptoms [36]. It was reported that there is significant genotypic variation among maize cultivars in Brazil [37].

4. Zn Efficiency Strategies in Crop Plants

Zn is a critical nutrient for plants [4] and certain plant species and varieties have developed strategies for securing an adequate supply or maximizing utilization of Zn. Zn-efficient crops and plant varieties are able to achieve sustainable growth and production as well as yield, especially in alkali soils, and could therefore be used to address the Zn deficiency problem. However, it is necessary to better understand the mechanisms of Zn efficiency, as well as natural variation in Zn efficiency traits in food crop plants. Although natural variation in Zn efficiency has been extensively studied in wheat, beans, rice, and chickpeas, the underlying physiological and genetic mechanisms are still not well understood [4–7]. Zn efficiency is a complex trait with two major mechanisms at a number of levels (Figure 1b). Furthermore, Zn efficiency could be explained with other potential mechanisms, as well as the combined effect of more than one mechanism.

4.1. Plant Zn efficiency Mechanism Candidate 1—Zn Uptake Systems and Transporters of Zn

In the uptake process, Zn^{2+} ions travel through the root epidermis, cortex, endodermis, pericycle, and xylem and are then translocated to the stem, leaves, phloem, and seeds [38]. In the past three decades, many attempts have been made to reveal the mechanisms of Zn-efficient plants in response to low-Zn stress in order to determine effective crop breeding strategies [39]. There have been various Zn efficiency mechanisms proposed for food crops in the literature; however, considerable experimental evidence comes from root uptake studies [4,9,15,39,40]. A number of recent uptake studies in crop plants found no strong correlation between root Zn^{2+} influx and Zn efficiency, especially in wheat [6,11,40]. This indicates that Zn efficiency in higher plants is likely not a root-focused trait but a shoot-focused trait. Furthermore, this was supported by the findings that the availability of Zn in soil solution may be an important cause of low-Zn stress compared to total Zn in the soil [15].

Zn uptake can be facilitated by root hairs that increase the availability of Zn from the soil [2]. It is well known that soil type and pH are important determinants of how much Zn is available for crop plants to use [2,9]. Soil pH is important for Zn because it can form insoluble complexes, especially in alkaline (high pH and high $CaCO_3$) soils [41]. Zn deficiency is also common in sandy soils with low total Zn availability [2]. Furthermore, biological factors such as phytosiderophores could affect Zn availability by exudation. As an example, Rengel and Graham [42] found that Zn deficiency caused Fe deficiency may be the major factor that leads to phytosiderophore release by Zn-efficient wheat varieties.

The uptake of Zn into the root follows a biphasic pattern of the high affinity transport system (HATS) and low affinity transport system (LATS) before remobilization [4]. While the LATS mechanism functions when Zn is at high concentrations, the HATS mechanism functions at low external Zn concentrations [25,43]. In wheat, our previous studies demonstrated that Km values were 0.6 to 2nM for HATS and 2 to 5 μM for LATS [4,25]. Milner et al. [43] further suggested a widespread role of the high affinity pathway within plants.

Zn is transported across the root plasma membrane into root cells by transporter proteins such as ZRT-IRT-like protein (ZIP) family, HMA (heavy metal ATPase) family (P-type ATPase), MTP (metal tolerance protein) family cation diffusion facilitators (CDFs), vacuolar iron transporter (VIT) family, and plant cadmium resistance family (PCR) proteins [44,45]. There are transporter genes such as NAS2, NAS4, ZIP4, and IRT3 that act as free Zn^{2+} sensors in the Arabidopsis genome [46]. There are Zn transporters such as MTP and HMA that are affected by Zn deficiency [47]. Zn transporter genes have been shown to have their expression regulated by transcription factors, such as bZIP19 and bZIP3, depending on cytoplasmic free Zn changes [48]. Other transporters involved in Zn uptake include OsHMA2 (in pericycle), OsZIP9, and OsZIP7 (in xylem) [49]. Additionally, it was hypothesized that phytosiderophores, which are organic substances produced by plants, including nicotinamine, mugeniec acid, and avenic acid, may promote Zn uptake, especially in rice in waterlogged soils with high Fe and low Zn levels [50]. Other Zn transporter families included P-type ATPase (metal transporting ATPases), cation diffusion facilitators (CDFs), CAX (cation exchangers) proteins, and natural resistance-associated macrophage protein (NRAMP) [38]. Future research on the characterization of Zn transporter proteins will help to understand how crop plants tolerate low-Zn soils.

4.2. Plant Zn Efficiency Mechanism Candidate 2—Shoot Internal Zn Utilization

Zn is regarded as the only metal that is involved in all enzyme classes, including lyases, transferases, isomerases, oxidoreductases, and hydrolases [4], which subsequently may affect Zn efficiency. Moreover, it was reported that Zn deficiency caused the inhibition of carbonic anhydrase in crop plants [4,15]. Therefore, it is required for the efficient functioning of more than 300 enzymes [4,5]. The use of more Zn-efficient crops will help to maintain crop yields in the future. It has been suggested that Zn efficiency points to the existence of a shoot-coordinated pathway [32,51,52]. One of the complex Zn efficiency mechanisms is the internal biochemical utilization of Zn in the shoot system. Considering the fact that Zn-efficient and Zn-inefficient crop plants have similar leaf Zn concentrations, Zn-efficient varieties have to be using their greater internal utilization efficiency mechanisms. There are several key enzymes that require Zn as part of their essential components [9]. As a result, considerable recent experimental evidence has been presented that plant shoot internal Zn utilization is based on enzymes requiring Zn [4,5,32,53,54]. It has been proposed that greater activities of carbonic anhydrase and Cu/Zn superoxide dismutase enzymes may be responsible for the increased utilization of cytoplasmic Zn in Zn-efficient wheat genotypes compared with inefficient genotypes [53]. Finally, this was further supported by higher expression of the genes for Zn-requiring enzymes, including Cu/Zn superoxide dismutase [4,51]. A study was carried out with wheat that reported that physiological Zn utilization plays an important role in Zn efficiency and grain Zn concentration was correlated with superoxide dismutase and carbonic anhydrase activities [55]. Additional future research will help further our understanding of Zn efficiency by discovering novel genes on shoot internal Zn utilization with regard to Zn enzymes in crop plants.

4.3. Other Mechanisms

Additional Zn efficiency mechanisms may be operating in crop plants (e.g., root system architecture or seed Zn) and future studies are needed to identify and characterize these [56]. Furthermore, it has been reported that soil conditions, together with the environmental conditions of geographic locations, can impact micronutrient contents, such as Zn in seeds [4,8,33]. For example, Zn concentration in plant parts such as seeds is an important parameter for human nutrition. Previous research reported seed Zn content QTLs (quantitative trait loci) in wheat [57], rice [58], maize [59], and beans [60] that can be used in the marker-assisted selection and breeding of Zn-biofortified crop varieties. There are 22 QTLs of concentration of Zn, copper (Cu), and cadmium (Cd) identified in brown rice [61]. There are two major QTLs of Zn efficiency and seed Zn accumulation identified in wheat [62]. Moreover, there are grain Zn and iron (Fe) QTLs on chromosome 1, 4, 7, and 11 in rice [63]. This will increase our understanding of plant Zn efficiency physiology and molecular genetics and contribute greatly to improving crop tolerance to low-Zn soils around the world.

5. Conclusions, Future Challenges, and Perspectives

Zn impacts not only plant growth and function but also human nutrition since plants are a dominant part of diets. Our understanding of the impact of Zn in living organisms continues to advance in Zn-efficient crop varieties that can cope with low-Zn stress in soils. A comprehensive understanding of plant Zn efficiency strategies, cellular mechanisms, and genes can facilitate opportunities for increasing agricultural sustainability, improving human nutrition, and reducing synthetic fertilizer usage. In turn, Zn efficiency could enhance crop production and nutritional quality for the increasing population of the 21st century.

There is a need for more research and some of the suggested research approaches to further explore Zn efficiency may include the following: (1) Identifying the target genes and pathways for Zn efficiency in plants; (2) investigating potential genome editing technologies (CRISPR-Cas9) [64]; (3) developing new methods to advance Zn efficiency phenotyping for food crops in the field; (4) metabolomic profiling of Zn efficiency responses under low-Zn stress in crop plants; and (5) genome-wide association studies (GWASs) to detect the genetic basis of Zn efficiency and seed Zn accumulation under low Zn stress environments.

Funding: No external funding was received for this study.

Acknowledgments: G.H. is grateful to colleagues researching plant zinc efficiency as well as Hacisalihoglu G. lab members at Florida A & M University (USA). Our sincere apologies to the authors whose work could not be mentioned due to limited space. We would like to thank "Plants" and Assistant Editor Leo Zou for their excellent help with the "Unraveling the Mechanisms of Zn Efficiency in Crop Plants: From Lab to Field Applications" Special Issue.

Conflicts of Interest: The author declares no conflict of interest.

References

1.   FAO. FAO Statistical Databases. Available online: http://apps.fao.org/ (accessed on 20 September 2020).
2.   Welch, R.M.; Graham, R.D. Agriculture: The real nexus for enhancing bioavailable micronutrients in food crops. *J. Trace Elem. Med. Biol.* **2005**, *18*, 299–307. [CrossRef]
3.   Wessels, K.R.; Brown, K.H. Estimating the global prevalence of zinc deficiency: Results based on zinc availability in national foods supplies and prevalence of stunting. *PLoS ONE* **2012**, *7*, e50568. [CrossRef] [PubMed]
4.   Marschner, H. *Mineral Nutrition of Higher Plants*; Academic Press: London, UK, 1995; p. 889.
5.   Alloway, B.J. Soil factors associated with zinc deficiency in crops and humans. *Environ. Geochem. Health* **2009**, *31*, 537–548. [CrossRef] [PubMed]
6.   Hacisalihoglu, G.; Kochian, L.V. How do some plants tolerate low levels of soil zinc? Mechanisms of zinc efficiency in crop plants. *New Phytol.* **2004**, *159*, 341–350. [CrossRef]

7. Hacisalihoglu, G.; Blair, M. Current advances in zinc in soils and plants: Implications for zinc efficiency and biofortification studies. *Achiev. Sustain. Crop Nutr.* **2020**, *76*, 337–353.

8. Hacisalihoglu, G.; Hart, J.J.; Kochian, L.V. High and low affinity Zn transport systems and their possible role in Zn efficiency in bread wheat. *Plant Physiol.* **2001**, *125*, 456–463. [CrossRef]

9. Cakmak, I.; Torun, B.; Erenoglu, B.; Oztürk, L.; Marschner, H.; Kalayci, M.; Ekiz, H.; Yilmaz, A. Morphological and physiological differences in the response of cereals to zinc deficiency. *Euphytica* **1998**, *100*, 349–357. [CrossRef]

10. Hacisalihoglu, G.; Ozturk, L.; Cakmak, I.; Welch, R.M.; Kochian, L.V. Genotypic variation in common bean in response to Zn deficiency in calcareous soil. *Plant Soil* **2004**, *259*, 71–83. [CrossRef]

11. Broadley, M.R.; White, P.J.; Hammond, J.P.; Zelko, I.; Lux, A. Zinc in plants. *New Phytol.* **2007**, *173*, 677–702. [CrossRef]

12. Grusak, M.A.; DellaPenna, D. Improving the nutrient composition of plants to enhance human nutrition and health. *Annu. Rev. Plant Biol.* **1999**, *50*, 133–161. [CrossRef]

13. Cakmak, I. Enrichment of cereal grains with zinc: Agronomic or genetic biofortification. *Plant Soil* **2008**, *302*, 1–17. [CrossRef]

14. Sommer, A.L.; Lipman, C.B. Evidence on indispensable nature of zinc and boron for higher green plants. *Plant Physiol.* **1926**, *1*, 231–249. [CrossRef]

15. Kochian, L.V. Zinc absorption from hydroponic solutions by plant roots. In *Zinc in Soils and Plants*; Kluwer Academic Publishers: Berlin, Germany, 1993; pp. 45–57.

16. Neilsen, G.H.; Neilsen, D. *Tree Fruit Zinc Nutrition*; GoodFruit Grower: Yakima, WA, USA, 1994; pp. 85–93.

17. Koo, R.C.J. Fertilization and Irrigation Effects on Fruit Quality. *Factors Affect. Fruit Qual. Citrus Short Course Proc.* **1988**, *97*, 35–42.

18. Khan, H.R.; McDonald, G.K.; Rengel, Z. Chickpea genotypes differ in their sensitivity to Zn deficiency. *Plant Soil* **1998**, *198*, 11–18. [CrossRef]

19. Ullah, A.; Farooq, M.; Rehman, A.; Hussain, M.; Siddique, K.H.M. Zinc nutrition in chickpea (*Cicer arietinum*): A review. *Crop Pasture Sci.* **2020**, *71*, 199–218. [CrossRef]

20. Fageria, N.K. Screening method of lowland rice genotypes for zinc uptake efficiency. *Sci. Agric.* **2001**, *58*, 623–626. [CrossRef]

21. Naik, S.M.; Raman, A.K.; Nagamallika, M.; Venkateshwarlu, C.; Singh, S.P.; Kumar, S.; Singh, S.K.; Tomizuddin, A.; Das, S.P.; Prasad, K.; et al. Genotype × environment interactions for grain iron and zinc content in rice. *J. Sci. Food Agric.* **2020**, *100*, 4150–4164. [CrossRef]

22. Shukla, U.C.; Raj, H. Zinc response in corn as influenced by genetic variability. *Agron. J.* **1976**, *68*, 20–22. [CrossRef]

23. Shukla, U.C.; Arora, S.K.; Singh, Z.; Prasad, K.G.; Safaya, N.M. Differential susceptibility of some sorghum genotypes to zinc deficiency in soil. *Plant Soil* **1973**, *39*, 423–427. [CrossRef]

24. Graham, R.D.; Rengel, Z. Genotypic variation in zinc uptake and utilization by plants. In *Zinc in Soils and Plants*; Robson, A.D., Ed.; Kluwer Academic Publishers: Dordrecht, The Netherlands, 1993; pp. 107–118.

25. Genc, Y.; McDonald, G.K.; Rengel, Z.; Graham, R.D. Genotypic variation in the response of barley to Zn deficiency. In *Plant Nutrition-Molecular Biology and Genetics*; Gissel-Nielsen, G., Ed.; Kluwer Publishers: Dordrecht, The Netherlands, 1999; pp. 205–221.

26. Behera, S.K.; Shukla, A.K.; Tiwari, P.K.; Tripathi, A.; Singh, P.; Trivedi, V.; Patra, A.K.; Das, S. Classification of Pigeonpea (*Cajanus cajan* (L.) Millsp.) Genotypes for Zinc Efficiency. *Plants* **2020**, *9*, 952. [CrossRef]

27. Genc, Y.; Humphries, J.M.; Lyons, G.H.; Graham, R.D. Exploiting genotypic variation in plant nutrient accumulation to alleviate micronutrient deficiency in populations. *J. Trace Elem. Med. Biol.* **2005**, *18*, 319–324. [CrossRef] [PubMed]

28. Gao, X.; Zou, C.; Zhang, F.; vander Zee, E.; Hoffland, E. Tolerance to zinc deficiency in rice correlates with zinc uptake and translocation. *Plant Soil* **2005**, *278*, 253–261. [CrossRef]

29. Cakmak, O.; Ozturk, L.; Karanlik, S.; Ozkan, H.; Kaya, Z.; Cakmak, I. Tolerance of 65 durum wheat genotypes to zinc deficiency in a calcareous soil. *J. Plant Nutr.* **2001**, *24*, 1831–1847. [CrossRef]

30. Lu, X.C.; Cui, J.; Tian, X.H.; Ogunniyi, J.E.; Gale, W.J.; Zhao, A.Q. Effects of zinc fertilization on zinc dynamics in potentially zinc-deficient calcareous soil. *Agron. J.* **2012**, *104*, 963–969. [CrossRef]

31. Siqueira, O.J.F. *Response of Soybeans and Wheat to Limestone Application on Acid Soils in RioGrande do Sul, Brazil*; Digital Repository Iowa State University: Ames, IA, USA, 1977; p. 224.

32. Genc, Y.; McDonald, G.K.; Graham, R.D. Contribution of different mechanisms to zinc efficiency in bread wheat during early vegetative stage. *Plant Soil* **2006**, *281*, 353–367. [CrossRef]

33. Blair, M.; Astudillo, C.; Grusak, M.; Graham, R.; Beebe, S. Inheritance of seed iron and zinc concentrations in common bean (*Phaseolus vulgaris* L.). *Mol. Breed.* **2009**, *23*, 197–207. [CrossRef]

34. QuickStats. USDA-NASS, State Agriculture Overview: Arkansas. 2019. Available online: https://www.nass.usda.gov/Quick_Stats/Ag_Overview/stateOverview.php?state=ARKANSAS (accessed on 20 September 2020).

35. Norman, R.J.; Wilson, C.E.; Slaton, N.A. Soil fertilization and mineral nutrition in US mechanized rice culture. In *Rice: Origin, History, Technology, and Production*; Wiley: Hoboken, NJ, USA, 2003; pp. 331–412.

36. Maze, P. Influences respective des elements de la solution minérale sur le development du mais. *Ann. Inst. Pasteur* **1914**, *28*, 21–68.

37. Furlani, A.M.C.; Furlani, P.R.; Meda, A.R.; Durate, A.P. Efficiency of maize cultivars for zinc uptake and use. *Sci. Agric.* **2005**, *62*, 264–273. [CrossRef]

38. Haslett, B.S.; Reid, R.J.; Rengel, Z. Zinc mobility in wheat: Uptake and distribution of zinc applied to leaves or roots. *Ann. Bot.* **2001**, *87*, 379–386. [CrossRef]

39. Welch, R.M. Micronutrient nutrition of plants. *Crit. Rev. Plant Sci.* **1993**, *14*, 49–87. [CrossRef]

40. Rengel, Z. Physiological mechanisms underlying differential nutrient efficiency of crop genotypes. In *Mineral Nutrition of Crops*; Food Products Press: Binghamton, NY, USA, 1999; pp. 231–261.

41. Barber, S. *Soil Nutrient Bioavailability: A Mechanistic Approach*; John Wiley & Sons: New York, NY, USA, 1984.

42. Rengel, Z.; Graham, R.D. Uptake of zinc from chelate-buffered nutrient solutions by wheat genotypes differing in zinc efficiency. *J. Exp. Bot.* **1996**, *47*, 217–226. [CrossRef]

43. Milner, M.J.; Craft, E.; Yamaji, N.; Ma, J.F.; Kochian, L.V. Characterization of the high affinity Zn transporter from *Noccaea caerulescens*, NcZNT1, and dissection of its promoter for its role in Zn uptake and hyperaccumulation. *New Phytol.* **2012**, *195*, 113–123. [CrossRef] [PubMed]

44. Tiong, J.; McDonald, G.K.; Genc, Y.; Pedas, P.; Hayes, J.E.; Toubia, J.; Langridge, P.; Huang, C.Y. HvZIP7 mediates zinc accumulation in barley (*Hordeum vulgare*) at moderately high zinc supply. *New Phytol.* **2014**, *201*, 131–143. [CrossRef]

45. Guerinot, M.L. The ZIP family of metal transporters. *Biochim. Biophys. Acta* **2000**, *1465*, 190–198. [CrossRef]

46. Assuncao, A.G.L.; Herrero, E.; Lin, Y.F.; Huettel, B.; Talukdar, S.; Smaczniak, C.; Immink, G.H.; van Eldik, M.; Fiers, M.; Schat, H.; et al. *Arabidopsis thaliana* transcription factors bZIP19 and bZIP23 regulate the adaptation to zinc deficiency. *Proc. Natl. Acad. Sci. USA* **2010**, *107*, 10296–10301. [CrossRef]

47. Fujiwara, T.; Kawachi, M.; Sato, Y.; Mori, H.; Kutsuna, N.; Hasezawa, S.; Maeshima, M. A high molecular mass zinc transporter MTP12 forms a functional heteromeric complex with MTP5 in the Golgi in *Arabidopsis thaliana*. *FEBS J.* **2015**, *282*, 1965–1979. [CrossRef]

48. Kramer, U.; Talke, I.N.; Hanikenne, M. Transition metal transport. *FEBS Lett.* **2007**, *581*, 2263–2272. [CrossRef]

49. Huang, S.; Sasaki, A.; Yamaji, N.; Okada, H.; Mitani-Ueno, N.; Ma, J.F. The ZIP Transporter Family Member OsZIP9 Contributes to Root Zinc Uptake in Rice under Zinc-Limited Conditions. *Plant Physiol.* **2020**, *183*, 1224–1234. [CrossRef]

50. Zhang, X.K.; Zhang, F.S.; Mao, D.R. Effect of iron plaque outside roots on nutrient uptake by rice (*Oryza sativa* L.): Zinc uptake by Fe-deficient rice. *Plant Soil* **1998**, *202*, 33–39. [CrossRef]

51. Hacisalihoglu, G.; Hart, J.J.; Vallejos, C.E.; Kochian, L.V. The Role of shoot-localized processes in the mechanism of Zn efficiency in common bean. *Planta* **2004**, *218*, 704–711. [CrossRef]

52. Frei, M.; Wang, Y.; Ismail, A.M.; Wissuwa, M. Biochemical factors conferring shoot tolerance to oxidative stress in rice grown in low zinc soil. *Funct. Plant Biol.* **2010**, *37*, 74–84. [CrossRef]

53. Hacisalihoglu, G.; Hart, J.J.; Cakmak, I.; Wang, Y.; Kochian, L.V. Zinc Efficiency is correlated with enhanced expression and activities of of Cu/Zn-SOD and carbonic anhydrase in wheat. *Plant Physiol.* **2003**, *131*, 595–602. [CrossRef]

54. Cakmak, I.; Ozturk, L.; Eker, S.; Torun, B.; Kalfa, H.I.; Yilmaz, A. Concentration of Zn and activity of copper/zinc superoxide dismutase in leaves of rye and wheat cultivars differing in sensitivity to zinc deficiency. *J. Plant Physiol.* **1997**, *151*, 91–95. [CrossRef]

55. Singh, P.; Shukla, A.K.; Behera, S.K.; Tiwari, P.K. Zinc applicationenhances super oxide dismutase and carbonic anhydrase activities in zinc efficient and inefficient wheat genotypes. *J. Soil. Sci. Plant Nutr.* **2019**, *19*, 477–487. [CrossRef]

56. Blair, M.W.; Izquierdo, P. Use of the advanced backcross-QTL method to transfer seed mineral accumulation nutrition traits from wild to Andean cultivated common beans. *Theor. Appl. Genet.* **2012**, *125*, 1015–1031. [CrossRef] [PubMed]

57. Shi, R.; Li, H.; Tong, Y.; Jing, R.; Zhang, F.; Zou, C. Identification of quantitative trait locus of zinc and phosphorus density in wheat (*Triticum aestivum* L.) grain. *Plant Soil* **2008**, *306*, 95–104. [CrossRef]

58. Stangoulis, J.C.R.; Huynh, B.L.; Welch, R.M.; Choi, E.Y.; Graham, R.D. Quantitative trait loci for phytate in rice grain and their relationship with grain micronutrient content. *Euphytica* **2007**, *154*, 289–294. [CrossRef]

59. Simic, D.; Mladenovic Drinic, S.; Zdunic, Z.; Jambrovic, A.; Ledencan, T.; Brkic, J.; Brkic, A.; Brkic, I. Quantitative trait Loci for biofortification traits in maize grain. *J. Hered.* **2012**, *103*, 47–54. [CrossRef]

60. Gelin, J.R.; Forster, S.; Grafton, K.F.; McClean, P.; Rojas-Cifuentes, G.A. Analysis of seed-zinc and other nutrients in a recombinant inbred population of navy bean (*Phaseolus vulgaris* L.). *Crop Sci.* **2007**, *47*, 1361–1366. [CrossRef]

61. Huang, F.; Wei, X.; He, J.; Sheng, Z.; Shao, G.; Wang, J.; Tang, S.; Xia, S.; Xiao, Y.; Hu, P. Mapping of quantitative trait loci associated with concentrations of five trace metal elements in rice (*Oryza sativa*). *Int. J. Biol.* **2018**, *20*, 554–560.

62. Velu, G.; Tutus, Y.; Gomez-Becerra, H.F.; Hao, Y.; Demir, L.; Kara, R. QTL mapping for grain zinc and iron concentrations and zinc efficiency in a tetraploid and hexaploid wheat mapping populations. *Plant Soil* **2017**, *411*, 81–99. [CrossRef]

63. Jeong, O.Y.; Bombay, M.; Ancheta, M.B.; Lee, J.H. QTL for the iron and zinc contents of the milled grains of a doubled-haploid rice (*Oryza sativa* L.) population grown over two seasons. *J. Crop Sci. Biotechnol.* **2020**. [CrossRef]

64. Doudna, J.A.; Charpentier, E. Genome editing. The new frontier of genome engineering with CRISPR-Cas9. *Science* **2014**, *346*. [CrossRef]

Publisher's Note: MDPI stays neutral with regard to jurisdictional claims in published maps and institutional affiliations.

Article

Comparative Transcriptome Analysis of Iron and Zinc Deficiency in Maize (*Zea mays* L.)

Mallana Gowdra Mallikarjuna [1,*], Nepolean Thirunavukkarasu [1,†], Rinku Sharma [1,‡], Kaliyugam Shiriga [1,§], Firoz Hossain [1], Jayant S Bhat [2], Amitha CR Mithra [3], Soma Sunder Marla [4], Kanchikeri Math Manjaiah [5], AR Rao [6] and Hari Shanker Gupta [1,*]

[1] Maize Research Lab, Division of Genetics, ICAR-Indian Agricultural Research Institute, New Delhi 110012, India; tnepolean@gmail.com (N.T.); rinkusharma882012@gmail.com (R.S.); kaliyugs@gmail.com (K.S.); fh_gpb@yahoo.com (F.H.)

[2] IARI-Regional Research Centre, Dharwad, Karnataka 580005, India; jsbhat73@gmail.com

[3] ICAR-National Institute for Plant Biotechnology, New Delhi 110012, India; amithamithra.nrcpb@gmail.com

[4] ICAR-National Bureau of Plant Genetic Resources, New Delhi 110012, India; ssmarl@yahoo.com

[5] Division of Soil Science and Agricultural Chemistry, ICAR-Indian Agricultural Research Institute, New Delhi 110012, India; manjaiah.math@gmail.com

[6] ICAR-Indian Agricultural Statistics Research Institute, New Delhi 110012, India; ar.rao@icar.gov.in

* Correspondence: mgrpatil@gmail.com (M.G.M.); hsgupta.53@gmail.com (H.S.G.)

† Present address: ICAR-Indian Institute of Millet Research, Hyderabad, Telangana 500030, India.

‡ Present address: Department of Life Sciences, Shiv Nadar University, Gautam Buddha Nagar, Uttar Pradesh 201314, India.

§ Present address: Nu-genes seeds company Pvt. Ltd., Hyderabad, Telangana 500003, India.

Received: 13 November 2020; Accepted: 30 November 2020; Published: 21 December 2020

Abstract: Globally, one-third of the population is affected by iron (Fe) and zinc (Zn) deficiency, which is severe in developing and underdeveloped countries where cereal-based diets predominate. The genetic biofortification approach is the most sustainable and one of the cost-effective ways to address Fe and Zn malnutrition. Maize is a major source of nutrition in sub-Saharan Africa, South Asia and Latin America. Understanding systems' biology and the identification of genes involved in Fe and Zn homeostasis facilitate the development of Fe- and Zn-enriched maize. We conducted a genome-wide transcriptome assay in maize inbred SKV616, under −Zn, −Fe and −Fe−Zn stresses. The results revealed the differential expression of several genes related to the mugineic acid pathway, metal transporters, photosynthesis, phytohormone and carbohydrate metabolism. We report here Fe and Zn deficiency-mediated changes in the transcriptome, root length, stomatal conductance, transpiration rate and reduced rate of photosynthesis. Furthermore, the presence of multiple regulatory elements and/or the co-factor nature of Fe and Zn in enzymes indicate their association with the differential expression and opposite regulation of several key gene(s). The differentially expressed candidate genes in the present investigation would help in breeding for Fe and Zn efficient and kernel Fe- and Zn-rich maize cultivars through gene editing, transgenics and molecular breeding.

Keywords: functional genomics; homeostasis; hormonal regulation; iron; maize; malnutrition; photosynthesis; zinc

---

1. Introduction

Iron (Fe) and zinc (Zn) are essential elements for all living organisms, including plants and animals. Fe and Zn also act as co-factors of numerous enzymes and in turn play vital roles in various physiological process, *viz.*, photosynthesis, respiration, electron transport, protein metabolism, chlorophyll synthesis and hormonal regulations [1,2]. Therefore, any deficiency in Fe and Zn affects the economic yield in crops and thereafter manifests in the form of micronutrient malnutrition in humans [3]. The development of Fe- and Zn-efficient cultivars is one of the effective approaches to sustain crop production and to alleviate widespread micronutrient-malnutrition. Hence, understanding the functional genomics of Fe and Zn homeostasis and identification of target genes and pathways in major staple crops will help in the genetic biofortification of crop plants for Fe and Zn.

Plants adapt a complex network of homeostatic mechanisms to regulate Fe and Zn uptake, transport and accumulation [4]. The uptake of Fe ions in plants occurs through two important strategies *viz.*, the reduction-based strategy (strategy-I) and chelation-based strategy (strategy-II). The strategy-I is present in all the plants except those from the Poaceae family. Under Fe deficiency, the strategy-I plants release the protons into the rhizosphere by H^+-ATPases and makes the Fe more soluble by lowering the soil pH. Subsequently, the NADPH-dependent ferric chelate reductase (FRO2), reduces Fe^{3+} to Fe^{2+} form and which can then be transported into the root epidermis by the divalent metal transporter, iron regulated transporter 1 (IRT1). On the other hand, plants of Poaceae family mostly follow a mugineic acid (MA) pathway-based chelation strategy (Strategy-II) to uptake the Fe from soil [5]. It has been reported that in the mugineic acid pathway, nicotianamine synthase (NAS) [6], nicotianamine aminotranferase (NAAT) [7] and deoxymugineic acid synthase (DMAS) [8] mediate the synthesis of deoxymugineic acid (DMA) from S-adenosyl-methionine via a series of reactions [9]. The grass plants release derivatives of deoxymugineic acid (DMA) called phytosiderophores (PS) which make the complex with the ferric ions (Fe^{3+}). The efflux of PS is facilitated by transporters of the mugineic acid family phytosiderophores (TOM1 and TOM2) whereas the influx of PS–Fe^{3+} is mediated through yellow stripe like-1 (YS1) transporters [10,11]. Among Poaceae members, rice possesses both strategy I and II for the uptake of Fe [2–4]. Interestingly, in maize, the recent finding showed the presence of strategy-I genes for the uptake of Fe [12].

Plants absorb the Zn, mainly in divalent cationic form (Zn^{2+}) [13]. The Zn uptake mechanisms in plants are equipped with a dual-transporter system, *viz.*, high-affinity transport system (HATS) and low-affinity transport system (LATS) [14–16]. The LATS mechanism operates when the Zn is in high concentration and the HATS operates when the Zn concentration is low [16]. Several transporters *viz.*, ZRT-IRT-like protein (ZIP) family [17], HMA (heavy metal ATPase) family [18], MTP (metal tolerance protein) family [19], vacuolar iron transporter (VIT) family, and plant cadmium resistance family (PCR) proteins mediate the Zn transport across the root plasma membrane into root cells [14,20]. Additionally, in plants many researchers also reported the chelation-based Zn uptake mechanism [20,21].

Various transporters and chelation agents *viz.*, oligopeptide transporters (OPT) and yellow stripe like (YSL) [22,23], ZIP [24–27], ferric reductase defective protein (FRD) transporters [28], DMA [29], nicotianamine (NA) [22,30], and citrate [31] have been reported to be involved in the mobilization of Fe and Zn ions. Furthermore, genome-wide transcriptome assays have been employed to understand the expression pattern of genes and pathways associated with Fe and Zn deficiencies in rice [32], *Arabidopsis* [33], maize [34], and several other crops [35]. The deficiency of Fe and Zn is known to activate distinct functional gene modules such as the 'transportome' which, encompasses genes encoding metal transporters, root system modifications, primary metabolic pathways and hormonal metabolism [33]. Under Fe deficiency, Li et al. [34] reported the induced regulation of genes associated with plant hormones, protein kinases and phosphatase in maize roots; whereas, Zanin et al. [36] showed the presence of strategy-I components, which is most prominent in dicot plant. Like Fe, Zn deficiency results in the higher expression of *ZIP* and *NAS* genes in maize [37]. The chlorophyll biosynthesis and rate of photosynthesis are severely affected through the altered expression of genes associated with chloroplast biosynthesis and photosynthesis under Fe and Zn deficiency [9,32,38]. Studies by Garnica et al. [38] in wheat and García et al. [39] in *Arabidopsis* revealed the coordinated action of indole acetic acid (IAA), ethylene and nitrate oxide (NO) in the signalling networks of Fe deficiency. Phytohormones *viz.*, abscisic acid (ABA), and salicylic acid (SA) regulate the expression pattern of *ZmNAS* in maize seedlings [40]. Similarly, ethylene increases the transcripts abundance of genes *viz.*, bHLH (*BASIC HELIX-LOOP-HELIX*), IRO2 (*IRON-RELATED*

TRANSCRIPTION FACTOR 2), NAS1 (NICOTIANAMINE SYNTHASE 1), NAS2 (NICOTIANAMINE SYNTHASE 2), YSL15 (YELLOW STRIPE LIKE 15) and *IRT1 (IRON REGULATED TRANSPORTER 1)* associated with Fe^{2+} and Fe^{3+}-phytosiderophore uptake systems in rice [41]. In *Arabidopsis*, NO enhances the expression of genes for Fe-acquisition and ethylene synthesis in roots [42,43] and the interaction between the auxin and NO modulates the root growth under Fe deficiency in rice [44]. Apart from root growth, Fe deficiency induced auxin signalling, which results in photosynthesis inhibition and defective shoot growth in rice seedlings [45].

The comparative analysis of nutrients and their interactions are crucial for the simultaneous improvement of multi-nutrient use efficiency and enrichment in crops. Few of the interaction studies at whole-genome transcriptome strata are available in crops for Fe and P [32] and Zn and P [46]. However, there are hardly any reports on the genome-wide expression studies under Fe and Zn deficiency interactions, in both root and shoot tissues of crops in general and maize in specific. Therefore, the present investigation was designed to study the transcriptome response to Fe, and Zn deficiencies individually and together to understand Fe and Zn homeostasis and metabolism in maize.

2. Results

2.1. Morpho-Physiological Evidence for Fe and Zn Interaction

Maize seedlings when tested under Fe and Zn deficiencies individually and together (−Zn, −Fe and −Fe−Zn) along with control (+Fe+Zn). The stress treatments started showing Fe and Zn deficiency symptoms from the fourth and fifth days after transplanting (DAT), respectively. However, the typical stress symptoms were prominent among the stress treatments at 10 DAT (Figure 1). Seedlings showed chlorosis under Fe deficiency (−Fe), interveinal chlorosis in the lower half of the top leaves under Zn deficiency (−Zn), and severe chlorosis coupled with slight whitish blotches near the base of top leaves under combined Fe and Zn deficiencies (−Fe−Zn). Furthermore, reduced root length and induced root hair formation were observed under stress treatments (Figure 1).

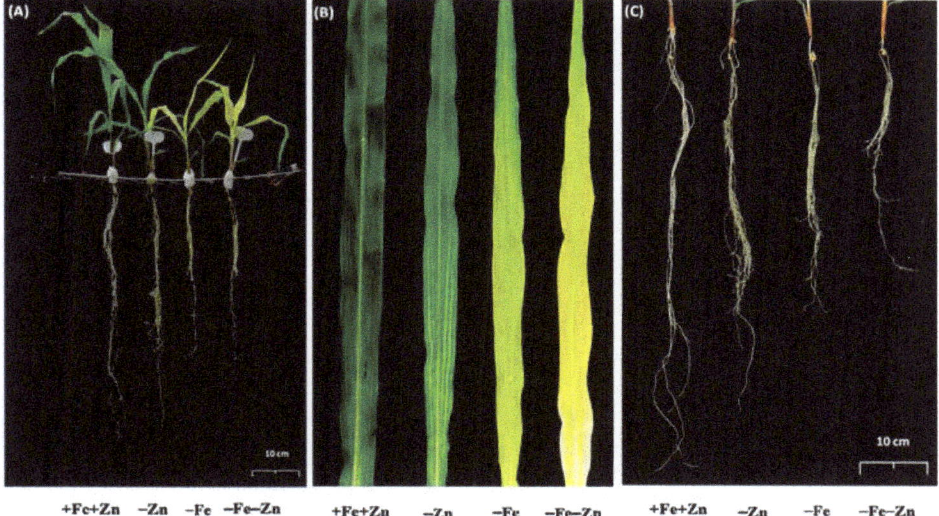

Figure 1. Phenotypic expression of 19 days-old maize SKV616 seedlings in response to 12 days exposure to Zn and Fe deficiencies (−Zn, −Fe and −Fe−Zn): (**A**) whole plant; (**B**) leaves; and (**C**) roots.

Plants in Fe (−Fe), and Fe and Zn (−Fe−Zn) deficiencies started showing chlorosis from the fourth DAT; in contrast, the chlorophyll content of leaves increased significantly in the seedlings of the control treatment (+Fe+Zn) (Figure S1). However, no significant variation in the chlorophyll content was observed between –Fe and –Fe–Zn deficiencies (Figure 2A). Zn starved the seedlings in comparison to the control which did not show a significant reduction in stomatal conductance and transpiration rate. In contrast, as compared to the control and Zn deficiency (–Zn), the Fe (–Fe) and Fe and Zn (–Fe–Zn) deficiencies recorded a significantly reduced transpiration rate and stomatal conductance (Figure 2B,C).

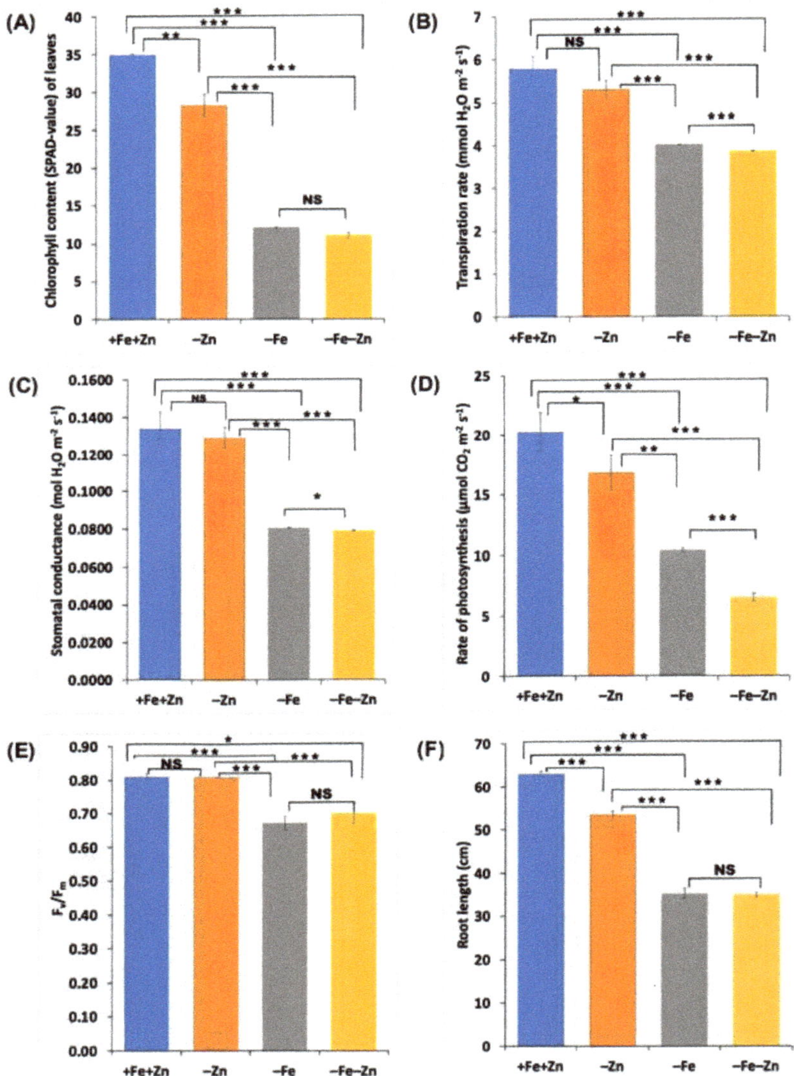

Figure 2. The morpho-physiological response of maize seedlings to –Zn, –Fe and –Fe–Zn deficiencies at 12 days after transplanting (DAT): (**A**) change in the chlorophyll content (soil plant analysis development (SPAD) value) of the leaves in response to the (**B**) transpiration rate; (**C**) stomatal conductance; (**D**) photosynthesis rate; (**E**) quantum efficiency of PS II photochemistry (Fv/Fm); and (**F**) root length (*, **, *** significant at $p < 0.05$, $p < 0.01$ and $p < 0.001$, respectively; NS: non-significant).

Additionally, the reduced rate of photosynthesis was observed in all the stress treatments as compared to the control (Figure 2D). The –Fe and –Fe–Zn stresses showed a significant reduction in the maximum photochemical efficiency of PSII (F_V/F_M), whereas no significant decrease was observed under –Zn stress (Figure 2E). Furthermore, –Fe and –Fe–Zn stressed seedlings showed a similar level of photochemical efficiency of PSII (Figure 2E).

Roots are the absorption points for nutrients and showed variation in the root length under nutrient stresses. The reduction in root lengths was significant for all the stress treatments. However, –Fe and –Fe–Zn treatments did not show any difference in root length. Therefore, Fe deficiency (–Fe and –Fe–Zn) is the major contributor for root length reduction as compared to Zn deficiency (–Zn) (Figure 2F). All the morpho-physiological traits showed a high degree of correlation ($p < 0.01$ and $p < 0.001$) owing to interdependency on each other at physiological level. The stomatal conductance and transpiration rate, and chlorophyll content and root length showed the maximum correlation coefficients (r = 0.99, $p < 0.001$) (Figure S2).

2.2. Maize Transcriptome Profiles in Response to Fe and Zn Deficiencies

Root and shoot tissues collected from hydroponically grown maize seedlings under treatments and control were analysed for a genome-wide transcriptome assay using the Affymetrix GeneChip maize genome arrays (Affymetrix Inc., Santa Clara, California, USA). The transcriptome snapshots of stress treatments were compared against control seedlings. Genome-wide transcriptome analysis under –Zn, –Fe and –Fe–Zn stresses identified differential expression of 1349, 1920 and 6467 genes, respectively, in root and 1466, 4224 and 6360 genes in the shoot, respectively (Figure 3). In both the root and shoot, –Fe–Zn resulted in a higher number of differentially expressed genes (DEGs) as compared to individual stress (–Zn or –Fe). The roots of –Fe–Zn treatment showed a relatively higher number of DEGs (6467) as compared to the shoot (6360) (Figure 3). Conversely, the overall transcriptome snapshot showed a greater number of DEGs in the shoot (8200) as compared to the root (7316).

2.3. Gene Ontology Assignments and KEGG Enrichment Analyses

Gene ontology (GO) assignments *viz.*, the cellular component, molecular function, and the biological process were used to classify the annotated DEGs based on the biological process, cellular localisation and molecular functions. The top 15 sub-components in each of the categories were depicted as bar graphs (Figures 4 and 5) and all the subcomponents of the categories with a *p*-value for false discovery rate (FDR) enrichment is mentioned in Table S1. Under the biological process category, a co-factor metabolic process and response to abiotic stimulus were common across the stress treatments in both root and shoot tissues (Table S1).

In the cellular component category, Zn and Fe stresses in the root shared the sub-components *viz.*, the extracellular region, cell wall, external encapsulating structure, apoplast and protein storage vacuole membrane. Similarly, in the shoot, all the stress treatments showed the cellular components subcategories *viz.*, plastid, chloroplast, cytosol, plastid part, organelle envelope, envelope, photosynthetic membrane and chloroplast thylakoid (Table S1).

Furthermore, the molecular function terms *viz.*, cation binding, metal ion binding, oxidoreductase activity and the co-factor binding across the stress treatments and tissues (Figures 4 and 5, Table S1). The numbers of GO terms under cation binding, metal ion binding, oxidoreductase activity and co-factor binding were high in –Fe–Zn stress as compared to the individual –Fe and –Zn stresses in both the root and shoot tissues (Table S1). Which confirms that the co-occurrence of Fe and Zn stresses (–Fe–Zn) causes more physiological disturbance than individual stresses. Additionally, a greater number of DEGs in the shoot as compared to the root tissue suggests the presence of diverse physiological and metabolic activities in shoot. Moreover, the significant number of DEGs associated with these mineral homeostasis GO terms indicates the metabolic and physiological demands for Fe and Zn ions under stressed conditions.

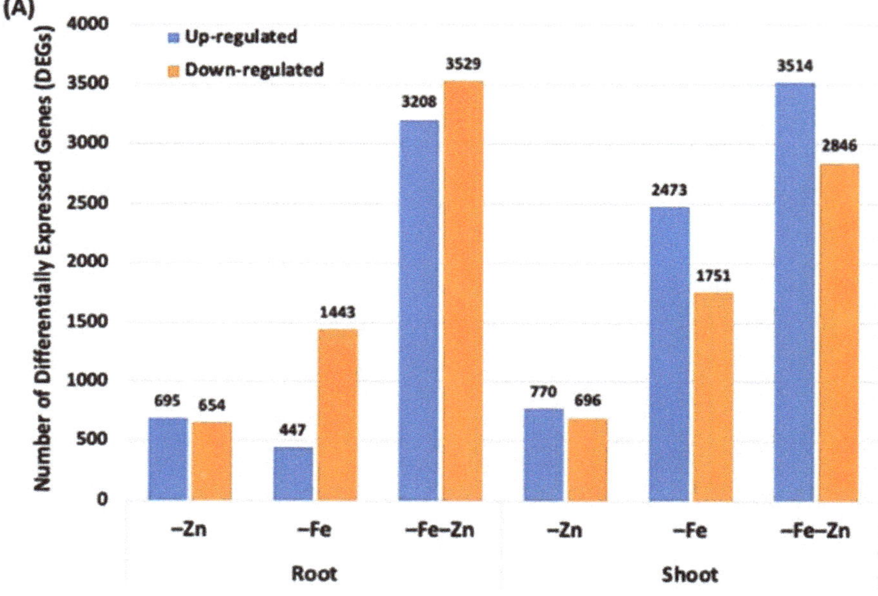

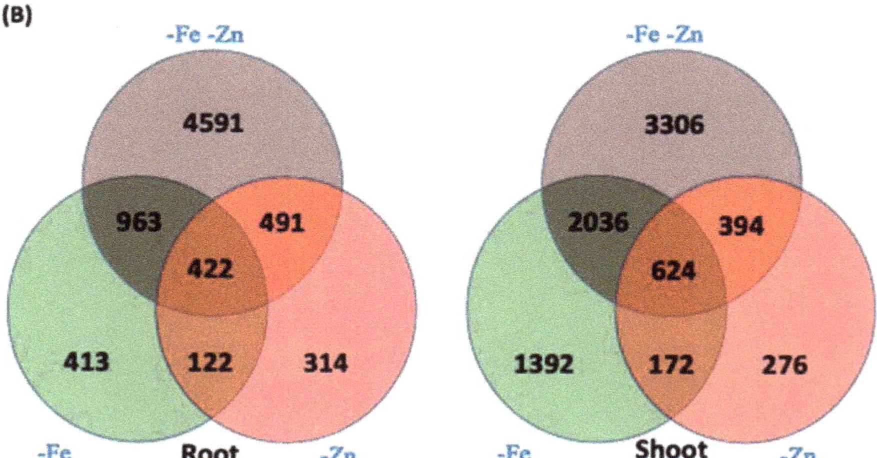

Figure 3. The overview of the spatial transcriptome responses to Fe and Zn stresses in the maize inbred line of SKV616 seedlings. The genes showing > 2-fold expression under stress treatments and $p < 0.05$ are considered as differentially expressed genes (DEGs) in stress treatments as compared to the control: (**A**) total number of upregulated and downregulated genes in response to −Zn, −Fe and −Fe−Zn stresses in the root and shoot; and (**B**) the Venn diagram depicting the stress-specific and common DEGs in response to −Zn, −Fe and −Zn−Fe stresses in the root and shoot tissues.

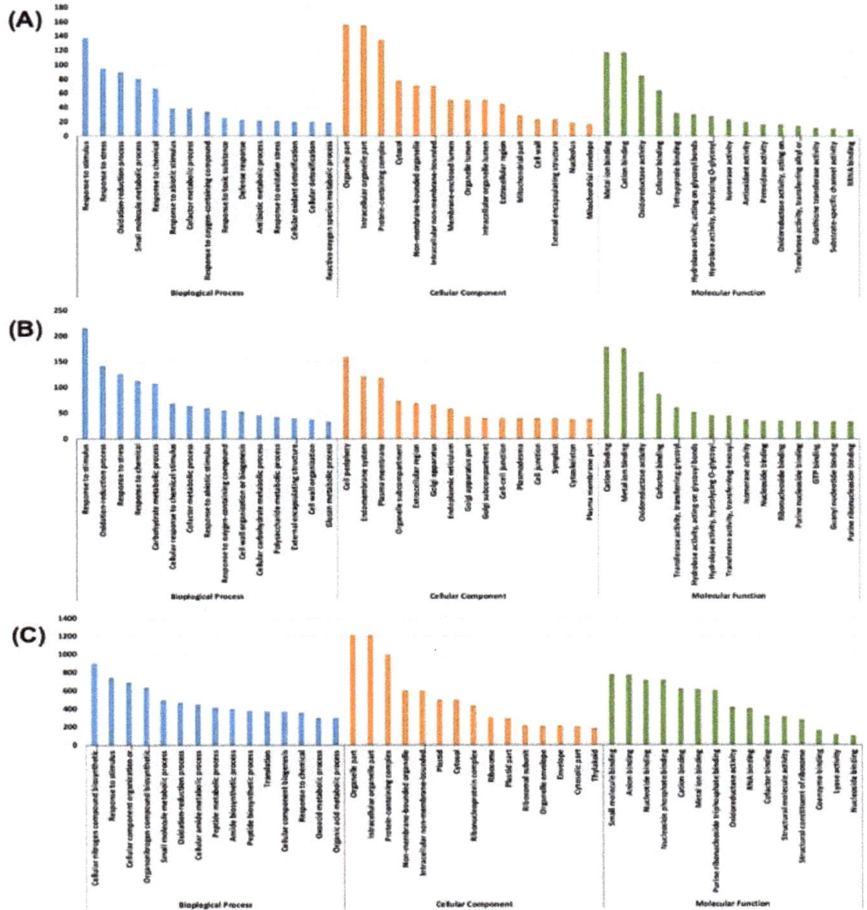

Figure 4. Functional gene ontology annotations of differentially expressed genes (DEGs) under Fe and Zn stresses in the root: (**A**) –Zn; (**B**) –Fe; and (**C**) –Fe–Zn. For graphical representation, we have the top 15 significant terms under each category *viz.*, biological process (blue), cellular component (orange), and molecular function (green). The complete list of terms along with significance of false discovery rate (FDR) < 0.05 are mentioned in Table S1.

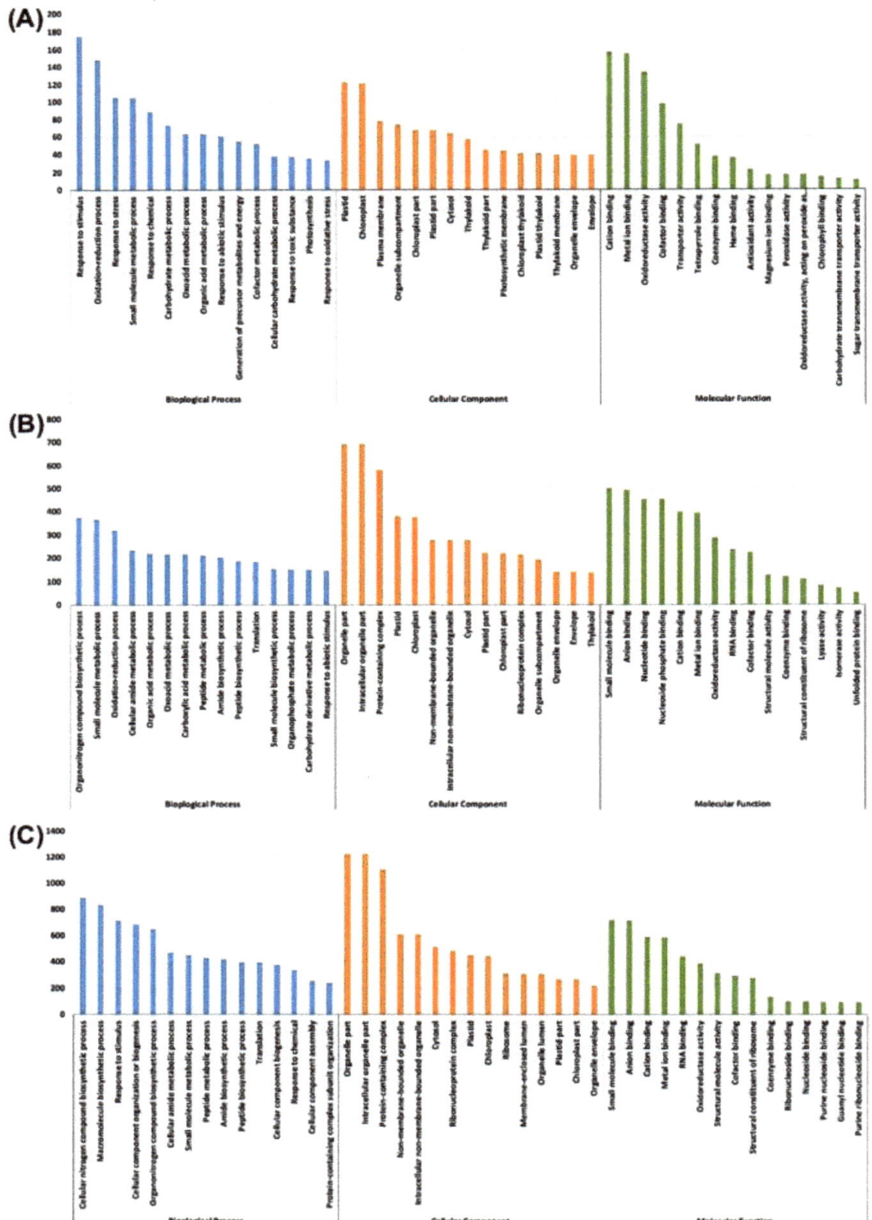

Figure 5. Functional gene ontology annotations of differentially expressed genes (DEGs) under Fe and Zn stresses in the shoot: (**A**)–Zn; (**B**)–Fe; and (**C**)–Fe–Zn. For graphical representation, we have the top 15 significant terms under each category *viz.*, biological process (blue), cellular component (orange), and molecular function (green). The complete list of terms along with the significance of FDR < 0.05 are mentioned in the Table S1.

The Kyoto Encyclopedia of Genes and Genomes (KEGG) enrichment analysis categorised DEGs into 15, 22 and 30 functional categories in the root under –Zn, –Fe and –Fe–Zn stresses, respectively (Table S2). Similarly, in shoot 28, 30 and 30 functional categories were identified for –Zn, –Fe and –Fe–Zn stress treatments, correspondingly (Table S2). The KEGG enrichment analysis and hierarchical clustering of pathways were undertaken to visualize the relatedness in the enrichment results. In this hierarchical clustering tree, related GO terms were grouped based on the number of common genes. The size of the solid circle corresponds to the enrichment FDR (Figure 6). The KEGG-enriched DEGs in the root are majorly related to themes such as amino acid, carbon and carbohydrate metabolism. Similarly, in the shoot, the major themes fall under amino acid metabolism, carbon and carbohydrate metabolism, photosynthesis and nucleic acid metabolism (Figure 6).

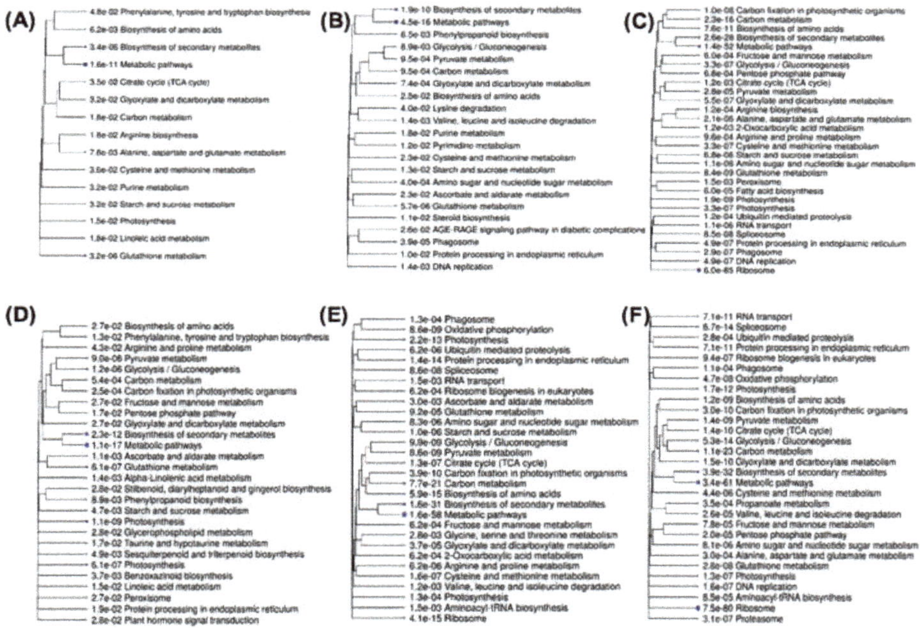

Figure 6. The clustering of KEGG-enriched functional categories of DEGs in the root and shoot under Fe and Zn stresses: (**A**) –Zn in root; (**B**) –Fe in root; (**C**) –Fe–Zn in the root; (**D**). –Zn in the shoot; (**E**) –Fe in shoot; and (**F**) –Fe–Zn in shoot.

2.4. Genes Expression in Response to Fe and Zn Deficiency

Fe and Zn are the vital micronutrients for various metabolisms and the survival of plants. During Fe and Zn deficiency, several genes are known to regulate Fe and Zn homeostasis, phytohormonal regulation and morpho-physiological adaptation. The present investigation recorded the DEGs associated with the mugineic acid pathway, transporters, phytohormone metabolism, photosynthesis and carbohydrate metabolism (Table 1).

Table 1. Differential expression of genes associated with the mugineic acid pathway, transporters, phytohormones, photosynthesis and carbohydrate metabolism in response to –Zn, –Fe and –Zn–Fe stresses in the root and shoot tissues. A minimum of a 2-fold change with $p < 0.05$ were considered for further analysis. The grey colour values indicate a non-significant *p-value*. Positive and negative values indicate increases and decreases in gene expression, respectively.

S. No.	Gene Models	Probe ID	Fold Change						Annotation
			Root			Shoot			
			–Zn	–Fe	–Fe–Zn	–Zn	–Fe	–Fe–Zn	
Transporters and Mugineic Acid Pathway									
1	GRMZM2G142870	Zm.17901.1.S1_at	2.09	1.18	15.46	1.93	–0.74	–1.21	ABC transporter C family member 14
2	GRMZM2G015295	Zm.10189.1.A1_at	2.10	–1.23	–9.12	1.25	7.51	30.81	Adenosylhomocysteinase
3	GRMZM5G843141	ZmAffx.1313.1.S1_s_at	1.13	2.65	–1.77	2.06	2.80	2.83	ATP synthase subunit alpha (atp1-a2)
4	GRMZM2G036908	Zm.15922.1.S1_at	–1.13	–2.38	10.63	1.53	–2.48	–6.57	Cation transmembrane transporter
5	GRMZM5G862882	Zm.18006.1.A1_at	2.14	3.71	7.69	2.03	–1.37	–6.30	Cation transmembrane transporter
6	GRMZM2G064023	Zm.5927.2.A1_a_at	3.04	1.03	1.92	–1.26	6.89	5.41	Citrate synthase 2
7	GRMZM2G157263	Zm.3964.1.A1_a_at	2.35	2.45	2.83	1.44	2.71	3.46	Ferric-chelate reductase
8	GRMZM2G123486	Zm.4621.1.A1_at	–4.04	–3.91	–3.58	1.17	–1.07	2.00	Heavy metal transport/detoxification superfamily protein
9	GRMZM2G178190	Zm.13647.1.S1_at	–1.23	5.07	7.22	–3.76	2.10	–3.66	Metal ion transmembrane transporter activity
10	GRMZM2G122437	Zm.6067.2.A1_at	–2.54	–2.33	–14.22	1.59	–1.34	9.14	Metal ion transporter
11	GRMZM2G099340	Zm.1356.2.S1_a_at	–1.40	–3.90	7.20	2.74	2.15	2.99	Metallothionein-like protein type 2
12	GRMZM2G131473	Zm.4989.1.A1_at	1.85	–1.05	1.48	–1.23	6.14	8.11	Methionine aminopeptidase
13	GRMZM2G015401	Zm.198.1.S1_at	2.78	–1.45	–1.93	1.08	3.46	10.28	Mitochondrial phosphate transporter
14	GRMZM2G034956	Zm.8336.1.S1_x_at	1.02	4.98	–1.29	1.06	–1.01	1.97	Nicotianamine synthase 1
15	GRMZM2G050108	Zm.9637.1.A1_at	2.07	–2.31	20.78	1.35	–1.18	–100.67	Nicotianamine synthase 3
16	GRMZM2G069198	Zm.12619.1.A1_at	3.15	1.56	7.28	2.34	3.38	1.67	NRAMP transporter1
17	GRMZM5G827496	Zm.17744.1.A1_at	–2.58	1.07	10.36	1.20	3.93	–14.55	NRT1/ PTR family 3.1
18	GRMZM2G567452	Zm.10825.1.A2_at	1.34	1.14	1.42	2.91	2.54	1.56	O-methyltransferase
19	GRMZM2G148800	Zm.19137.1.A1_at	–1.04	6.72	25.37	–48.72	–0.75	–126.24	Oligopeptide transmembrane transporter
20	GRMZM2G092125	Zm.605.1.S1_a_at	2.07	–1.85	–4.81	–1.02	7.21	3.86	Plasma membrane intrinsic protein 2
21	GRMZM2G014914	Zm.14436.1.A1_at	–1.70	–1.45	–1.61	1.37	18.02	8.85	Plasma membrane intrinsic protein 2
22	GRMZM2G099628	Zm.15800.5.A1_a_at	1.83	–2.40	–2.35	–1.06	15.83	26.41	Probable methionine-tRNA ligase

Table 1. *Cont.*

S. No.	Gene Models	Probe ID	Fold Change						Annotation
			Root			Shoot			
			−Zn	−Fe	−Fe–Zn	−Zn	−Fe	−Fe–Zn	
23	GRMZM2G104418	Zm.12750.2.A1_x_at	1.16	−1.67	−1.25	1.05	19.23	8.65	Proton-transporting V-type ATPase, V0 domain
24	GRMZM2G070605	Zm.14951.4.S1_at	1.99	2.26	2.77	−1.33	6.68	2.82	S-adenosylmethionine decarboxylase proenzyme
25	GRMZM2G054123	Zm.5785.1.S1_at	1.15	−1.17	−1.57	−1.32	3.92	9.42	S-adenosylmethionine synthetase 1
26	–	Zm.9197.1.A1_at	−2.23	−1.61	−2.50	−1.16	6.86	3.34	Tonoplast intrinsic protein 1
27	GRMZM2G027098	Zm.613.1.A1_at	−2.26	−6.54	−4.77	3.91	1.70	4.49	Tonoplast intrinsic protein 2
28	GRMZM2G056908	Zm.614.1.A1_at	−2.27	−9.93	−636.10	1.62	−1.29	120.97	Tonoplast intrinsic protein 2
29	GRMZM2G070360	Zm.3314.2.A1_a_at	1.91	1.44	2.89	1.18	21.50	6.22	V-type proton ATPase subunit E3
30	GRMZM2G094497	Zm.6859.1.A1_at	1.50	1.05	1.00	−1.23	9.84	2.00	Vacuolar ATP synthase subunit B
31	GRMZM2G421857	Zm.6945.1.A1_at	2.01	−1.22	7.50	−2.40	19.02	2.91	Vacuolar proton pump 3
32	GRMZM2G128995	Zm.4630.2.A1_a_at	1.48	−1.32	1.43	1.24	11.89	7.92	Vacuolar proton-transporting V-type ATPase, V1 domain
33	GRMZM2G126860	Zm.12751.1.S1_at	1.12	−1.15	1.93	−1.13	5.05	2.05	Vacuolar sorting protein 4b
34	GRMZM2G067546	Zm.3321.1.A1_a_at	1.33	1.23	4.31	1.89	8.57	4.03	Vacuolar sorting receptor homolog 1
35	GRMZM2G156599	Zm.582.1.S1_at	−2.12	8.55	35.45	−10.90	2.09	−38.26	Yellow stripe 1
36	GRMZM2G015955	Zm.13452.2.A1_at	1.64	1.26	10.59	3.62	1.19	1.39	Zinc transporter 4
37	GRMZM2G064382	Zm.18511.1.S1_at	5.53	1.15	3.85	8.72	−1.82	−4.10	ZRT-IRT-like protein 5
Phytohormonal Metabolism									
1	GRMZM2G074267	Zm.7858.1.A1_at	−1.25	−3.48	−5.35	1.26	−1.50	3.40	Auxin efflux carrier component PIN1bA: J
2	GRMZM2G427451	Zm.16735.1.A1_at	1.57	−1.96	−99.75	−1.30	4.24	112.22	Auxin induced in root cultures 12
3	GRMZM2G035465	Zm.2321.2.A1_a_at	2.08	1.32	9.13	3.87	1.03	−1.35	Auxin-responsive protein
4	GRMZM2G077356	Zm.4896.1.A1_at	1.30	−2.17	−9.12	2.59	1.30	9.07	Auxin-responsive Aux/IAA family member IAA 13
5	GRMZM2G147243	Zm.2321.3.A1_a_at	1.80	1.19	10.43	1.40	41.19	11.45	Auxin-responsive Aux/IAA family member IAA 17
6	GRMZM2G115354	Zm.7611.2.A1_at	2.06	−1.01	2.13	1.40	1.57	−1.17	Auxin-responsive Aux/IAA family member IAA 24

Table 1. *Cont.*

S. No.	Gene Models	Probe ID	Fold Change						Annotation
			Root			Shoot			
			−Zn	−Fe	−Fe−Zn	−Zn	−Fe	−Fe−Zn	
7	GRMZM2G141903	Zm.5181.2.A1_a_at	2.39	−1.23	5.46	1.15	10.64	6.42	CK2 protein kinase alpha 2
8	GRMZM2G050997	Zm.9754.1.A1_at	2.39	−1.12	21.32	−1.10	1.67	−32.91	Cytokinin oxidase 2
9	GRMZM2G167220	Zm.16971.1.A1_at	1.00	2.01	11.57	−3.13	−1.09	−23.68	Cytokinin oxidase 3
10	GRMZM2G167220	Zm.16971.1.A1_s_at	1.87	5.12	23.03	−2.88	−1.12	−35.98	Cytokinin oxidase 3
11	GRMZM2G041699	Zm.15533.1.S1_at	−1.54	−2.01	−1.93	−4.77	−2.15	−5.99	Cytokinin-O-glucosyltransferase 2
12	GRMZM2G065694	Zm.13139.1.A1_a_at	−1.11	−2.27	3.40	1.17	−1.58	−4.02	EREBP-4 like protein
13	GRMZM2G025062	Zm.10181.1.A1_at	−4.27	−8.86	−8.92	1.85	−1.23	−1.07	ERF-like protein
14	GRMZM2G102601	Zm.14806.1.A1_at	1.26	−2.12	1.43	1.46	5.50	3.43	Ethylene receptor
15	GRMZM2G052667	Zm.6775.1.A1_at	1.46	2.86	3.61	2.51	1.26	−2.48	Ethylene response factor
16	GRMZM2G053503	Zm.11441.1.S1_at	−1.43	−2.61	−4.86	−4.02	−4.16	−5.74	Ethylene-responsive factor-like protein 1
17	GRMZM5G831102	Zm.8468.1.A1_at	−1.35	−5.78	−8.20	5.00	−1.96	15.97	Gibberellin receptor GID1L2
18	GRMZM2G165901	Zm.2157.6.S1_x_at	−1.25	−1.50	−17.02	−1.34	13.02	8.67	Responsive to abscisic acid 15
Photosynthesis and Carbohydrate Metabolism									
1	GRMZM2G448174	ZmAffx.1483.1.S1_at	1.81	1.28	−3.47	−1.27	15.00	2.92	Apocytochrome f precursor
2	GRMZM2G385635	ZmAffx.1092.1.A1_at	−1.47	4.19	−2.34	1.29	2.32	4.10	ATPase beta chain
3	GRMZM2G336448	Zm.19239.1.A1_at	1.13	4.22	2.12	−4.79	−1.33	−8.81	Carbohydrate transporter/sugar porter/transporter
4	GRMZM2G029219	Zm.3331.1.A1_at	−1.49	−2.28	24.02	2.81	5.37	−2.63	Carbohydrate transporter/sugar porter/transporter
5	GRMZM2G121878	Zm.1079.1.A1_a_at	6.24	1.16	804.18	3.14	13.05	−129.47	Carbonic anhydrase
6	GRMZM2G424832	Zm.520.1.S1_x_at	−1.49	−2.11	−4.06	3.91	−1.66	2.04	Cellulose synthase 4
7	GRMZM2G018241	Zm.523.1.S1_at	−1.46	−2.08	−9.33	2.24	−1.28	3.77	Cellulose synthase 9
8	GRMZM2G142898	ZmAffx.5.1.S1_at	1.11	−1.77	−7.57	2.07	−1.84	2.85	Cellulose synthase catalytic subunit 12
9	GRMZM2G047513	Zm.7289.1.A1_at	1.00	1.14	5.87	−1.57	−4.78	−32.62	Chloroplast 30S ribosomal protein S10
10	–	ZmAffx.1219.1.S1_s_at	6.83	2.12	−2.13	9.49	7.00	18.67	Chlorophyll a-b binding protein 6A
11	GRMZM2G145460	Zm.15915.1.S1_at	−1.10	−1.42	−2.38	−1.78	−7.22	−5.92	Chloroplast SRP54 receptor 1
12	GRMZM2G064302	Zm.14043.3.S1_a_at	5.18	2.35	−2.88	1.48	46.52	35.20	Enolase 1

Table 1. *Cont.*

S. No.	Gene Models	Probe ID	Fold Change						Annotation
			Root			Shoot			
			−Zn	−Fe	−Fe–Zn	−Zn	−Fe	−Fe–Zn	
13	GRMZM2G048371	Zm.410.1.A1_at	1.39	−1.02	−2.32	−2.00	7.79	24.38	Enolase 2
14	GRMZM2G053458	Zm.4.1.S1_at	1.35	1.71	−2.32	−1.21	25.82	17.88	Ferredoxin 3
15	GRMZM2G106190	Zm.136.1.S1_at	−1.24	−8.66	−2.28	3.47	−1.50	3.73	Ferredoxin 6, chloroplastic
16	GRMZM2G113325	Zm.1691.1.S1_a_at	−1.38	2.35	20.44	1.12	1.11	−9.52	Ferrochelatase
17	GRMZM2G051004	Zm.8992.2.A1_a_at	2.86	−1.56	−4.31	1.06	3.41	34.02	Glyceraldehyde-3-phosphate dehydrogenase
18	GRMZM2G415359	Zm.5727.1.S1_a_at	2.37	−1.01	1.36	2.37	21.14	15.47	Malate dehydrogenase 5
19	GRMZM2G142873	Zm.8962.1.A1_at	2.55	−1.15	3.49	1.40	4.29	1.20	Opaque endosperm 5
20	GRMZM2G065083	Zm.3318.1.S1_at	1.42	−2.86	1.79	−1.06	15.22	2.71	Phosphohexose isomerase 1
21	GRMZM2G382914	Zm.6780.1.A1_a_at	−1.14	−1.79	−1.06	−1.22	104.09	36.76	Phosphoglycerate kinase
22	GRMZM2G382914	Zm.6780.3.A1_x_at	−1.25	−1.41	−1.07	−1.34	25.90	11.82	Phosphoglycerate kinase
23	GRMZM2G076276	Zm.15933.1.A1_at	1.89	−1.29	−1.18	3.89	1.21	7.50	Predicted polygalacturonate 4-alpha-galacturonosyltransferase
24	GRMZM2G165357	Zm.730.3.A1_x_at	1.70	−1.53	−4.90	1.49	3.50	11.60	Predicted UDP-glucuronic acid decarboxylase 1
25	GRMZM2G059151	Zm.2422.1.A1_at	1.17	−1.49	−2.16	2.07	−1.53	3.21	Pyrophosphate–fructose 6-phosphate 1-phosphotransferase subunit beta
26	GRMZM2G113033	Zm.6763.4.S1_a_at	1.56	2.39	−1.09	1.87	520.30	27.36	Ribulose bisphosphate carboxylase small subunit 2
27	GRMZM2G095252	Zm.3461.1.A1_a_at	1.32	−1.19	1.14	−1.15	104.80	8.52	RuBisCO large subunit-binding protein subunit beta
28	GRMZM5G875238	Zm.26.1.A1_at	−1.24	2.96	308.21	9.08	3.65	−31.29	Sucrose phosphate synthase 1
29	GRMZM2G140107	Zm.18317.1.A1_at	1.32	2.44	68.25	−1.20	2.11	−80.91	Sucrose-phosphate synthase-like
30	GRMZM2G134256	Zm.2199.1.A1_a_at	2.15	−1.08	2.35	2.03	5.68	3.10	Transaldolase 2
31	GRMZM2G030784	Zm.3889.1.A1_at	1.91	1.38	−1.22	1.36	6.97	10.31	Triosephosphate isomerase
32	GRMZM2G073725	Zm.752.1.S1_a_at	2.15	−1.47	−2.33	−1.25	25.38	23.42	UDP-arabinopyranose mutase 3
33	GRMZM2G328500	Zm.4986.1.S1_at	1.14	-3.00	−4.23	−1.80	5.47	11.70	UDP-glucose 6-dehydrogenase
34	GRMZM5G862540	Zm.4986.2.S1_at	1.09	−2.44	−5.64	−1.40	3.00	4.10	UDP-glucose 6-dehydrogenase
35	GRMZM2G042179	Zm.12322.1.A1_at	−1.17	−3.17	−1.08	2.18	2.30	3.02	UDP-glucuronic acid 4-epimerase
36	GRMZM2G067707	Zm.4793.1.S1_a_at	2.99	1.35	1.49	1.32	22.03	24.99	Ubiquinol-cytochrome c reductase complex
37	GRMZM5G833389	Zm.5548.1.A1_x_at	2.37	−1.32	−6.95	−1.04	12.38	13.28	2,3-bisphosphoglycerate-independent phosphoglycerate mutase

2.4.1. Mugineic Acid Pathway and Transporter Genes

The stress treatments (−Zn, −Fe and −Fe–Zn) showed a differential expression of mugineic acid pathway genes in maize seedlings (Figure S3). S-adenosyl-L-homocysteine hydrolysing enzyme *ADENOSYL HOMOCYSTEINASE* (*GRMZM2G015295*; −Fe: 7.51-fold and −Fe–Zn: 30.81-fold) showed the upregulated expression under Fe deficiency in the shoot. O-methyltransferases mediate the conversion of homocysteine to methionine and it serves as a precursor for derivatives of the mugineic acid pathway. *METHYL TRANSFERASES* (*GRMZM2G567452*) showed the upregulation under Fe and Zn deficiency treatments. The *SAMS1* (*GRMZM2G054123*) is known to mediate the conversion of methionine to S-adenosylmethionine (SAM) showed an increased transcript abundance under −Fe (3.92-fold) and −Fe–Zn (−Fe–Zn: 9.42-fold) deficiencies in the shoot. Furthermore, the roots showed a differential expression of *NAS3* (*GRMZM2G054123*) in response to the stress treatments. The upregulation of *YELLOW STRIPE 1* (*ZmYS1*; *GRMZM2G156599*) was observed under Fe deficiency in both roots (8.55-fold) and shoot (2.09-fold). However, in response to the −Fe–Zn treatment, the opposite regulation of *ZmYS1* was observed in the root (35.45-fold) and shoot (−38.26 fold). The expression pattern of *ZmYS1* suggests its higher affinity towards PS-Fe^{3+} and the prominent activity in roots.

In addition to mugineic acid pathway genes, Fe and Zn deficiencies resulted in the differential expression of several transporters. The upregulation of membrane-localised natural resistance associated macrophage protein (NRAMP), *NRAMP TRANSPORTER 1* (GRMZM2G069198) have been observed in both root and shoot. In plants, vacuoles serve as a sequestering site for minerals, including Fe and Zn. Various tonoplast-located transporters mediate Fe and Zn homeostasis in plants. Vacuolar transporters such as V-type ATPases, vacuolar proton pumps (VPPs), tonoplast intrinsic proteins (TIPs) were differentially expressed under the Fe and Zn deficiency. Two TIPs *viz.*, *Zm.9197.1.A1_at* and *GRMZM2G027098,* were specifically upregulated in response to Fe deficiency (−Fe and −Fe–Zn) and Zn (−Zn and −Fe–Zn) deficiencies, respectively. Similarly, vacuolar-type H^+-ATPase (*V-TYPES ATPASE*; *GRMZM2G070360*) showed Fe responsive expression in the shoot. Interestingly, *VPP3* (*GRMZM2G421857*) showed Zn (−Zn and −Fe–Zn) and Fe (−Fe and −Fe–Zn) specific upregulation in the root and shoot, respectively. Ferric chelate reductases (FRO) are the critical component of strategy-I of Fe homeostasis, which showed the upregulated (*GRMZM2G157263*) expression in both roots and shoot under all the stress treatments. Citrate is one of the potent chelators for the mineral transportation in xylem [47]. *CITRATE SYNTHASE 2* (*GRMZM2G064023*) showed a 6.89-fold and 5.41-fold increased expression under −Fe and −Fe–Zn stresses, respectively, in the roots. Among the ABC transporters, the expression of *ABC TRANSPORTER C FAMILY MEMBER 14* (*GRMZM2G142870*) increased the expression by 2.09-fold and 15.46-fold in the root under the −Zn and −Fe–Zn stresses, respectively. The ZIP-like transporter (*ZIP5;* GRMZM2G064382) showed Zn stress-specific upregulation (−Zn: 5.53-fold; −Fe–Zn: 3.85-fold) in the root, whereas the metal ion transmembrane transporter activity (*GRMZM2G178190*) was observed under Fe deficiencies only. Fe deficiency in the shoot (−Fe and −Fe–Zn) resulted in the upregulation of *MITOCHONDRIAL PHOSPHATE TRANSPORTER* (*GRMZM2G015401*; −Fe: 3.46-fold and −Fe–Zn: 10.28-fold).

2.4.2. Phytohormones

Phytohormone metabolism shows the dynamic responses to Fe and Zn availability, which enable the plants to adapt to Fe and Zn deficiencies. In the present investigation, Fe and Zn deficiencies have altered the expression of various genes associated with various phytohormone metabolism *viz.*, ethylene, auxin, gibberellins, cytokinin (Figure S4). The roots under Zn (−Zn and −Fe–Zn) and Fe (−Fe and −Fe–Zn) deficiencies upregulated the *CYTOKININ OXIDASE 2* (*GRMZM2G050997*) and *CYTOKININ OXIDASE 3* (*GRMZM2G167220*) expressions, respectively. The *CYTOKININ-O-GLUCOSYLTRANSFERASE 2* (*GRMZM2G041699*) showed downregulation across the stress treatments.

The upregulation of *Aux/IAA* genes *IAA13* (*GRMZM2G077356*) and *IAA24* (*GRMZM2G115354*) in the shoot (−Zn: 2.59-fold; −Fe–Zn: 9.07-fold) and root (−Zn: 2.06-fold; −Fe–Zn: 2.13-fold), respectively, was observed specifically to Zn deficiency. Similarly, *IAA17* (*GRMZM2G147243*) showed Zn-deficiency induced expression in root tissue (–Zn: 41.19-fold; –Fe–Zn: 11.45-fold). The pin-formed (PIN) protein *PIN1b* (*GRMZM2G074267*) crucial for the polar transport of auxins showed the downregulated expression under –Fe (–3.48) and –Fe–Zn (–5.35) in the root. The repression or non-significant expression of AIR12 (GRMZM2G427451) was observed in the root under –Fe and –Fe–Zn stresses.

Ethylene is another important class of hormones mediating plants' adaptation under Fe and Zn deficiencies. The –Fe and –Fe–Zn stresses resulted in the high expression of *ETHYLENE RECEPTOR* (*GRMZM2G102601*) in the shoot (–Fe: 5.50-fold; –Fe–Zn: 3.43-fold). On the contrary, *ETHYLENE RESPONSE FACTOR* (*GRMZM2G052667*) showed an upregulation in the roots under –Fe (2.86-fold) and –Fe–Zn (3.61-fold) stresses. Ethylene responsive factors (ERF) mediate stress signals. The repression of ERFs has been recorded in response to Fe and Zn deficiencies in a tissue-specific manner. *ERF-LIKE PROTEIN* (*GRMZM2G025062*) was downregulated only in roots, whereas *ETHYLENE-RESPONSIVE FACTOR-LIKE PROTEIN 1* (*GRMZM2G053503*) was downregulated in both the root and shoot. Furthermore, Zn stresses in the shoot (–Zn and –Fe–Zn) upregulated the gibberellin receptor gene *GID1L2* (*GRMZM5G831102*; –Zn: 5.00-fold; –Fe–Zn: 15.97-fold).

2.4.3. Photosynthesis and Carbohydrate Metabolism

Several photosynthesis and carbohydrate metabolism-related genes showed differential expression under Fe and Zn stresses (Figure S5). Under –Fe (2.35-fold) and –Fe–Zn (20.44-fold) stresses, *FERROCHELATASE* (GRMZM2G113325) showed the upregulated expression in roots. In photosynthesis, ferredoxins transfer the electrons from photo-reduced photosystem-I to ferredoxin NADP$^+$ reductase in which NADPH is produced for CO_2 assimilation. *FERREDOXIN 3* (*GRMZM2G053458*) showed the enhanced expression under –Fe (25.82-fold) and –Fe–Zn (17.88-fold) stress in the shoot. Similarly, the cytochrome complexes mediate the electron transfer between PSII and PSI. Fe stresses in the shoot (–Fe, –Fe–Zn) resulted in the increased accumulation of *APOCYTOCHROME F PRECURSOR* transcripts (GRMZM2G448174) in the shoot. Carbonic anhydrases (CAs) are Zn metalloenzymes that catalyse the interconversion of CO_2 and HCO_3^- in plants. The root under –Zn and –Fe–Zn deficiencies and shoot under –Zn and –Fe deficiencies upregulated the *CARBONIC ANHYDRASE* (*GRMZM2G121878*), whereas –Fe–Zn stress in the shoot showed the downregulation of *CARBONIC ANHYDRASE* (*GRMZM2G121878*). The activity of carbonic anhydrase mediates the supply of CO_2 to Rubisco (Ribulose bisphosphate carboxylase). The –Fe and –Fe–Zn deficiencies resulted in higher *RIBULOSE BISPHOSPHATE CARBOXYLASE SMALL SUBUNIT 2* (*GRMZM2G113033*) transcripts accumulation. However, as compared to combined Fe and Zn deficiency (–Fe–Zn), Fe deficiency (–Fe) showed very high expressions. The chlorophyll *a/b*-binding proteins are the apoproteins of the light-harvesting complex of photosystem II. In the present investigation, the *CHLOROPHYLL A-B BINDING PROTEIN 6A* (*ZmAffx.1219.1.S1_s_at*) was upregulated under all the stresses in the shoot. The various proteins associated with photosynthesis co-translationally targeted the chloroplast via signal recognition particles. The transcripts of *CHLOROPLAST SIGNAL RECOGNITION PARTICLE SUBUNIT* (*GRMZM2G145460*; *cpSRP54*) were repressed across the stresses.

Fe stress resulted in a higher expression of DEGs associated with carbohydrate metabolism. The Fe deficiency (–Fe, –Fe–Zn) showed a consistently greater accumulation of transcripts of the glycolysis pathway in the shoot. The majority of glycolytic enzymes *viz.*, *PHOSPHOGLYCERATE KINASE* (*GRMZM2G382914*), *ENOLASES* (*GRMZM2G064302, GRMZM2G048371*), *PHOSPHOHEXOSE ISOMERASE* (*GRMZM2G065083*), *TRIOSEPHOSPHATE ISOMERASE* (*GRMZM2G030784*), *GLYCERALDEHYDE-3-PHOSPHATE DEHYDROGENASE* (*GRMZM2G051004*) showed an enhanced expression in the shoots under –Fe and –Fe–Zn deficiencies. However, Zn deficiency did not show any consistent expression of glycolysis-associated transcripts. Furthermore, the roots enhanced the transcripts of *CARBOHYDRATE TRANSPORTER* (*GRMZM2G336448*) in response to Fe deficiency (–Fe: 4.22-fold; –Fe–Zn: 2.12-fold). On the other hand, genes associated with polysaccharide synthesis and cell wall biogenesis *viz.*, *UDP-GLUCOSE 6-DEHYDROGENASE* (*GRMZM2G328500; GRMZM5G862540*), *UDP-GLUCURONIC ACID DECARBOXYLASE 1* (predicted; *GRMZM2G165357*) and *POLYGALACTURONATE 4-ALPHA-GALACTURONOSYLTRANSFERASE* (*GRMZM2G076276*) showed increased transcripts accumulation in the shoot under –Fe and –Fe–Zn stresses. The Zn stresses in the shoot (–Zn and –Fe–Zn) resulted in the higher expression of *CELLULOSE SYNTHASES* in the shoot (*GRMZM2G424832, GRMZM2G018241, GRMZM2G142898*).

2.5. Regulation and Interaction of Fe and Zn Transporters in Maize

The gene-regulatory network (GRN) is constructed to visualize the regulatory relationships of miRNA and the transcription factors with their downstream differentially expressed transporter and mugineic acid pathway genes under Fe and Zn deficiency. GRN revealed a total of 454 interactions of transporters with miRNA and transcription factors (TFs) (Tables S3 and S4; Figure 7). The highest number of total degrees (31) were observed for the *ZINC TRANSPORTER 4* (*GRMZM2G015955*), *METALLOTHEIONINE LIKE PROTEIN* (*GRMZM2G099340*) and *METHIONINE t-RNA LIGASE* (*GRMZM2G099628*). However, the *ZmYS1* (*GRMZM2G156599*) and *VACUOLAR PROTON-TRANSPORTING V-TYPE ATPASE, V1 DOMAIN* (*GRMZM2G128995*) showed the highest degrees (23) with miRNAs and TFs, respectively. Therefore, these genes are the potential hub nodes in the regulation of the Fe and Zn homeostasis in maize. TF and miRNA and their interactions are adding the regulatory complexity to transporters genes. The regulatory network showed the highest gene–miRNA interactions with miRNAs of the zma-miR395 family (36) and the highest gene–TF interactions with ERFs (104). Hence, zma-miR395 and ERFs are the major regulatory elements in transporter and mugineic acid pathway regulatory network of maize. Furthermore, transporters' regulatory network in the present investigation showed the regulation of genes with a common expression pattern by a common transporter under Fe and Zn deficiency although this may not be a universal phenomenon (Figure 7; Table S3). For instance, six transporter genes (*GRMZM2G015295, GRMZM2G067546, GRMZM2G070605, GRMZM2G099340, GRMZM2G099628, GRMZM2G128995*) showing Fe specific upregulation showed a common TF-mediated regulation through EREB142 (*GRMZM2G010100*). Conversely, many of the transporter genes' expression is regulated by several miRNA and TFs.

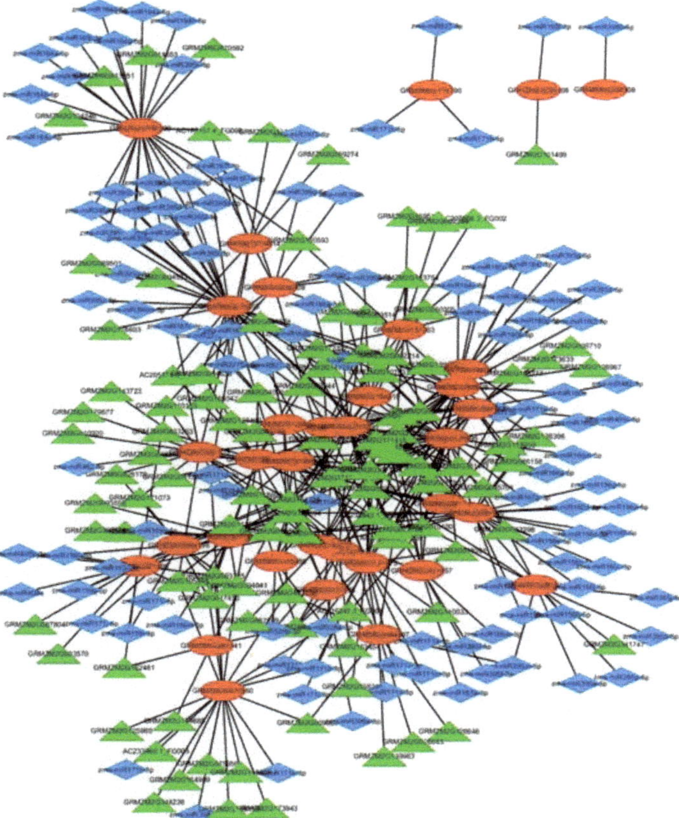

Figure 7. The transcription factors (TFs) and miRNAs mediated the regulation of differentially expressed transporters under Fe and Zn deficiency in maize: the regulatory network is characterized by 148 miRNA–target gene and 306 TF–target genes interactions. The green triangles represent the TFs, the red ovals represent the target transporters and mugineic acid pathway genes, and the blue diamonds represent miRNAs.

3. Discussion

3.1. Fe and Zn Deficiencies Shifted the Expression Pattern of the Transcriptome in Root and Shoot

The current investigation presents a genome-wide transcriptome response of individual and simultaneous Fe and Zn deficiencies in both the root and shoot tissues. The experiments were performed with the aim of understanding the changes in the global transcripts' level under –Zn, –Fe and –Fe–Zn stresses, and to decipher the expression of transcripts associated with Fe and Zn homeostasis, phytohormone metabolism, photosynthesis and carbohydrate metabolism. Genome-wide expression assay revealed a higher number of DEGs in the shoot (8200) as compared to root (7316) for –Fe and –Zn stresses indicating greater physiological demand for Fe and Zn minerals in the shoot over the root. The shoot controls most of the essential physiological activities *viz.*, photosynthesis, chlorophyll biosynthesis, carbohydrate and cytochrome biosynthesis, and that of several plant pigments. Moreover, several enzymes employed in the shoot physiological activities required Fe and/or Zn as co-factor(s) or their integral component(s) [48,49]. On the other hand, high metabolic and physiological pressure on the root transportome and membrane-bound proteins to meet the Fe and Zn ions under

–Fe–Zn stress could have resulted in contrasting trends of a higher number of DEGs in the roots as compared to shoot. The results of the present investigations are in contrast with those of Zheng et al. [32], who reported a higher number of DEGs in the root as compared to the shoot under Fe and P interactions in rice. The functional classification of DEGs in the current investigation revealed many of the GO terms associated with ion uptake and homeostasis *viz.*, cation binding, metal ion binding, oxidoreductase activity etc. The deficiency of Fe and Zn alters the expression of genes associated with Fe and Zn homeostasis to regulate the plant physiological and metabolic requirement of Fe and Zn ions [2,3,23,26,27].

3.2. Fe and Zn Stresses Alter the Root Length via Differential Expression of Phytohormonal Genes

The plant root system is a vital component for the uptake and transport of nutrients from the soil and also exhibits considerable plasticity in response to endogenous and environmental signals [50]. The notable changes in the root length of maize inbred line SKV-616 under –Zn, –Fe and –Fe–Zn stresses were associated with the differential expressions of several genes of hormonal metabolism (Figure 1C, Table 1). Owing to Zn stresses (–Zn, –Fe–Zn), the upregulated expression of *IAA17* (*GRMZM2G147243*) and *IAA24* (*GRMZM2G115354*) was observed in the root tissue. *IAA17* prevents the initiation of lateral roots by preventing the binding of AXR3 to *auxin-responsive element* (*ARE*) [51]. Similarly, the *IAA3* gain of function in rice causes growth defects in crown roots [52,53]. Furthermore, *AtIAA8* is an ortholog of *ZmIAA24* in *Arabidopsis* which has also been reported as a negative regulator of lateral root formation [54]. In addition to the regulation of root phenotype, auxins also serve as signalling molecules in mineral deficiency in plants. The enhanced expression of *IAA13* (*GRMZM2G077356*) and *IAA17* (*GRMZM2G147243*) in –Zn and –Fe stress-specific manner in the shoot indicate their probable role in a shoot-to-root auxin signalling under Zn and Fe deficiency, respectively. In wheat, the application of IAA to shoots of Fe-sufficient plants resulted in the enhanced release of PS in roots [38]. Furthermore, the opposite regulation pattern *IAA* genes could be associated with variation in the spatial responses. In contrast to auxins, cytokinins play an inhibitory role in the lateral root initiation and development [55,56]. The maize seedlings showed Zn and Fe stress-specific expression of *CYTOKININ OXIDASES* (*CKXs*) in root tissues. Remarkably, *CKX3* (*GRMZM2G167220*) showed a higher upregulation in response to Fe stresses (–Fe, –Fe–Zn) as compared to *CKX2* (*GRMZM2G050997*) under Zn stresses (–Zn, –Fe–Zn). In *Arabidopsis,* the overexpression of cytokinin-degrading enzyme encoding gene *CKX* in the lateral root primordia (LRP) results in reduced cytokinin levels and the elimination of cytokinin signalling [57,58], thereafter which enhances the lateral root density by reducing the distance between the adjacent LRP cells [59,60].

The synthesis of ethylene is perceived by membrane-bound ethylene receptors that initiate the ethylene signalling pathway. The higher expression of *ETHYLENE RESPONSE FACTOR* (*GRMZM2G052667*) in –Fe and –Fe–Zn stresses of the root appear to have contributed towards reduced root length. In maize, ethylene-responsive factors (*ERF1*) play a vital role in stress signalling pathways [61]. Similarly, in tomato, the overexpression of ethylene-responsive factors *LeERF2/TERF2* showed the increased synthesis of ethylene [62]. Furthermore, the inverse association between higher ethylene biosynthesis and primary root growth through decreased cell proliferation and elongation in root apical meristem was observed in *Arabidopsis* [63,64].

Gibberellins increase primary root length through the cell elongation and proliferation of root meristem and reduced levels of GAs result in shorter primary roots [65]. GA signalling in plants involves the GA receptor *GIBBERELLIN INSENSITIVE DWARF1* (*GID1*), GA signalling repressor DELLA proteins and ubiquitin ligase complex [66]. The binding of GID1 to bioactive GA, causing the degradation of DELLA proteins, which subsequently initiates the GA signalling in plants. In the absence of bioactive GA, DELLA proteins repress the GA responses [67]. Hence, the downregulation of *GID1L2* (*GRMZM5G831102*) under –Fe and –Fe–Zn stresses in the root indicate a direct association between the reduced root length and of GA signalling in maize. Previous findings also reported an association between the reduced root growth and endogenous GA level under Zn and Fe deficiency in maize [68] and rice [67], respectively.

3.3. Fe and Zn Deficiency Affect Photosynthesis and Carbohydrate Metabolism Through Altered Expression of Key Genes

In the present investigation, the enhanced expression of *FERROCHELATASES* (*GRMZM2G113325*), *APOCYTOCHROME F* (*GRMZM2G448174*), *CHLOROPHYLL A-B BINDING PROTEIN 6A* (*ZmAffx.1219.1.S1_s_at*) were observed under Fe deficiency stresses (–Fe and –Fe–Zn). APOCYTOCHROME F (GRMZM2G448174) is the precursor of cytochrome b6-f complex, which facilitates electron transfer between photosystem II (PSII) and PSI [69]. Similarly, ferredoxins involved in electron transport from the photo-reduced photosystem-I to $NADP^+$ reductases. The chlorophyll a-b binding protein 6A (ZmAffx.1219.1.S1_s_at) acts as a component of the light-harvesting complex (LHC) which is involved in the formation of a full-size NAD(P)H dehydrogenase-PSI super complex (NDH-PSI) that triggers cyclic and chlororespiratory electron transport in chloroplast thylakoids, especially under stress conditions [70]. Therefore, the Fe and Zn deficiency-induced overexpression of *APOCYTOCHROME F*, *CHLOROPHYLL A-B BINDING PROTEIN 6A* and *FERREDOXIN 3* affects photosynthesis by disrupting the cellular redox poise and electron transfer mechanisms [69,71].

The increased expression of *CARBONIC ANHYDRASE* (CA; *GRMZM2G121878*) during Zn deficiency (–Zn and –Fe–Zn) requires Zn^{2+} ions as co-factors for its functional activity. CA catalyses the first biochemical step of the carbon fixation in C_4 plants. As a co-factor in CA, Zn creates a proton H^+ and a nucleophilic hydroxide ion. The nucleophilic water molecules attack the carbonyl group of carbon dioxide to convert it into bicarbonate. Additionally, the H^+ ions reduce the pH, and inactivate the enzymes associated with the hydrolysis of starch and glucose, and subsequently reduce the osmotic gradient of cells and stomatal closure [72].

Rubisco is the most prominent protein on the Earth and serves as the primary engine of carbon assimilation through photosynthesis. Fe deficiency resulted in higher accumulation of RuBisCO subunit transcripts (*GRMZM2G113033*, *GRMZM2G095252*) in the shoot. However, previous reports in soybean and sugar beet pointed the reduced as well as increased RuBisCO protein content under Fe deficiency, respectively [73,74]. Fe and Zn deficiency-induced the disruption in the protein synthesis machinery, which could be the reason for reduced RuBisCO proteins' accumulation. In barley and rice, the accumulation of RuBisCO was higher in the older leaves of Fe-deficient plants than in the older control leaves, which suggests that RuBisCO accumulation is not decreased significantly by Fe deficiency. Therefore, there is a non-significant association between chlorophyll concentration and RuBisCO accumulation [75,76]. Overall, the photosynthetic machinery is known to be highly sensitive to Fe deficiency as compared to Zn owing to the role of Fe ions as co-factors in the key components of the photosynthetic machinery. There are two or three Fe atoms per PSII, 12 Fe atoms per PSI, five Fe atoms per cytochrome complex b6–f (Cyt b6–f) and two Fe atoms per ferredoxin molecule as reported by Briat et al. [77] and Krohling et al. [71].

The upregulation of glycolytic pathways genes *viz.*, PHOSPHOGLYCERATE KINASE (GRMZM2G382914), ENOLASES (GRMZM2G064302, GRMZM2G048371), PHOSPHOHEXOSE ISOMERASE (GRMZM2G065083), TRIOSEPHOSPHATE ISOMERASE (GRMZM2G030784), and GLYCERALDEHYDE-3-PHOSPHATE DEHYDROGENASE (GRMZM2G051004) under Fe deficiency is associated with Fe deficiency-associated metabolic pressure. The Fe deficiency increases the demand for reducing power, and ATP owing to the enhanced expression of Fe acquisition mechanisms and increased organic acids content [76]. Fe deficiency enhanced the activity of the glycolysis pathway to meet the reducing ATP and the enhanced carboxylation. However, the uptake and transport of Fe demand the chelators and organic acids from the carbohydrate metabolism as compared to free ionic movement of Zn and which could be one of the reasons for the greater response of the glycolytic pathway to Fe deficiency as compared to Zn deficiency in plants.

The Zn deficiency in the present investigation resulted in the enhanced expression of cellulose synthase enzymes (CESAs) (*GRMZM2G424832, GRMZM2G018241, GRMZM2G142898*) to meet the Zn requirement for cellulose synthesis. The flexibility of primary cell walls in plants is necessary to adapt the plants under stress conditions to withstand the turgor pressure. Cellulose is the main constituent of the cell wall and is directly synthesized at the plasma membrane [78]. The CESAs carry the terminal Zn-binding domain, which oligomerizes the CESAs under oxidative stress [78,79]. Likewise, Fe deficiency treatments (–Fe, –Fe–Zn) also enhanced the accumulation of the transcripts of other non-cellulosic constituents of cell walls *viz.*, pectins (*POLYGALACTURONATE 4-ALPHA-GALACTURONOSYLTRANSFERASE: GRMZM2G076276*), xylans (*PREDICTED UDP-GLUCURONIC ACID DECARBOXYLASE 1: GRMZM2G165357*), arabinoses (*UDP-ARABINOPYRANOSE MUTASE 3: GRMZM2G073725*).

Sucrose-phosphate synthases (*GRMZM5G875238, GRMZM2G140107*) involved in sucrose biosynthesis showed higher upregulation under Fe deficiency. The Fe deficiency in *Arabidopsis* increases sucrose accumulation in the roots and subsequently promotes the auxin signalling cascades [80]. Therefore, the enhanced accumulation of sucrose may act as one of the junctions between carbohydrate metabolism and phytohormonal signalling under Fe and Zn deficiency in plants.

3.4. Differential Expression of Transporters under Fe and Zn Stresses in Maize

The release of Fe and Zn chelators are crucial in root tissues. The expression pattern of *ADENOSYL HOMOCYSTEINASE* (*GRMZM2G015295*), *O-METHYL TRANSFERASES* (*GRMZM2G567452*), *SAMS* (*GRMZM2G054123*) and *ZmYS1* (*GRMZM2G156599*) suggest higher sensitivity to Fe deficiency. Furthermore, the upregulation of *NAS3* (*GRMZM2G050108*) was observed under –Zn and –Fe–Zn treatments, whereas the *NAS1* (*GRMZM2G034956*) expression was induced under –Fe deficiency only. Interestingly, *NAS3* expression was highly repressed in the shoot under –Fe–Zn treatment suggesting the possible spatial regulation of *NAS3*. The upregulation of the mugineic acid pathway genes is an adaptive strategy in grasses to overcome the Fe and Zn starvation [81,82]. Furthermore, the activity of transporter and chelator genes during Fe and Zn deficiencies occurs in a stress-specific spatiotemporal manner. Nicotianamine is the potent chelator and mediates the uptake and transport of Fe and Zn in plants [8]. The dual regulation of *NAS* genes in response to Fe deficiency has also been reported in maize [83]. In addition to uptake and transport, *NAS* genes also known to be engaged in the accumulation of Fe into the seed [84]. Hence, NA could be one of the significant contributors to the long-distance transport of Fe and Zn in maize. The root tissue showed an enhanced expression of *NRAMP TRANSPORTER 1* (*GRMZM2G069198*) under Zn deficiency. However, for Fe deficiency, the expression was prominent in the shoot. Similarly enhanced upregulation was also observed in maize under Fe deficiency [36]. In *Arabidopsis* and rice, *NRAMP1* is a key transporter for Fe homeostasis [85,86]. Additionally, *NRAMP1* showed to translocate both divalent and trivalent ions in peanut [87] and *Arabidopsits* [88].

Among Fe and Zn chelating complexes, the upregulated expression of *CITRATE SYNTHASE 2* (*GRMZM2G064023*) was observed in response to Zn (–Zn, –Fe–Zn) and Fe (–Fe, –Fe–Zn) deficiencies in the root and shoot, respectively, whereas the expression levels were very high under Fe deficiency in the shoot (–Fe). Both strategy-I and II plants hold Fe–citrate complexes in the xylem sap [89,90]. Fe starvation has been reported to increase the citric acid level in the xylem to enhance Fe mobility in plants [91,92]. In maize, Fe^{3+}–citrate and Fe^{3+}–phytosiderophore are the Fe form in the xylem sap [93]. Similarly, the xylem of *Thlaspi caerulescen* showed ~21% of total cellular Zn as Zn-citrate complex [94].

The reduction of Fe^{3+} to more soluble Fe^{2+} ions by membrane bound *FERRIC CHELATE REDUCTASE 2* (*GRMZM2G157263*) is crucial for the subsequent Fe transport as Fe^{2+}-NA and electron transfer mechanism. Oligopeptide transporter family proteins can transport minerals in the form of metal–NA complexes. *OPT* (*GRMZM2G148800*) showed an upregulated expression response to Fe deficiency (–Fe, –Fe–Zn) in the root. However, the downregulation was observed in the shoot under Zn deficiency (–Zn, –Fe–Zn). The accumulation of *ZINC-TRANSPORTER 4* (*GRMZM2G015955*) transcripts under Zn deficiency in the root and shoot indicates the major function in Zn homeostasis. Similarly, the transcripts of *ZIP5* (*GRMZM2G064382*) were enhanced under Zn deficiency in the roots (–Zn, –Fe–Zn), whereas anti-directional regulation was observed in the shoot. Several authors reported a role of ZIP transporters in the homeostasis of divalent ions *viz.*, Fe, Zn, Mg [17,25–27,95]. In maize, Li et al. [96] showed that the overexpression of *ZmZIP5* enhances the Zn and Fe accumulation. Similarly, the overexpression of *ZIP5* and *ZIP9* genes in rice showed an increased accumulation of Zn over the control plants [25–27].

Vacuoles in the cell help to regulate Fe and Zn flux. Transporters located in the vacuole membrane play an important role in long-distance transport in addition to intracellular metal homeostasis [97]. During Fe and Zn deficiencies, the enhanced expression of vacuole transporters is necessary for the mobilisation of stored Fe and Zn ions to meet the metabolic demands [98]. The enhanced expression of *TIPs* in the shoot showed a possible association of *Zm.9197.1.A1_at* and *GRMZM2G027098* in regulating Fe and Zn homeostasis, respectively. Studies showed that the enhanced expression of *ZmTIP1* facilitates the rapid intracellular osmotic equilibration and permits quick water flow through the vacuoles [99]. The *VACUOLAR PROTON PUMP 3* (*GRMZM2G421857*) transcripts increased in response to Zn and Fe deficiencies in the root and shoot, respectively; and Fe deficiency enhanced the expression of *V-TYPE PROTON ATPASE SUBUNIT E3* (*GRMZM2G070360*), *VACUOLAR ATP SYNTHASE SUBUNIT B* (*GRMZM2G094497*), and *VACUOLAR PROTON-TRANSPORTING V-TYPE ATPASE, V1 DOMAIN* (*GRMZM2G128995*) in the shoot. The specialized functions of vacuoles depend on the tissue and cell type and their developmental stage. All vacuoles seem to hold the majority of the proton pump transporters and differ in their function, depending on the type of vacuole in which they reside [100].

3.5. Hormonal Signalling and Homeostasis Networks Involved in the Fe and Zn Deficiency Regulatory Network

Transcriptome response to Fe and Zn interaction helps in understanding the association and crosstalk of Fe and Zn metabolism in maize. Very few attempts have been made to decipher Fe and Zn interactions in crops at the transcriptome strata. At the transcriptome level, the Fe and Zn stresses acted additively as the number of genes differentially expressed were quite high in the combined Fe and Zn stress (–Fe–Zn) as compared to the individual Fe (–Fe) or Zn (–Zn) stress. However, the direction of the individual gene(s) expression pattern varied with the genes. Many of the key genes associated with Fe and Zn metabolisms were highly downregulated and showed an opposite regulation pattern under both Fe and Zn (–Fe–Zn) deficiencies in the shoot (Table 1). It is known that Fe and Zn act as co-factors of enzymes associated with various metabolic processes and the expression of key genes [48,49]. The simultaneous deficiency of both Fe and Zn could hinder the deficiency-responsive enhanced expression of key genes through altering the respective regulatory enzymes expression. Furthermore, the exertion of high regulatory and metabolic pressure during Fe and Zn stresses to sustain the key metabolisms in the shoot could have resulted in the downregulation and opposite regulation of various key genes. As a support to the findings, the inverse regulation of auxin metabolism by Fe and Zn deficiency was reported in bean [49] as well as in rice [101].

The expression of transporter genes under Fe and Zn deficiency is regulated by various regulatory miRNA and TFs. The binding of miRNA to respective transcripts suppresses the expression of target genes. The maximum interactions was observed for zm-miR395 family miRNAs. The zm-miR395, is a regulator of key proteins *viz.*, low-affinity sulphate transporter (*AST68*) and ATP sulfurylases (*APS1*, *APS3* and *APS4*) associated with sulphur homeostasis [102] and Fe–S clusters in plants. Fe is present in the centre of Fe–S clusters which act as electron acceptors and donors in several cellular processes including photosynthesis, respiration, sulphate assimilation and ethylene biosynthesis [77]. Furthermore, the binding of Zn ions to scaffold assembly protein ISU1 is more crucial for the stability of Fe–S cluster [103]. Fe deficiency results in the downregulation of miR395 in *Arabidopsis* [104]. The Fe and Zn deficiency-induced downregulation of miR395 is associated with the upregulation of its target genes *viz.*, *ZIP4* (*GRMZM2G015955*), *ZmYS1* (*GRMZM2G156599*), *VACUOLAR SORTING RECEPTOR*

HOMOLOG 1 (GRMZM2G067546), VACUOLAR PROTON PUMP 3 (GRMZM2G421857), CITRATE SYNTHASE 2 (GRMZM2G064023).

Similarly, among all the transcription factors, ERFs showed the highest interactions (104) with transporters and mugineic acid pathway genes. In the present and previous investigations, Fe and Zn deficiency resulted in differential expression ERFs [34]. Ethylene acts as an important signalling molecule in the regulatory networks of transporter and hormones. The regulatory network showed gene–ERF interactions with *ZINC TRANSPORTER 4 (GRMZM2G015955), ZRT-IRT-LIKE PROTEIN 5 (GRMZM2G064382), METAL ION TRANSPORTER (GRMZM2G122437), VACUOLAR PROTON-TRANSPORTING V-TYPE ATPASE, V1 DOMAIN (GRMZM2G128995), ABC TRANSPORTER C FAMILY MEMBER 14 (GRMZM2G142870), OLIGOPEPTIDE TRANSMEMBRANE TRANSPORTER (GRMZM2G148800), FERRIC-CHELATE REDUCTASE (GRMZM2G157263), VACUOLAR PROTON PUMP 3 (GRMZM2G421857)* and *MITOCHONDRIAL PHOSPHATE TRANSPORTER (GRMZM2G015401).* Many of the ERFs are known to acts as repressor proteins for transporters and ethylene biosynthesis through feedback inhibition [105]. Therefore, the downregulation of ERFs (GRMZM2G025062, GRMZM2G053503) could be associated with the enhanced regulation of transporters and ethylene synthesis under Fe and Zn deficiencies. Furthermore, in rice, the upregulated expression of Fe and Zn transporters was observed under enhanced ethylene synthesis [41]. Additionally, the ethylene is known to alter the auxin signalling cascades in plants [106]. The sensitivity and spatiotemporal expression of regulatory elements could also play a role in the dual regulation of transporter genes. Furthermore, the regulation of genes associated with different adaptive physiological pathways by miRNA and TFs enhances the system's complexity. In addition to regulatory elements, regulatory enzyme systems also increase the interactions in plants. Fe and Zn deficiency has been reported to alter the hormonal signalling through the accumulation of sugars [80]. These complex crosstalks and interactions through various pathways need to be addressed through reverse genetic approaches.

4. Materials and Methods

4.1. Plant Material and Stress Treatment

Genotype SKV-616, a quality protein maize (QPM with superior protein quality) inbred line was selected to carry out the genome-wide expression assay owing to its ability to accumulate a moderately high level of Fe and Zn concentration [107]. The seeds of a maize inbred SKV-616 were surface sterilised with 1% NaClO solution for 5 min followed by six rinses with Milli-Q water (18.2 MΩ). Seeds were germinated on a Milli-Q water (18.2 MΩ)-soaked filter paper roll. The seedlings were germinated and grown for seven days in the darkroom at 25 °C. The uniform seedlings were transferred to a nutrient solution containing 0.7 mM K_2SO_4, 0.1 mM KCl, 0.1 mM KH_2PO_4, 2.0 mM $Ca(NO_3)_2$, 0.5 mM $MgSO_4$, 10 μM H_3BO_3, 0.5 μM $MnSO_4$, 0.2 μM $CuSO_4$, 0.5 μM $ZnSO_4$, 0.05 μM Na_2MoO_4, and 0.1 mM Fe^{3+}-EDTA. The pH of the nutrient solution was maintained at 5.5 throughout the experiment with 1 M HCl [108]. Fe^{3+}-EDTA and $ZnSO_4$ were excluded from the media to induce Fe (–Fe +Zn) and Zn (+Fe–Zn) deficiencies, respectively. For induction of simultaneous Fe and Zn stresses (–Fe–Zn), both Fe^{3+}-EDTA and $ZnSO_4$ were removed from the complete solution. After the seeds germinated, the plants were transferred to the respective treatment solution and grown until the clear expression of Fe and Zn deficiency symptoms (12 days).

4.2. Morpho-Physiological Characterisation

All the morphophysiological parameters were recorded on five plants grown in hydroponic culture with three data points per seedling. The chlorophyll content was measured with soil plant analysis development (SPAD) chlorophyll meter (SPAD-502 Plus chlorophyll meter, Konica Minolta Sensing Inc., Osaka, Japan). The readings were taken on the top, middle and base of the fully expanded top leaf in each seedling. The same leaf was used for measuring the photosynthesis rate, transpiration rate and stomatal conductance with a portable photosynthesis system (LI-6400, Nebraska, USA). The maximal quantum efficiency of the photo-system II (PS II) photochemistry (*Fv/Fm*) was recorded on 30 min dark-adapted leaves using a portable photosynthesis system with fluorescence attachment (LI-6400, LI-COR, Nebraska, USA). The root length was measured manually with the measuring scale. The statistical analysis was performed with a two-sample t test with the assumption of unequal variances with all possible comparisons among the control and treatments.

4.3. Genome-Wide Expression Assay Using Affymetrix GeneChip Maize Genome Array

The microarray experiment was designed as a two-factor experiment with the four possible treatments among Fe and Zn combination *viz.*, 1) +Fe+Zn, 2) –Zn, 3) –Fe and 4) –Fe–Zn. The total RNA was isolated in triplicates separately from the control and stress-treated root and shoot tissues of 19 day-old seedlings with a stress period of 12 days (50 mg each) using the RNeasy Mini Kit (Qiagen, Hilden, North Rhine-Westphalia, Germany) as per manufacturer's guidelines. The RNase free DNAse treatment with a Thermo Scientific kit (La layette, USA) was given to each RNA sample to eliminate the residual DNA present in RNA samples. Each DNAse-treated RNA sample obtained was examined for quantity using a NanoDrop 1000 spectrophotometer (Thermo Scientific, Wilmington, Delaware, USA). RNA samples with a A260/280 ratio of 1.8 to 2.1 in were considered for further analysis. Affymetrix GeneChip maize genome arrays (Affymetrix Inc., Santa Clara, California, USA) were used in three replications for each treatment to ensure the reproducibility and quality of the chip hybridisation using standard Affymetrix protocols (3'IVT protocol on GeneChip Fluidics Station 450; scanning on Affymetrix GSC3000). For the microarray experiment, ~300 ng of total RNA was biotin-labelled for GeneChip analysis and 10 µg of purified fragmented cRNA was used for hybridisation. All expression data were derived based on the Minimum Information About a Microarray Experiment (MIAME) guidelines [109].

4.4. Microarray Data Analyses

The GeneChip Operating Software (GCOS, Affymetrix GeneChip operating software with autoloader ver. 1.4, manual) was used to generate the CEL file for each of the scanned microarray chips. The raw CEL files containing probe hybridisation intensities from 24 chips were imported into the R platform using the *affy* package [110]. The GeneChip Robust Multiarray Average (*GCRMA*) algorithm was used for background correction, normalisation, and probe set summarisation [111]. The *arrayQualityMetrics* package was employed to generate microarray quality metrics reports [112]. The linear modelling of the microarray data and the identification of DEGs was performed with the *limma* package [113]. The *limma* computes t-statistics and log-odds of differential expression by the empirical Bayes shrinkage of the standard errors toward a common value. Probe sets having a *p*-value of < 0.05 and a 2-fold expression change were considered as differentially expressed under mineral stress treatments compared to the control. The gene model IDs for the probes were retrieved from the Gramene database [114] for subsequent analyses. Where the gene model IDs are not available, the Affymetrix probe IDs were used to mention the genes.

4.5. Gene Ontology, KEGG Enrichment and Gene-Regulatory Network (GRN) Analyses

Gene ontology (GO) and KEGG enrichment analyses were performed using ShinyGO v 0.61 [115]. The function categories of the genes were selected with the enrichment FDR (false discover rate) of *p* < 0.05. The transcription factors (TFs) regulating the expression of transporters (DEGs) were retrieved from the PlantRegMap database and the miRNAs regulating the expression of stress responsive genes at post-transcriptional level were predicted by the psRNATarget tool [116,117]. The complete Fe and Zn transporters regulatory network was realised using Cytoscape [118] and furthermore, the network was analysed to find important nodes (genes) that played a major role in the flow of information within the biological system and helping plants' adaptation during Fe and Zn deficiencies.

4.6. Validation and Expression Correlation of DEGs

A total of 16 expression data points was validated using four genes through the qRT-PCR assay (Agilent Technologies, Santa Clara, CA, USA) (Table S5; Figure S2). The first-strand cDNA was synthesised from 250 ng of total RNA using an Affinity Script qRT-PCR cDNA synthesis kit (Stratagene, Agilent Technologies, Santa Clara, CA, USA). With the help of the IDT server [119], gene-specific primers were designed (Table S5). The qRT-PCR reaction was performed using Stratagene MX3005P (Agilent Technologies, Santa Clara, CA, USA) with the following PCR conditions: 10 min at 95 °C (preheating), followed by 40 cycles of amplification with denaturation for 30 s at 60 °C, primer annealing for 1 min at 58 °C and primer extension for 30 s at 72 °C [120,121].

5. Conclusions

The current investigation uncovered Fe and Zn-deficiency-responsive transcriptome signatures of transporters, phytohormone and carbohydrate metabolisms in maize. The Fe and Zn deficiencies altered the morpho-physiological and molecular responses through the differential expression of genes associated with phytohormonal regulations, transporters and photosynthesis. The result of the present investigation also revealed the dual regulation and anti-directional expression of key transporter genes suggesting something more than a direct/inverse association between Fe and Zn at transcriptome levels through regulatory proteins. Furthermore, the GRN analysis revealed the interactions among genes associated with hormone signalling and Fe and Zn transporters. The various DEGs regulating Fe and Zn homeostasis could be employed as candidate genes for enhancing Fe and Zn efficiency in maize through marker-assisted breeding, genetic engineering and genome editing approaches. Furthermore, this investigation sets the stage for the use of reverse genetic tools to analyse the stress-hormone signalling pathways and the interaction of various metabolic processes in the maize under Fe and Zn deficiencies and their interactions.

Supplementary Materials: The following are available online at http://www.mdpi.com/2223-7747/9/12/1812/s1, Figure S1. Initial and final chlorophyll content in the leaves of SKV616 maize inbred under complete (+Fe+Zn) hydroponic solution (*significant at $p < 0.5$); Figure S2. Correlation coefficients among morpho-physiological parameters under Fe and Zn stresses in maize. (CC: chlorophyll content (SPAD-value); PR: photosynthesis rate; TR: transpiration rate; SC: stomatal conductance; Fv/Fm: quantum efficiency of PS II photochemistry; and RL: Root length; **, *** significant at $p < 0.01$ and $p < 0.001$, respectively); Figure S3. Heatmap of differentially expressed transporters and mugineic acid pathways genes. The z-scores are computed for all genes that are differentially expressed with $p < 0.05$ and > 2-fold expression; Figure S4. Heatmap of differentially expressed phytohormonal metabolism genes. The z-scores are computed for all genes that are differentially expressed with $p < 0.05$ and > 2-fold expression; Figure S5. Heatmap of differentially expressed photosynthesis and carbohydrate metabolism genes. The z-scores are computed for all genes that are differentially expressed with $p < 0.05$ and > 2-fold expression; Table S1. Gene ontology terms for differentially expressed genes in the root and shoot under the –Zn, –Fe and –Fe–Zn treatments; Table S2. KEGG-enriched functional categories of differentially expressed genes in the root and shoot under –Zn, –Fe and –Fe–Zn stress treatments; Table S3. The predicted miRNA–gene and TF–gene interactions used to construct the gene regulatory network for the transporter and mugineic acid pathway genes; Table S4. The GRN features of the transporter and mugineic acid pathway genes; Table S5. Primer* sequences of DEGs selected from microarray analyses for qRT-PCR validation; Figure S6. Validation of DEGs from microarray analysis through qRT-PCR. The x axis represents the probes and the y axis represents the fold change expression values of genes. Error bars in the column represent the standard error. The letters 'S' and 'R' in the stress name on the x axis refer to the shoot and root, respectively. The reactions were performed using the Stratagene MX3005P (Agilent Technologies, Santa Clara, California, USA) Real-Time PCR system with the following PCR conditions: 10 min, at 95 °C (preheating), followed by 40 cycles of amplification with denaturation for 30 s at 60 °C, primer annealing for 1 min at 58 °C and primer extension for 30 s at 72 °C.

Author Contributions: H.S.G., N.T. and M.G.M. conceived and designed the experiments; M.G.M., K.S., N.T., F.H., J.S.B., and K.M.M. performed the experiments; M.G.M., R.S., A.C.R.M., S.S.M. and A.R.R. analysed the data; M.G.M., H.S.G. and N.T. contributed to manuscript preparation. All authors have read and agreed to the published version of the manuscript.

Funding: The experiment was funded by the National Agricultural Innovation Project (NAIP, Component IV), the ICAR Network Project on Transgenics in Crop Plants (Maize Functional Genomics Component) and DST grant (DST/INSPIRE/2011/IF110504). Bioinformatic analyses were supported by a network project on "Computational Biology and Agricultural Bioinformatics (Agril.Edn.14(44)/2014-A&P)". The funding agencies had no role in the study design, data collection and analysis, the decision to publish, or the preparation of the manuscript.

Acknowledgments: We are thankful to the National Phytotron Facility, IARI, New Delhi, for extending the glasshouse facility for the experiment, and to the Division of Plant Pathology for providing GeneChip Fluidics service and Madan Pal, from the Division of Plant Physiology, for extending the IRGA instrument facility.

Conflicts of Interest: The authors declare that the research was conducted in the absence of any commercial or financial relationships that could be construed as a potential conflict of interest.

Data Availability: The raw microarray expression data files of the present study have been deposited in GEO (Accession No. GSE122581; Sample No. GSM3474993 to GSM3475016). All other supporting data are included as supplementary files.

References

1. Mallikarjuna, M.G.; Thirunavukkarasu, N.; Hossain, F.; Bhat, J.S.; Jha, S.K.; Rathore, A.; Agrawal, P.K.; Pattanayak, A.; Reddy, S.S.; Gularia, S.K.; et al. Stability performance of inductively coupled plasma mass spectrometry-phenotyped kernel minerals concentration and grain yield in maize in different agro-climatic zones. *PLoS ONE* **2015**, *10*, e0139067. [CrossRef]

2. Kobayashi, T.; Nozoye, T.; Nishizawa, N.K. Iron transport and its regulation in plants. *Free Radic. Biol. Med.* **2019**, *133*, 11–20. [CrossRef] [PubMed]

3. Gupta, H.S.; Hossain, F.; Nepolean, T.; Vignesh, M.; Mallikarjuna, M.G. Understanding genetic and molecular bases of Fe and Zn accumulation towards development of micronutrient-enriched maize. In *Nutrient Use Efficiency: From Basics to Advances*; Rakshit, A., Singh, H., Sen, A., Eds.; Springer India: New Delhi, India, 2015; pp. 255–282. ISBN 9788132221692.

4. Clemens, S. Molecular mechanisms of plant metal tolerance and homeostasis. *Planta* **2001**, *212*, 475–486. [CrossRef] [PubMed]

5. Römheld, V.; Marschner, H. Evidence for a specific uptake system for iron phytosiderophores in roots of grasses. *Plant Physiol.* **1986**, *80*, 175–180. [CrossRef] [PubMed]

6. Higuchi, K.; Suzuki, K.; Nakanishi, H.; Yamaguchi, H.; Nishizawa, N.K.; Mori, S. Cloning of nicotianamine synthase genes, novel genes involved in the biosynthesis of phytosiderophores. *Plant Physiol.* **1999**, *119*, 471–479. [CrossRef] [PubMed]

7. Takahashi, M.; Yamaguchi, H.; Nakanishi, H.; Shioiri, T.; Nishizawa, N.K.; Mori, S. Cloning two genes for nicotianamine aminotransferase, a critical enzyme in iron acquisition (strategy II) in graminaceous plants. *Plant Physiol.* **1999**, *121*, 947–956. [CrossRef] [PubMed]

8. Bashir, K.; Inoue, H.; Nagasaka, S.; Takahashi, M.; Nakanishi, H.; Mori, S.; Nishizawa, N.K. Cloning and characterization of deoxymugineic acid synthase genes from graminaceous plants. *J. Biol. Chem.* **2006**, *281*, 32395–32402. [CrossRef]

9. Kawakami, Y.; Bhullar, N.K. Molecular processes in iron and zinc homeostasis and their modulation for biofortification in rice. *J. Integr. Plant Biol.* **2018**, *60*, 1181–1198. [CrossRef]

10. Nozoye, T.; Nagasaka, S.; Kobayashi, T.; Takahashi, M.; Sato, Y.; Sato, Y.; Uozumi, N.; Nakanishi, H.; Nishizawa, N.K. Phytosiderophore efflux transporters are crucial for iron acquisition in graminaceous plants. *J. Biol. Chem.* **2011**, *286*, 5446–5454. [CrossRef]

11. Nozoye, T.; Nagasaka, S.; Kobayashi, T.; Sato, Y.; Uozumi, N.; Nakanishi, H.; Nishizawa, N.K. The phytosiderophore efflux transporter TOM2 is involved in metal transport in rice. *J. Biol. Chem.* **2015**, *290*, 27688–27699. [CrossRef]

12. Li, S.; Zhou, X.; Chen, J.; Chen, R. Is there a strategy I iron uptake mechanism in maize? *Plant Signal. Behav.* **2018**, *13*, 1–4. [CrossRef] [PubMed]

13. Ramesh, S.A. Differential metal selectivity and gene expression of two zinc transporters from rice. *Plant Physiol.* **2003**, *133*, 126–134. [CrossRef] [PubMed]

14. Hacisalihoglu, G. Zinc (Zn): The last nutrient in the alphabet and shedding light on zn efficiency for the future of crop production under suboptimal zn. *Plants* **2020**, *9*, 1471. [CrossRef] [PubMed]

15. Marschner, H. *Mineral Nutrition of Higher Plants*, 2nd ed.; Academic Press: Cambridge, MA, USA, 1995; Volume 66, ISBN 9780124735439.

16. Hacisalihoglu, G.; Hart, J.J.; Kochian, L.V. High- and low-affinity zinc transport systems and their possible role in zinc efficiency in bread wheat. *Plant Physiol.* **2001**, *125*, 456–463. [CrossRef] [PubMed]

17. Guerinot, M. Lou The ZIP family of metal transporters. *Biochim. Biophys. Acta Biomembr.* **2000**, *1465*, 190–198. [CrossRef]

18. Takahashi, R.; Bashir, K.; Ishimaru, Y.; Nishizawa, N.K.; Nakanishi, H. The role of heavy-metal ATPases, HMAs, in zinc and cadmium transport in rice. *Plant Signal. Behav.* **2012**, *7*, 1605–1607. [CrossRef]

19. Ricachenevsky, F.K.; Menguer, P.K.; Sperotto, R.A.; Williams, L.E.; Fett, J.P. Roles of plant metal tolerance proteins (MTP) in metal storage and potential use in biofortification strategies. *Front. Plant Sci.* **2013**, *4*, 1–16. [CrossRef]

20. Gupta, N.; Ram, H.; Kumar, B. Mechanism of Zinc absorption in plants: Uptake, transport, translocation and accumulation. *Rev. Environ. Sci. Biotechnol.* **2016**, *15*, 89–109. [CrossRef]

21. Zhang, F.S.; Römheld, V.; Marschner, H.; Zhang, F.S. Diurnal rhythm of release of phytosiderophores and uptake rate of zinc in iron-deficient wheat. *Soil Sci. Plant Nutr.* **1991**, *37*, 671–678. [CrossRef]

22. Waters, B.M.; Chu, H.-H.; DiDonato, R.J.; Roberts, L.A.; Eisley, R.B.; Lahner, B.; Salt, D.E.; Walker, E.L. Mutations in arabidopsis *Yellow Stripe-Like1* and *Yellow Stripe-Like3* reveal their roles in metal ion homeostasis and loading of metal ions in seeds. *Plant Physiol.* **2006**, *141*, 1446–1458. [CrossRef]

23. Mallikarjuna, M.G.; Nepolean, T.; Mittal, S.; Hossain, F.; Bhat, J.S.; Manjaiah, K.M.; Marla, S.S.; Mithra, A.C.; Agrawal, P.K.; Rao, A.R.; et al. In-silico characterisation and comparative mapping of yellow stripe like transporters in five grass species. *Indian J. Agric. Sci.* **2016**, *86*, 721–727.

24. Lin, Y.F.; Liang, H.M.; Yang, S.Y.; Boch, A.; Clemens, S.; Chen, C.C.; Wu, J.F.; Huang, J.L.; Yeh, K.C. Arabidopsis IRT3 is a zinc-regulated and plasma membrane localized zinc/iron transporter. *New Phytol.* **2009**, *182*, 392–404. [CrossRef] [PubMed]

25. Yang, M.; Li, Y.; Liu, Z.; Tian, J.; Liang, L.; Qiu, Y.; Wang, G.; Du, Q.; Cheng, D.; Cai, H.; et al. A high activity zinc transporter OsZIP9 mediates zinc uptake in rice. *Plant J.* **2020**, *103*, 1695–1709. [CrossRef] [PubMed]

26. Tan, L.; Qu, M.; Zhu, Y.; Peng, C.; Wang, J.; Gao, D.; Chen, C. ZINC TRANSPORTER 5 and ZINC TRANSPORTER 9 function synergistically in zinc/cadmium uptake. *Plant Physiol.* **2020**, *183*, 1235–1249. [CrossRef] [PubMed]

27. Huang, S.; Sasaki, A.; Yamaji, N.; Okada, H.; Mitani-Ueno, N.; Ma, J.F. The ZIP transporter family member OsZIP9 contributes to root zinc uptake in rice under zinc-limited conditions. *Plant Physiol.* **2020**, *183*, 1224–1234. [CrossRef]

28. Rogers, E.E.; Guerinot, M.L. FRD3, a member of the multidrug and toxin efflux family, controls iron deficiency responses in arabidopsis. *Plant Cell* **2002**, *14*, 1787–1799. [CrossRef]

29. Inoue, H.; Takahashi, M.; Kobayashi, T.; Suzuki, M.; Nakanishi, H.; Mori, S.; Nishizawa, N.K. Identification and localisation of the rice nicotianamine aminotransferase gene *OsNAAT1* expression suggests the site of phytosiderophore synthesis in rice. *Plant Mol. Biol.* **2008**, *66*, 193–203. [CrossRef]

30. Curie, C.; Cassin, G.; Couch, D.; Divol, F.; Higuchi, K.; Le Jean, M.; Misson, J.; Schikora, A.; Czernic, P.; Mari, S. Metal movement within the plant: Contribution of nicotianamine and yellow stripe 1-like transporters. *Ann. Bot.* **2008**, *103*, 1–11. [CrossRef]

31. Brown, J.C.; Chaney, R.L. Effect of iron on the transport of citrate into the xylem of soybeans and tomatoes. *Plant Physiol.* **1971**, *47*, 836–840. [CrossRef]

32. Zheng, L.; Huang, F.; Narsai, R.; Wu, J.; Giraud, E.; He, F.; Cheng, L.; Wang, F.; Wu, P.; Whelan, J.; et al. Physiological and transcriptome analysis of iron and phosphorus interaction in rice seedlings. *Plant Physiol.* **2009**, *151*, 262–274. [CrossRef]

33. Yang, T.J.W.; Lin, W.D.; Schmidt, W. Transcriptional profiling of the arabidopsis iron deficiency response reveals conserved transition metal homeostasis networks. *Plant Physiol.* **2010**, *152*, 2130–2141. [CrossRef] [PubMed]

34. Li, Y.; Wang, N.; Zhao, F.; Song, X.; Yin, Z.; Huang, R.; Zhang, C. Changes in the transcriptomic profiles of maize roots in response to iron-deficiency stress. *Plant Mol. Biol.* **2014**, *85*, 349–363. [CrossRef] [PubMed]

35. Zamboni, A.; Zanin, L.; Tomasi, N.; Avesani, L.; Pinton, R.; Varanini, Z.; Cesco, S. Early transcriptomic response to Fe supply in Fe-deficient tomato plants is strongly influenced by the nature of the chelating agent. *BMC Genomics* **2016**, *17*, 1–17. [CrossRef] [PubMed]

36. Zanin, L.; Venuti, S.; Zamboni, A.; Varanini, Z.; Tomasi, N.; Pinton, R. Transcriptional and physiological analyses of Fe deficiency response in maize reveal the presence of strategy I components and Fe/P interactions. *BMC Genomics* **2017**, *18*, 1–15. [CrossRef]

37. Mager, S.; Schönberger, B.; Ludewig, U. The transcriptome of zinc deficient maize roots and its relationship to DNA methylation loss. *BMC Plant Biol.* **2018**, *18*, 1–16. [CrossRef]

38. Garnica, M.; Bacaicoa, E.; Mora, V.; San Francisco, S.; Baigorri, R.; Zamarreño, A.M.; Garcia-Mina, J.M. Shoot iron status and auxin are involved in iron deficiency-induced phytosiderophores release in wheat. *BMC Plant Biol.* **2018**, *18*, 1–14. [CrossRef]

39. García, M.J.; Suárez, V.; Romera, F.J.; Alcántara, E.; Pérez-Vicente, R. A new model involving ethylene, nitric oxide and Fe to explain the regulation of Fe-acquisition genes in strategy I plants. *Plant Physiol. Biochem.* **2011**, *49*, 537–544. [CrossRef]

40. Zhou, M.L.; Qi, L.P.; Pang, J.F.; Zhang, Q.; Lei, Z.; Tang, Y.X.; Zhu, X.M.; Shao, J.R.; Wu, Y.M. Nicotianamine synthase gene family as central components in heavy metal and phytohormone response in maize. *Funct. Integr. Genomics* **2013**, *13*, 229–239. [CrossRef]

41. Wu, J.; Wang, C.; Zheng, L.; Wang, L.; Chen, Y.; Whelan, J.; Shou, H. Ethylene is involved in the regulation of iron homeostasis by regulating the expression of iron-acquisition-related genes in *Oryza sativa*. *J. Exp. Bot.* **2011**, *62*, 667–674. [CrossRef]

42. Chen, W.W.; Yang, J.L.; Qin, C.; Jin, C.W.; Mo, J.H.; Ye, T.; Zheng, S.J. Nitric oxide acts downstream of auxin to trigger root ferric-chelate reductase activity in response to iron deficiency in arabidopsis. *Plant Physiol.* **2010**, *154*, 810–819. [CrossRef]

43. García, M.J.; Lucena, C.; Romera, F.J.; Alcántara, E.; Pérez-Vicente, R. Ethylene and nitric oxide involvement in the up-regulation of key genes related to iron acquisition and homeostasis in *Arabidopsis*. *J. Exp. Bot.* **2010**, *61*, 3885–3899. [CrossRef] [PubMed]

44. Sun, H.; Feng, F.; Liu, J.; Zhao, Q. The interaction between auxin and nitric oxide regulates root growth in response to iron deficiency in rice. *Front. Plant Sci.* **2017**, *8*, 1–14. [CrossRef] [PubMed]

45. Liu, K.; Yue, R.; Yuan, C.; Liu, J.; Zhang, L.; Sun, T.; Yang, Y.; Tie, S.; Shen, C. Auxin signaling is involved in iron deficiency-induced photosynthetic inhibition and shoot growth defect in rice (*Oryza sativa* L.). *J. Plant Biol.* **2015**, *58*, 391–401. [CrossRef]

46. Bouain, N.; Shahzad, Z.; Rouached, A.; Khan, G.A.; Berthomieu, P.; Abdelly, C.; Poirier, Y.; Rouached, H. Phosphate and zinc transport and signalling in plants: Toward a better understanding of their homeostasis interaction. *J. Exp. Bot.* **2014**, *65*, 5725–5741. [CrossRef] [PubMed]

47. Conte, S.S.; Walker, E.L. Transporters contributing to iron trafficking in plants. *Mol. Plant* **2011**, *4*, 464–476. [CrossRef] [PubMed]

48. Balk, J.; Schaedler, T.A. Iron cofactor asembly in plants. *Annu. Rev. Plant Biol.* **2014**, *65*, 125–153. [CrossRef] [PubMed]

49. Cakmak, I.; Marschner, H.; Bangerth, F. Effect of zinc nutritional status on growth, protein metabolism and levels of indole-3-acetic acid and other phytohormones in bean (*Phaseolus vulgaris* L.). *J. Exp. Bot.* **1989**, *40*, 405–412. [CrossRef]

50. Ivanchenko, M.G.; Muday, G.K.; Dubrovsky, J.G. Ethylene-auxin interactions regulate lateral root initiation and emergence in *Arabidopsis thaliana*. *Plant J.* **2008**, *55*, 335–347. [CrossRef]

51. Luo, J.; Zhou, J.J.; Zhang, J.Z. *Aux/IAA* gene family in plants: Molecular structure, regulation, and function. *Int. J. Mol. Sci.* **2018**, *19*, 259. [CrossRef]

52. Nakamura, A.; Umemura, I.; Gomi, K.; Hasegawa, Y.; Kitano, H.; Sazuka, T.; Matsuoka, M. Production and characterization of auxin-insensitive rice by overexpression of a mutagenized rice IAA protein. *Plant J.* **2006**, *46*, 297–306. [CrossRef]

53. Zhang, Z.; Li, J.; Tang, Z.; Sun, X.; Zhang, H.; Yu, J.; Yao, G.; Li, G.; Guo, H.; Li, J.; et al. Gnp4/LAX2, a RAWUL protein, interferes with the OsIAA3–OsARF25 interaction to regulate grain length via the auxin signaling pathway in rice. *J. Exp. Bot.* **2018**, *69*, 4723–4737. [CrossRef] [PubMed]

54. Arase, F.; Nishitani, H.; Egusa, M.; Nishimoto, N.; Sakurai, S.; Sakamoto, N.; Kaminaka, H. IAA8 involved in lateral root formation interacts with the TIR1 auxin receptor and ARF transcription factors in *Arabidopsis*. *PLoS ONE* **2012**, *7*. [CrossRef] [PubMed]

55. Jing, H.; Strader, L.C. Interplay of auxin and cytokinin in lateral root development. *Int. J. Mol. Sci.* **2019**, *20*, 486. [CrossRef] [PubMed]

56. Li, X.; Mo, X.; Shou, H.; Wu, P. Cytokinin-mediated cell cycling arrest of pericycle founder cells in lateral root initiation of arabidopsis. *Plant Cell Physiol.* **2006**, *47*, 1112–1123. [CrossRef] [PubMed]

57. Werner, T.; Motyka, V.; Strnad, M.; Schmülling, T. Regulation of plant growth by cytokinin. *Proc. Natl. Acad. Sci. USA* **2001**, *98*, 10487–10492. [CrossRef]

58. Werner, T.; Motyka, V.; Laucou, V.; Smets, R.; Van Onckelen, H.; Schmülling, T. Cytokinin-deficient transgenic arabidopsis plants show multiple developmental alterations indicating opposite functions of cytokinins in the regulation of shoot and root meristem activity. *Plant Cell* **2003**, *15*, 2532–2550. [CrossRef]

59. Chang, L.; Ramireddy, E.; Schmülling, T. Cytokinin as a positional cue regulating lateral root spacing in *Arabidopsis*. *J. Exp. Bot.* **2015**, *66*, 4759–4768. [CrossRef]

60. Mahonen, A.P. Cytokinin signaling and its inhibitor AHP6 regulate cell fate during cascular development. *Science* **2006**, *311*, 94–98. [CrossRef]

61. Shi, Q.; Dong, Y.; Qiao, D.; Wang, Q.; Ma, Z.; Zhang, F.; Zhou, Q.; Xu, H.; Deng, F.; Li, Y. Isolation and characterization of *ZmERF1* encoding ethylene responsive factor-like protein 1 in popcorn (*Zea mays* L.). *Plant Cell Tissue Organ Cult.* **2015**, *120*, 747–756. [CrossRef]

62. Zhang, Z.; Huang, R. Enhanced tolerance to freezing in tobacco and tomato overexpressing transcription factor *TERF2/LeERF2* is modulated by ethylene biosynthesis. *Plant Mol. Biol.* **2010**, *73*, 241–249. [CrossRef]

63. Street, I.H.; Aman, S.; Zubo, Y.; Ramzan, A.; Wang, X.; Shakeel, S.N.; Kieber, J.J.; Eric Schaller, G. Ethylene inhibits cell proliferation of the arabidopsis root meristem. *Plant Physiol.* **2015**, *169*, 338–350. [CrossRef] [PubMed]

64. Vaseva, I.I.; Qudeimat, E.; Potuschak, T.; Du, Y.; Genschik, P.; Vandenbussche, F.; Van Der Straeten, D. The plant hormone ethylene restricts *Arabidopsis* growth via the epidermis. *Proc. Natl. Acad. Sci. USA* **2018**, *115*, E4130–E4139. [CrossRef] [PubMed]

65. Li, J.; Zhao, Y.; Chu, H.; Wang, L.; Fu, Y.; Liu, P.; Upadhyaya, N.; Chen, C.; Mou, T.; Feng, Y.; et al. SHOEBOX modulates root meristem size in rice through dose-dependent effects of gibberellins on cell elongation and proliferation. *PLoS Genet.* **2015**, *11*, e1005464. [CrossRef] [PubMed]

66. Shimada, A.; Ueguchi-Tanaka, M.; Nakatsu, T.; Nakajima, M.; Naoe, Y.; Ohmiya, H.; Kato, H.; Matsuoka, M. Structural basis for gibberellin recognition by its receptor GID1. *Nature* **2008**, *456*, 520–523. [CrossRef] [PubMed]

67. Wang, B.; Wei, H.; Xue, Z.; Zhang, W.H. Gibberellins regulate iron deficiency-response by influencing iron transport and translocation in rice seedlings (*Oryza sativa*). *Ann. Bot.* **2017**, *119*, 945–956. [CrossRef] [PubMed]

68. Sekimoto, H.; Hoshi, M.; Nomura, T.; Yokota, T. Zinc deficiency affects the levels of endogenous gibberellins in *Zea mays* L. *Plant Cell Physiol.* **1997**, *38*, 1087–1090. [CrossRef]

69. Schöttler, M.A.; Tóth, S.Z.; Boulouis, A.; Kahlau, S. Photosynthetic complex stoichiometry dynamics in higher plants: Biogenesis, function, and turnover of ATP synthase and the cytochrome b6f complex. *J. Exp. Bot.* **2015**, *66*, 2373–2400. [CrossRef]

70. Grouneva, I.; Gollan, P.J.; Kangasjärvi, S.; Suorsa, M.; Tikkanen, M.; Aro, E.M. Phylogenetic viewpoints on regulation of light harvesting and electron transport in eukaryotic photosynthetic organisms. *Planta* **2013**, *237*, 399–412. [CrossRef]

71. Krohling, C.A.; Eutrópio, F.J.; Bertolazi, A.A.; Dobbss, L.B.; Campostrini, E.; Dias, T.; Ramos, A.C. Ecophysiology of iron homeostasis in plants. *Soil Sci. Plant Nutr.* **2016**, *62*, 39–47. [CrossRef]

72. Escudero-Almanza, D.J.; Ojeda-Barrios, D.L.; Hernández-Rodríguez, O.A.; Sánchez Chávez, E.; Ruíz-Anchondo, T.; Sida-Arreola, J.P. Carbonic anhydrase and zinc in plant physiology. *Chil. J. Agric. Res.* **2012**, *72*, 140–146. [CrossRef]

73. Muneer, S.; Ryong, B. Silicon decreases Fe deficiency responses by improving photosynthesis and maintaining composition of thylakoid multiprotein complex proteins in soybean plants (*Glycine max* L.). *J. Plant Growth Regul.* **2015**, *34*, 485–498. [CrossRef]

74. Andaluz, S.; López-Millán, A.-F.; De las Rivas, J.; Aro, E.-M.; Abadía, J.; Abadía, A. Proteomic profiles of thylakoid membranes and changes in response to iron deficiency. *Photosynth. Res.* **2006**, *89*, 141–155. [CrossRef] [PubMed]

75. Higuchi, K.; Saito, A.; Mikami, Y.; Miwa, E. Modulation of macronutrient metabolism in barley leaves under iron-deficient condition. *Soil Sci. Plant Nutr.* **2011**, *57*, 233–247. [CrossRef]

76. López-Millán, A.F.; Grusak, M.A.; Abadía, A.; Abadía, J. Iron deficiency in plants: An insight from proteomic approaches. *Front. Plant Sci.* **2013**, *4*, 1–7. [CrossRef] [PubMed]

77. Briat, J.F.; Curie, C.; Gaymard, F. Iron utilization and metabolism in plants. *Curr. Opin. Plant Biol.* **2007**, *10*, 276–282. [CrossRef] [PubMed]

78. Kesten, C.; Menna, A.; Sánchez-Rodríguez, C. Regulation of cellulose synthesis in response to stress. *Curr. Opin. Plant Biol.* **2017**, *40*, 106–113. [CrossRef]

79. Kurek, I.; Kawagoe, Y.; Jacob-Wilk, D.; Doblin, M.; Delmer, D. Dimerization of cotton fiber cellulose synthase catalytic subunits occurs via oxidation of the zinc-binding domains. *Proc. Natl. Acad. Sci. USA* **2002**, *99*, 11109–11114. [CrossRef]

80. Lin, X.Y.; Ye, Y.Q.; Fan, S.K.; Jin, C.W.; Zheng, S.J. Increased sucrose accumulation regulates iron-deficiency responses by promoting auxin signaling in Arabidopsis plants. *Plant Physiol.* **2016**, *170*, 907–920. [CrossRef]

81. Kobayashi, T. Understanding the complexity of iron sensing and signaling cascades in plants. *Plant Cell Physiol.* **2019**, *60*, 1440–1446. [CrossRef]

82. Nozoye, T. The nicotianamine synthase gene is a useful candidate for improving the nutritional qualities and Fe-deficiency tolerance of various crops. *Front. Plant Sci.* **2018**, *9*, 1–7. [CrossRef]

83. Mizuno, D.; Higuchi, K.; Sakamoto, T.; Nakanishi, H.; Mori, S.; Nishizawa, N.K. Three nicotianamine synthase genes isolated from maize are differentially regulated by iron nutritional status. *Plant Physiol.* **2003**, *132*, 1989–1997. [CrossRef] [PubMed]

84. Klatte, M.; Schuler, M.; Wirtz, M.; Fink-Straube, C.; Hell, R.; Bauer, P. The analysis of arabidopsis nicotianamine synthase mutants reveals functions for nicotianamine in seed iron loading and iron deficiency responses. *Plant Physiol.* **2009**, *150*, 257–271. [CrossRef] [PubMed]

85. Curie, C.; Alonso, J.M.; Le Jean, M.; Ecker, J.R.; Briat, J.F. Involvement of NRAMP1 from *Arabidopsis thaliana* in iron transport. *Biochem. J.* **2000**, *347*, 749–755. [CrossRef] [PubMed]

86. Castaings, L.; Caquot, A.; Loubet, S.; Curie, C. The high-affinity metal transporters NRAMP1 and IRT1 team up to take up iron under sufficient metal provision. *Sci. Rep.* **2016**, *6*, 1–11. [CrossRef] [PubMed]

87. Xiong, H.; Kobayashi, T.; Kakei, Y.; Senoura, T.; Nakazono, M.; Takahashi, H.; Nakanishi, H.; Shen, H.; Duan, P.; Guo, X.; et al. *AhNRAMP1* iron transporter is involved in iron acquisition in peanut. *J. Exp. Bot.* **2012**, *63*, 4437–4446. [CrossRef] [PubMed]

88. Tiwari, M.; Sharma, D.; Dwivedi, S.; Singh, M.; Tripathi, R.D.; Trivedi, P.K. Expression in *Arabidopsis* and cellular localization reveal involvement of rice NRAMP, OsNRAMP1, in arsenic transport and tolerance. *Plant Cell Environ.* **2014**, *37*, 140–152. [CrossRef]

89. Rellán-Álvarez, R.; Giner-Martínez-Sierra, J.; Orduna, J.; Orera, I.; Rodríguez-Castrilln, J.Á.; García-Alonso, J.I.; Abadía, J.; Álvarez-Fernández, A. Identification of a tri-Iron(III), tri-citrate complex in the xylem sap of iron-deficient tomato resupplied with iron: New insights into plant iron long-distance transport. *Plant Cell Physiol.* **2010**, *51*, 91–102. [CrossRef]

90. Saridis, G.; Chorianopoulou, S.N.; Ventouris, Y.E.; Bouranis, D.L.; Sigalas, P.P. An exploration of the roles of ferric iron chelation-strategy components in the leaves and roots of maize plants. *Plants* **2019**, *8*, 133. [CrossRef]

91. Brown, J.C. Fe and Ca uptake as related to root-sap and stem-exudate citrate in soybeans. *Physiol. Plant.* **1966**, *19*, 968–976. [CrossRef]

92. Lopez-Millan, A.F.; Morales, F.; Abadia, A.; Abadia, J. Effects of iron deficiency on the composition of the leaf apoplastic fluid and xylem sap in sugar beet. Implications for iron and carbon transport. *Plant Physiol.* **2000**, *124*, 873–884. [CrossRef]

93. Ariga, T.; Hazama, K.; Yanagisawa, S.; Yoneyama, T. Chemical forms of iron in xylem sap from graminaceous and non-graminaceous plants. *Soil Sci. Plant Nutr.* **2014**, *60*, 460–469. [CrossRef]

94. Salt, D.E.; Prince, R.C.; Baker, A.J.M.; Raskin, I.; Pickering, I.J. Zinc ligands in the metal hyperaccumulator *Thlaspi caerulescens* as determined using x-ray absorption spectroscopy. *Environ. Sci. Technol.* **1999**, *33*, 713–717. [CrossRef]

95. Yang, J.; Wang, M.; Li, W.; He, X.; Teng, W.; Ma, W.; Zhao, X.; Hu, M.; Li, H.; Zhang, Y.; et al. Reducing expression of a nitrate-responsive bZIP transcription factor increases grain yield and N use in wheat. *Plant Biotechnol. J.* **2019**, *1*, 1–11. [CrossRef] [PubMed]

96. Li, S.; Liu, X.; Zhou, X.; Li, Y.; Yang, W.; Chen, R. Improving zinc and iron accumulation in maize grains using the zinc and iron transporter ZmZIP5. *Plant Cell Physiol.* **2019**, *60*, 2077–2085. [CrossRef]

97. Thomine, S.; Lelièvre, F.; Debarbieux, E.; Schroeder, J.I.; Barbier-Brygoo, H. AtNRAMP3, a multispecific vacuolar metal transporter involved in plant responses to iron deficiency. *Plant J.* **2003**, *34*, 685–695. [CrossRef]

98. Lanquar, V.; Lelièvre, F.; Bolte, S.; Hamès, C.; Alcon, C.; Neumann, D.; Vansuyt, G.; Curie, C.; Schröder, A.; Krämer, U.; et al. Mobilization of vacuolar iron by AtNRAMP3 and AtNRAMP4 is essential for seed germination on low iron. *EMBO J.* **2005**, *24*, 4041–4051. [CrossRef]

99. Barrieu, F.; Chaumont, F.; Chrispeels, M.J. High expression of the tonoplast aquaporin *ZmTIP1* in epidermal and conducting tissues of maize. *Plant Physiol.* **1998**, *117*, 1153–1163. [CrossRef]

100. Martinoia, E.; Maeshima, M.; Neuhaus, H.E. Vacuolar transporters and their essential role in plant metabolism. *J. Exp. Bot.* **2007**, *58*, 83–102. [CrossRef]

101. Shen, C.; Yue, R.; Sun, T.; Zhang, L.; Yang, Y.; Wang, H. OsARF16, a transcription factor regulating auxin redistribution, is required for iron deficiency response in rice (*Oryza sativa* L.). *Plant Sci.* **2015**, *231*, 148–158. [CrossRef]

102. Jagadeeswaran, G.; Li, Y.F.; Sunkar, R. Redox signaling mediates the expression of a sulfate-deprivation-inducible microRNA395 in arabidopsis. *Plant J.* **2014**, *77*, 85–96. [CrossRef]

103. Lewis, B.E.; Mason, Z.; Rodrigues, A.V.; Nuth, M.; Dizin, E.; Cowan, J.A.; Stemmler, T.L. Unique roles of iron and zinc binding to the yeast Fe-S cluster scaffold assembly protein "Isu1. " *Metallomics* **2019**, *11*, 1820–1835. [CrossRef] [PubMed]

104. Kong, W.W.; Yang, Z.M. Identification of iron-deficiency responsive microRNA genes and *cis*-elements in *Arabidopsis*. *Plant Physiol. Biochem.* **2010**, *48*, 153–159. [CrossRef] [PubMed]

105. Thirugnanasambantham, K.; Durairaj, S.; Saravanan, S.; Karikalan, K.; Muralidaran, S.; Islam, V.I.H. Role of ethylene response transcription factor (ERF) and its regulation in response to stress encountered by plants. *Plant Mol. Biol. Report.* **2015**, *33*, 347–357. [CrossRef]

106. Parry, G.; Estelle, M. Auxin receptors: A new role for F-box proteins. *Curr. Opin. Cell Biol.* **2006**, *18*, 152–156. [CrossRef] [PubMed]

107. Mallikarjuna, M.G.; Nepolean, T.; Hossain, F.; Manjaiah, K.M.; Singh, A.M.; Gupta, H.S. Genetic variability and correlation of kernel micronutrients among exotic quality protein maize inbreds and their utility in breeding programme. *Indian J. Genet. Plant Breed.* **2014**, *74*, 166–173. [CrossRef]

108. Nozoye, T.; Nakanishi, H.; Nishizawa, N.K. Characterizing the crucial components of iron homeostasis in the maize mutants *ys1* and *ys3*. *PLoS ONE* **2013**, *8*, e62567. [CrossRef]

109. Brazma, A.; Hingamp, P.; Quackenbush, J.; Sherlock, G.; Spellman, P.; Stoeckert, C.; Aach, J.; Ansorge, W.; Ball, C.A.; Causton, H.C.; et al. Minimum information about a microarray experiment (MIAME) - Toward standards for microarray data. *Nat. Genet.* **2001**, *29*, 365–371. [CrossRef]

110. Gentile, A.; Dias, L.I.; Mattos, R.S.; Ferreira, T.H.; Menossi, M. MicroRNAs and drought responses in sugarcane. *Front. Plant Sci.* **2015**, *6*, 58. [CrossRef]

111. Wu, Z.; Irizarry, R.A.; Gentleman, R.; Martinez-Murillo, F.; Spencer, F. A model-based background adjustment for oligonucleotide expression arrays. *J. Am. Stat. Assoc.* **2004**, *99*, 909–917. [CrossRef]

112. Kauffmann, A.; Gentleman, R.; Huber, W. arrayQualityMetrics - A bioconductor package for quality assessment of microarray data. *Bioinformatics* **2009**, *25*, 415–416. [CrossRef]

113. Smyth, G.K. limma: Linear models for microarray data. In *Bioinformatics and Computational Biology Solutions Using R and Bioconductor*; Springer-Verlag: New York, NY, USA, 2005; pp. 397–420. ISBN 1431-8776.

114. Tello-Ruiz, M.K.; Naithani, S.; Stein, J.C.; Gupta, P.; Campbell, M.; Olson, A.; Wei, S.; Preece, J.; Geniza, M.J.; Jiao, Y.; et al. Gramene 2018: Unifying comparative genomics and pathway resources for plant research. *Nucleic Acids Res.* **2018**, *46*, D1181–D1189. [CrossRef] [PubMed]

115. Ge, S.X.; Jung, D.; Jung, D.; Yao, R. ShinyGO: A graphical gene-set enrichment tool for animals and plants. *Bioinformatics* **2020**, *36*, 2628–2629. [CrossRef] [PubMed]

116. Jin, J.; Tian, F.; Yang, D.C.; Meng, Y.Q.; Kong, L.; Luo, J.; Gao, G. PlantTFDB 4.0: Toward a central hub for transcription factors and regulatory interactions in plants. *Nucleic Acids Res.* **2017**, *45*, D1040–D1045. [CrossRef] [PubMed]

117. Dai, X.; Zhuang, Z.; Zhao, P.X. PsRNATarget: A plant small RNA target analysis server (2017 release). *Nucleic Acids Res.* **2018**, *46*, W49–W54. [CrossRef] [PubMed]

118. Shannon, P.; Markiel, A.; Ozier, O.; Baliga, N.S.; Wang, J.T.; Ramage, D.; Amin, N.; Schwikowski, B.; Idekar, T. Cytoscape: A software environment for integrated models of biomolecular interaction networks. *Genome Res.* **2003**, *13*, 2498–2504. [CrossRef] [PubMed]

119. IDT Server. Available online: https://www.idtdna.com (accessed on 10 April 2018).

120. Thirunavukkarasu, N.; Hossain, F.; Mohan, S.; Shiriga, K.; Mittal, S.; Sharma, R.; Singh, R.K.; Gupta, H.S. Genome-wide expression of transcriptomes and their co-expression pattern in subtropical maize (*Zea mays* L.) under waterlogging stress. *PLoS ONE* **2013**, *8*, e70433. [CrossRef]
121. Van Gioi, H.; Mallikarjuna, M.G.; Shikha, M.; Pooja, B.; Jha, S.K.; Dash, P.K.; Basappa, A.M.; Gadag, R.N.; Rao, A.R.; Nepolean, T. Variable level of dominance of candidate genes controlling drought functional traits in maize hybrids. *Front. Plant Sci.* **2017**, *8*. [CrossRef]

Publisher's Note: MDPI stays neutral with regard to jurisdictional claims in published maps and institutional affiliations.

Article

Zinc Plant Uptake as Result of Edaphic Factors Acting

Vyacheslav Sergeevich Anisimov *, Lydia Nikolaevna Anisimova and Andrey Ivanovich Sanzharov

Russian Institute of Radiology and Agroecology 1, Kievskoe sh., 109th km, Kaluga Region, 249032 Obninsk, Russia; lanisimovan@list.ru (L.N.A.); ais_55@mail.ru (A.I.S.)
* Correspondence: vsanisimov@list.ru

Abstract: The influence of soil characteristics on the lability and bioavailability of zinc at both background and phytotoxic concentrations in Albic Retisol soil (Loamic, Ochric) was studied using various methods. Ranges of insufficient, non-phytotoxic, and phytotoxic zinc concentrations in soil solutions were established in an experiment with an aqueous barley culture. It was experimentally revealed that for a wide range of non-toxic concentrations of Zn in the soil corresponding to the indicative type of plant response, there was constancy of the concentration ratio (CR) and concentration factor (CF) migration parameters. As a result, a new method for assessing the buffer capacity of soils with respect to Zn (PBC_{Zn}) is proposed. The transformation processes of the chemical forms and root uptake of native (natural) zinc contained in the Albic Retisol (Loamic, Ochric) through the aqueous culture of barley were studied using a cyclic lysimetric installation and radioactive ^{65}Zn tracer. The distribution patterns of $Zn(^{65}Zn)$ between different forms (chemical fractions) in the soil were established using the sequential fractionation scheme of BCR. The coefficients of distribution and concentration factors of natural Zn and ^{65}Zn, as well as accumulation and removal of the metal by plants were estimated. The values of the enrichment factor of natural (stable) Zn contained in sequentially extracted chemical fractions with the ^{65}Zn radioisotope were determined and the amount of the pool of labile zinc compounds in the studied soil was calculated.

Keywords: zinc; ^{65}Zn; soil; soil solution; barley; lability; specific activity; potential buffer capacity; forms; labile zinc pool

Citation: Anisimov, V.S.; Anisimova, L.N.; Sanzharov, A.I. Zinc Plant Uptake as Result of Edaphic Factors Acting. *Plants* **2021**, *10*, 2496. https://doi.org/10.3390/plants10112496

Academic Editor: Gokhan Hacisalihoglu

Received: 25 October 2021
Accepted: 15 November 2021
Published: 18 November 2021

Publisher's Note: MDPI stays neutral with regard to jurisdictional claims in published maps and institutional affiliations.

1. Introduction

Increases in the concentrations of heavy metals (HMs) in soils as a result of technogenic pollution lead to negative effects in agricultural ecosystems such as crop losses, deterioration in the quality of agricultural products, and decreases in soil microbiological activity. However, among a wide range of pollutants, Zn deserves special attention for a number of reasons

First of all, zinc is one of the 17 elements necessary for plants, with an average content of 0.002% (20 ppm) in dry vegetative mass [1]. At the same time, it is also an important trace element—zinc deficiency negatively affects the growth and maturation of plants [1,2], which leads to crop losses and even, in the most severe cases, plant death. Zinc deficiency is often caused not only by the low metal content in the soil but also by the influence of the type of soil that determines its availability to plants. [2,3]. Thus, it was noted in [4] that Zn deficiency is observed in all soils with a low availability of Zn, which primarily include calcareous and highly phosphated soils with high pH.

In general, about 50% of the world's soils contain insufficient amounts of Zn [5]. Zn deficiency has been observed in significant amounts of soil in Bangladesh, Brazil, Pakistan, the Philippines, Sudan, sub-Saharan Africa, India, Turkey, Western Australia, and China, as well as on the Great Plains and in the western regions of the United States [2,6].

At the same time, zinc is the most common among heavy metals (HMs) polluting soil as a result of anthropogenic impact [1,7–10]. The most significant sources of pollution are the use of fertilizers and sewage sludge in agricultural practice, emissions and discharges from the mining and metallurgical industries, and road transport [11,12]. At high concentrations of zinc in the soil, various cytotoxic effects are manifested and lead to decreases in the biomass and yield of agricultural plants.

Thus, the study of zinc behavior in soils is of particular interest. In addition, due to its high mobility in soils and lack of ability to change the degree of oxidation in the environment, Zn is a convenient object for studying HM migration.

Zn's mobility in soils and availability to plants strongly depend on the ratio of the different forms (chemical fractions) of the element in soils [5,8,12,13].

To date, a large number of methods for determining the fractional composition of HMs in soils have been developed. All of them use specific reagents to isolate individual groups of HM compounds in soils (exchangeable, easily soluble, and associated with organo–mineral complex compounds, organic matter, carbonates, Fe, Mn oxides, etc.). Using schemes of the sequential chemical fractionation of HMs [4,10,14–23], it is possible to assess their lability and bioavailability, the distribution by fractions of the main groups of compounds and minerals in soils (exchange associated with oxides of Al, Fe, Mn, carbonates, phosphates, sulfides, organic matter, and crystal matrixes of soil minerals), the lability and bioavailability of HMs under changing soil conditions (pH, redox potential, humidity, salinity, etc.).

However, all of the above methods only allow researchers to solve the main task with a modest degree of approximation—to estimate the real pool of labile and bioavailable HM compounds. In fact, a more accurate solution of this problem is only possible when using the isotopic dilution method [11,23–28].

This method is based on the law of ideal isotope exchange [29] of ions of applied radioactive or stable isotopes acting as tracers and ions of native (stable) isotopes of the HMs under study in a soil–soil solution–plant system. The isotopic dilution method allows one to calculate the amount of reserves (pools) in the soil: (a) of the total number of labile compounds of the studied HM ("*E*-value" (E_{HM})) and (b) the total number of biologically available compounds of HM ("*L*-value" (L_{HM})).

Additionally, the isotopic dilution method, being a source of valuable data on the total amount of stocks (pools) of labile and biologically available HM compounds in soil, does not allow for the obtainment of information regarding potential lability in a soil–soil solution system and the bioavailability of various forms of HMs. To solve the abovementioned problem, it is necessary to apply an integrated approach, including the joint use of methods of isotopic dilution and subsequent fractionation of soil into which a stable or radioactive HM tracer has been previously introduced, followed by the analysis of isotope ratios for individual chemical fractions.

An important aspect of the problem of interaction of technogenic HMs (including Zn) with soils is the latter's potential for the specific and non-specific sorption of heavy metals due to the presence of soil minerals, the high molecular weight organic compounds, occluding processes, co-deposition, and the action of other mechanisms. These processes lead to the immobilization of HMs in soils. Their total effect, which can be quantified, is manifested in the form of the buffer capacity of soils in relation to HMs [16,30–32].

The objectives of this work were:

(1) To clarify the issue with different models of plant behavior in conditions of Zn pollution and to evaluate the buffer capacity of the soil with respect to Zn using barley as a test plant.

(2) Determine the pool of labile compounds of native Zn using ^{65}Zn as a radioactive tracer and assess the potential lability and bioavailability of individual forms of the metal in the soil under conditions as close as possible to equilibrium. For this purpose, a special flow lysimetric installation of cyclic action was developed and tested in practice.

2. Results and Discussion

2.1. Experiment I

The studied Albic Retisol (Loamic, Ochric) soil was found to be coarsely textured, poor in available potassium, weakly acidic, contained a small amount of organic carbon, and had a low cation exchange capacity. However, the content of biologically available phosphorus was found to be high (the consequences of the abundant phosphating of agricultural land during the Soviet era). As such, it was an infertile low buffer soil that was not located in the best natural conditions.

More interesting is the behavior of zinc—one of the most important trace elements for living organisms and the most common pollutant –when it enters soil in increased quantities [33].

Before the experiment, a total content of zinc in the soil were determined in the samples (Table 1).Because the soil did not contain carbonates, the results obtained for the exchange of Ca^{2+} and Mg^{2+} (5.20 $\pm$ 0.06 and 0.40 $\pm$ 0.09 cmol(+) kg^{-1}, respectively) with the use of AAB-4.8 as an extractant did not significantly differ from the results obtained using neutral salts as extractants: 1 M NH_4Cl (5.70 $\pm$ 0.68 and 0.53 $\pm$ 0.05 cmol(+) kg^{-1}) and AAB-7.0 (5.21 $\pm$ 0.08 and 0.47 $\pm$ 0.10 cmol(+) kg^{-1}). Therefore, extraction with AAB-4.8 was used to estimate the total pool of exchange (non-specific bound) forms of alkali, alkaline earth elements and zinc. In addition, the exchange forms of HMs extracted by AAB-4.8 from soils in the extracting solution were capable of forming weak complexes of $HM(Ac)_n$ that prevented the hydrolysis and re-deposition of metals [32].

Table 1. Main characteristics of Albic Retisol (Loamic, Ochric) soil (mean $\pm$ standard deviation).

Parameter	Value
Mass fraction of particles (mm) in soil, %	
1–0.25	35.08
0.25–0.05	15.64
0.05–0.01	30.88
0.01–0.005	5.20
0.005–0.001	7.30
<0.002	8.75
<0.001	5.89
Exchangeable cation content, cmol (+) kg^{-1}	
Ca^{2+}	5.20 $\pm$ 0.06
Mg^{2+}	0.40 $\pm$ 0.09
K^+	0.15 $\pm$ 0.01
pH_{KCl}	5.05 $\pm$ 0.01
pH_{water}	6.04 $\pm$ 0.01
C_{org}, %	1.0 $\pm$ 0.01
Total acidity (TA), cmol (+) kg^{-1} soil	1.89 $\pm$ 0.02
Total exchangeable bases (S), cmol (+) kg^{-1}	5.3 $\pm$ 0.2
Labile P_2O_5, mg kg^{-1} (Kirsanov method)	126.9 $\pm$ 1.9
Mass fraction of total Zn in native soil, mg kg^{-1}	37.1 $\pm$ 2.8

Extraction using the same reagent to obtain non-specifically related compounds of various macro- and microelements was convenient in practical terms and created conditions for the in-depth study of the relationship between links in the soil–soil solution–plant migration chain. In addition, the validity of the determination of non-specifically bound (exchange) forms of HMs in soils by extracting them with neutral salt solutions at pH 7.0 is doubtful due to the tendency of multicharged transition element cations to hydrolyze with repeated precipitation [32,34].

The mass fraction of "labile" forms of zinc in the studied soil linearly increased with the amount of metal introduced (Figure 1). At the same time, the relative content of the labile form of Zn with the amount of metal introduced into the soil non-linearly increased in accordance with a power dependence. Thus, in the native soil, the proportion of "labile (or accessible to plants)" Zn from the total metal content was equal to 33%. With increases in the dose of Zn introduced into the soil, the proportion of its "labile" forms from the total amount of metal in the soil increased to 75%.

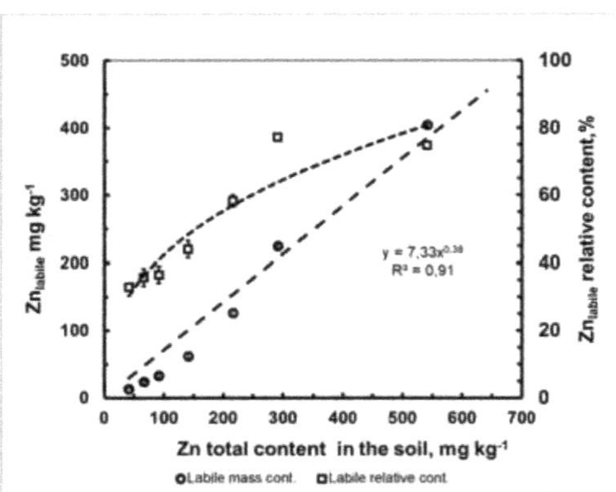

Figure 1. The change in the content of the labile form of Zn in Albic Retisol (Loamic, Ochric) soil (mean value ± standard deviation; $n = 3$). Data are from the vegetation experiment I.

The study of the features of Zn migration in the soil–barley system under the conditions of a vegetative experiment with increasing amounts of introduced Zn is important for the purposes of predicting the behavior of metals in agroecosystems, e.g., in the case of technogenic contamination with metal in a water-soluble form. With the help of such experiments, it was possible for us to obtain a general picture of the concentration dependence of $[Zn]_{plant} = f[Zn]_{soil}$ without investigating the essence of the involved migration mechanisms. In our case, the obtained dependence of zinc accumulation in barley straw on the total amount of metal in the soil presented the form of a straight line in a wide range of metal concentrations in the soil from 38 to 538 mg kg^{-1}, with a proportionality coefficient, called the "Concentration Ratio (CR)", equal to 3.8 (Figure 2a). This indicator depended on both the properties of the soil and the individual characteristics of zinc uptake by plants.

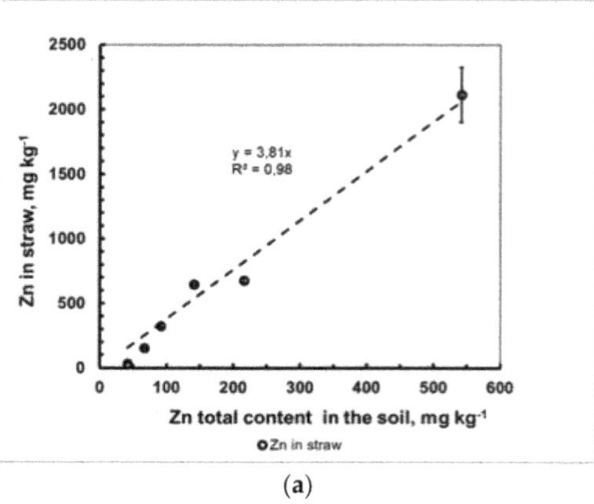

(a)

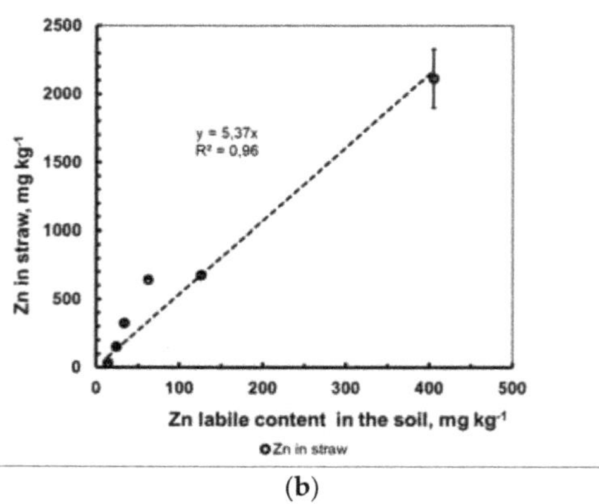

(b)

Figure 2. Dependence of the mass concentration of Zn in barley straw on the total amount (a) and labile form (b) of the metal in the soil (mean value ± standard deviation; $n = 3$).

So why does the $[Zn]_{plant} = f[Zn]_{soil}$ dependence present a straightforward character? This was firstly due to the rectilinear nature of the dependence $[Zn]_{soil,labile} = f[Zn]_{soil,total}$, $(R^2 = 0.98)$ (Figure 1) and secondly to the universal nature of the biological mechanisms of Zn absorption involved in a wide range of non-toxic metal concentrations in the soil. The relationship between the accumulation level Zn (and any other HMs) in the plant and its content in the soil can be considered a response of the plant. If there is a directly proportional relationship between the metal content in the soil and its accumulation in the biomass of plants, then this type of response is called "indicative" [35]. Attention is drawn to the fact that the concentration range of zinc is large, even in the low-fertility and low-buffer soil under study.

Figure 2b shows the dependence of the concentration of Zn in barley straw on the content of metal in the labile form in the soil. This relationship was found to be directly proportional, although slightly less pronounced than the previous case $(CR_{Zn,labile} = 5.4, R^2 = 0.96)$. Studying the dependence of $[Zn]_{plant} = f[Zn]_{soil,labile}$ allowed us to more objectively assess the contribution of other soil characteristics (such as pH and cationic composition) to the migration ability of HMs based only on the content of labile form of the metal in different soils. This was due to the fact that part of labile form (α) of the total soil metal content, as shown in the $[Zn]_{soil,labile} = \alpha[Zn]_{soil,total}$ expression, can differ for different soils many times with the same amount of $[Zn]_{soil,total}$ and, accordingly, make an additional contribution to the overall variance of the $[Zn]_{plant}$ dependent variable.

The $[Zn]_{grain} = f[Zn]_{soil,total}$ and $[Zn]_{grain} = f[Zn]_{soil,labile}$ dependencies presented, respectively, in Figure 3a,b, were obviously non-linear. They were satisfactorily approximated by a power function. The presence of such a dependence indicates the involvement of different mechanisms during the translocation of Zn from vegetative mass into the economically valuable part of the crop—grain. In addition, at a 250 mg kg^{-1} dose of introduced zinc (Zn_{250}), grain formation did not occur, and at a dose of Zn_{500}, even the development of generative organs was not observed and the plants themselves did not survive to the phase of the beginning of earing.

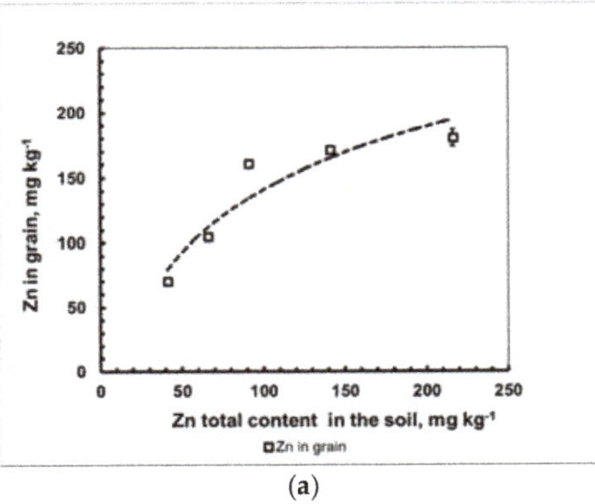

(a)

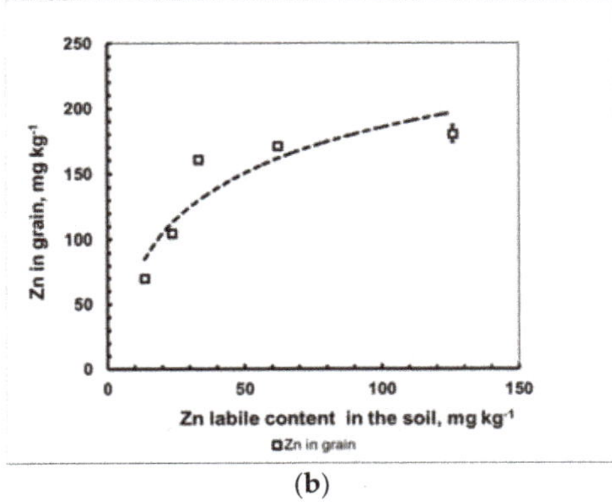

(b)

Figure 3. Dependence of the mass concentration of Zn in barley grain on the total amount (**a**) and labile form content (**b**) of the metal in the soil (mean value $\pm$ standard deviation; *n* = 3).

The concentration dependence between the amount of the labile form of Zn in the soil and the content of Zn in the phytomass of barley was most clearly demonstrated by the data obtained for relatively young 21-day-old plants (Figure 4a,b). The dependences were found to be directly proportional, and their angular coefficients (CR_{Zn}) were equal to 4.02 and 5.68, respectively; although these were close to the values shown in Figure 2 for straw, they were still 5–10% more, which indicated a slight decrease in the concentration of Zn in the straw compared to the dry biomass of the 21-day-old plants. Perhaps this was due to the effect of the "biological dilution" of Zn in the vegetative mass of barley in the phenophases following the tillering phase (21 days) and with leaf litter at maturity.

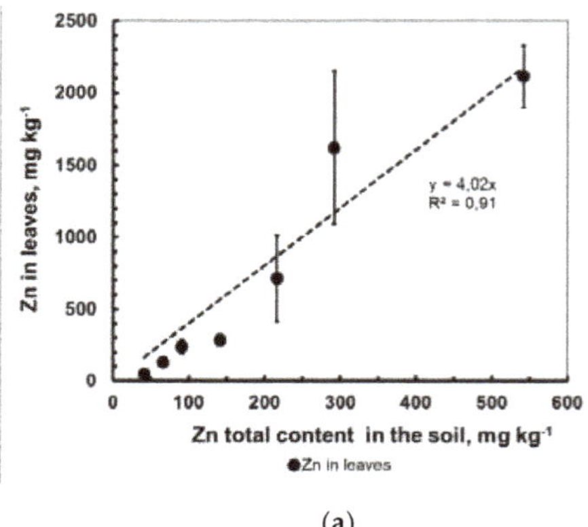

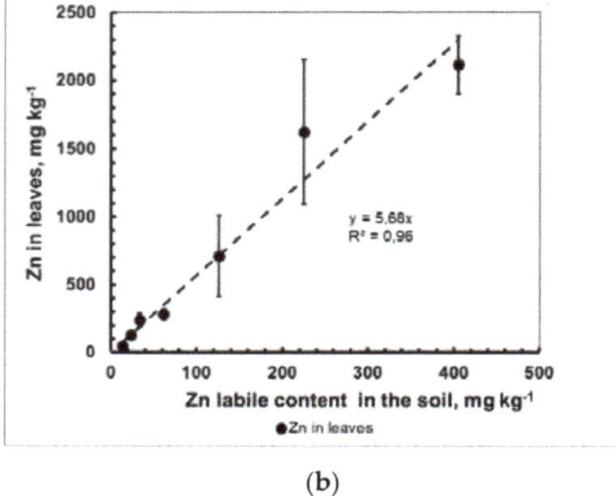

(a)   (b)

Figure 4. Dependence of the mass concentration of Zn in the phytomass of 21-day-old barley plants on the total amount (**a**) and labile form (**b**) of the metal in the soil (mean value $\pm$ standard deviation; n = 3).

Knowing the patterns of Zn behavior in the soil–plant system in a wide range of concentrations provided us to an effective tool for predicting the accumulation of metal in agricultural products under conditions of technogenic pollution. However, without a detailed study of the liquid phase of the soil–soil solution, knowledge of the role of various mechanisms in the accumulation of HMs by plants will be incomplete. Thus, due to the almost ubiquitous increased content of Zn in soil due to technogenesis, the "zero" point (without introducing Zn) in experiments with native agricultural soils will most likely be in the area of sufficient concentrations with an indicative type of plant response. In this case, it will not be possible to establish the patterns of Zn behavior in low concentrations zones, where the accumulative type of plant response is formed. Nutrient solutions representing an aqueous extract from the corresponding soils are ideal for clarifying migration patterns in areas of low concentrations of Zn in long-term vegetation experiments.

2.2. Experiment II

Vegetation experiments with a water culture of barley were carried out for a fairly long period of time—21 days. As a nutrient solution, a soil solution extracted from the studied soil was used with a narrow soil:solution ratio = 1:2. Increasing amounts of $Zn(NO_3)_2$ were added to various soil solution batches. No additional nutritious elements were added to the solution, except for a small amount of nitrogen in the form of $Ca(NO_3)_2$ (based on 0.2 mM N (NO_3^-)) due to the need to adjust the nitrate content according to each variant with a Zn concentration of 114 μmol L^{-1} (the metal concentration of 430 μmol L^{-1} turned out to be highly phytotoxic). The obtained results regarding the accumulation of zinc in plant roots and phytomass are presented in the form of a diagram in Figure 5a.

It is obvious that with a relatively low concentration of zinc in the soil solution ($<5.84 \times 10^{-3}$ mM), an accumulative type of plant response to the metal content was observed, characteristic of the lack of an element [35]. With a similar type of plant response, the kinetic curve of metal accumulation by roots from a nutrient solution (obtained using the water culture method) was satisfactorily described by a function resembling the power-law $y = a \times x^{1/2}$ in appearance but representing a superposition of two functions: asymptotic, described by the Michaelis–Menten enzymatic catalysis equation [36], and linear [6,37,38]. Moreover, the role of the latter was found to increase with increases in the concentration of Zn in the solution. This indicated the dominance of at least two transmembrane transfer systems in the region of very low and low concentrations of Zn in the substrate, differing in the degree of affinity to metal and using both highly specific carrier proteins with respect to Zn, mainly ZIP family, and other less specific ones, the number of which is limited in the cell [1,37,39]. Due to its participation in the process of the transmembrane transfer of metal into the roots of highly specific carrier proteins, we found an increased accumulation of Zn compared to the amount of metal that the plant would be able to assimilate as a result of electrochemical diffusion processes. The linear component was primarily due to the absorption of metal ions by plant apoplast and electrochemical diffusion [37,40].

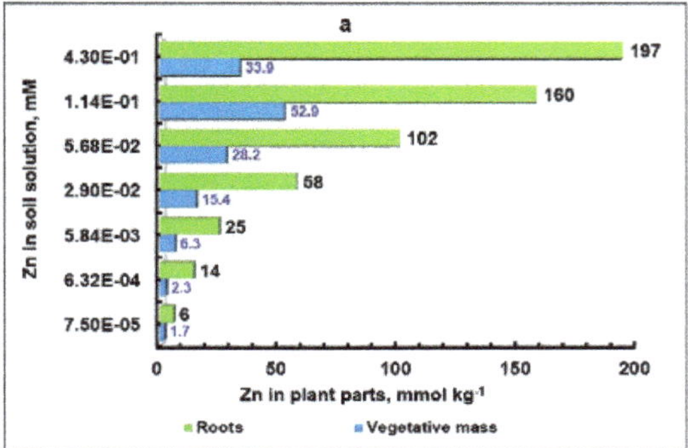
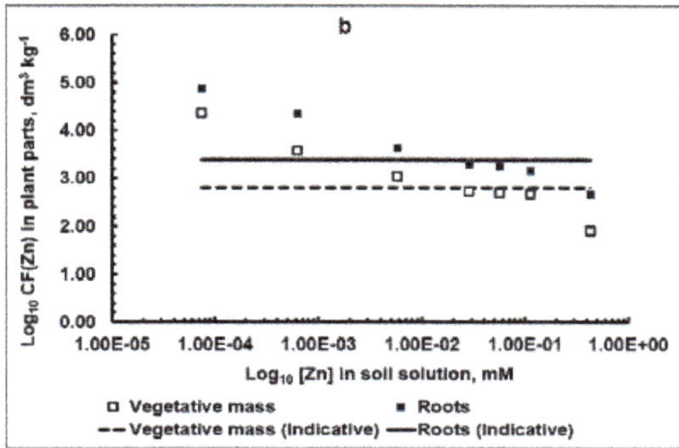

Figure 5. The relationship between the concentration of Zn in the soil solution and its content in different parts of 21-day-old barley plants (in terms of dry biomass) (**a**); values of $CF_{Zn,roots}$ and $CF_{Zn,VPs}$, expressed for clarity in logarithmic form (**b**). Straight lines denote constant values in the "indicative" region $CF_{Zn,roots} = 2375$ and $CF_{Zn,VPs} = 640$ dm^3 kg^{-1}.

Corresponding to the specified range of Zn concentrations in the soil solution, the ranges of its concentrations in barley vegetative and root phytomass were found to be $1.7 \div 6.3$ and $6 \div 25$ mmol kg^{-1} of dry mass, respectively.

At higher concentrations of Zn in a soil solution, in the range from 5.84×10^{-3} to 1.14×10^{-1} mM, the processes of the non-specific transmembrane transport of Zn ions entering the root surface begin to prevail as a result of a convective moisture flow and electrochemical diffusion processes [1,7,40] and the absorption of metal ions by apoplast. In our case, it can be argued that this range coincided with the range of non-toxic (normal, or optimal [35,41]) concentrations of Zn in the soil solution. As already mentioned, this type of plant response is called "indicative" (there is a directly proportional relationship between the metal content in the substrate and its accumulation in the biomass of plants). Corresponding to the specified range of Zn concentrations in the soil solution, the ranges of its concentrations in the VPs and roots of barley were $6.3 \div 52.9$ and $25 \div 160$ mmol kg^{-1} of dry mass, respectively, and the values of the corresponding constant coefficients of proportionality, called "Concentration Factors (CF)", were equal to $CF_{Zn,roots} = 2375$ and $CF_{Zn,VPs} = 640$ L kg^{-1}, respectively ($CF_{Zn} = $ [Zn]$_{plant}$/[Zn]$_{soil\ solution}$) (Figure 5b).

Finally, in the area of high concentrations of Zn in the soil solution (1.14×10^{-1} mM $\div 4.30 \times 10^{-1}$ mM), the phytotoxicity of Zn for barley was manifested. In this case, the physiological mechanisms regulating the uptake and translocation of Zn in plants were seriously disrupted. The type of plant response could be characterized as barrier restrictive [35]. At the same time, there was a strong oppression of barley plants and their premature death.

Accordingly, it should be noted that the abovementioned indicative type of plant response [35] in relation to zinc is the most common in real conditions with man-made soil contamination via metal.

2.3. Assessment of Zn Mobility in the Soil–Plant System and Determination of the Inactivating (Buffering) Ability of the Soil in Case of Contamination

As already mentioned, the migration ability of Zn in a soil–plant system is influenced by biological factors related to the physiological characteristics of plants [8,39,41–45], as well as the edaphic factors that determine the inactivating ability or buffering of soils in relation to Zn and other pollutants [34,35,40].

In the case of an indicative type of plant response, for example, for the aboveground biomass of 21-day-old barley plants (Figure 4b), a simple ratio will be observed:

$$d[Zn]_{plant}/d[Zn]_{exch} = [Zn]_{plant}/[Zn]_{exch} = CR_{Zn,exch} = const \qquad (1)$$

and (Figure 5)

$$d[Zn]_{plant}/d[Zn]_{soil\ solution} = [Zn]_{plant}/[Zn]_{soil\ solution} = CF_{Zn} = const \qquad (2)$$

where $[Zn]_{plant}$ is the mass (or molar) fraction of HMs in the dry aboveground biomass of the plant and $CR_{Zn,labile}$ is the concentration ratio of HMs in the plant in terms of the mass (or molar) content of its "labile (available to plants)" form in the soil. As follows from Figure 4b, the value of $CR_{Zn,exch}$ was equal to the tangent of the slope angle of the linear section of the above dependence: 5.68.

Earlier [34,46], we proposed a methodological approach for assessing the inactivating ability of soils with respect to HMs in conditions of technogenic pollution by using test plants. It was based on the fact that plants are essentially a universal integrating link in the soil–plant migration chains of different HMs. First of all, they uptake these metals for a long time and from a sufficiently large soil volume; secondly, they demonstrate similar patterns of behavior in conditions of HM pollution. The applied pollutant was zinc. Since then, this approach has undergone some corrections and additions. Its current version is presented below.

According to the definition given in [31] (p. 34), "the buffering of the system of compounds of microelements of the soil horizon in relation to any chemical element is understood as the ability to maintain the level of concentration of the element in the soil solution of this horizon at a constant level when the element level changes from the outside."

As a measure of the potential buffering capacity (*PBC*) of soils in relation to a chemical element (for example, potassium), F. Beckett [47] proposed the use of the quotient between the change in the concentration of an exchange-sorbed element in the soil (Q, capacity factor) and the activity of the cations of the element under study in a quasi-equilibrium extracting (soil) solution a_{Me}, normalized for the total activity of macronutrient cations a_{Ca+Mg} (I, intensity factor) in a wide range of concentrations of the element in the soil, with the exception of extremely low concentrations at which deviations from the law of ion exchange are observed. In a formal form, a similar dependence for Zn^{2+} ions can be written as the following equation:

$$PBC_{Zn} = \frac{Q}{I} = \frac{\Delta[Zn]_{exch}}{\Delta AR} \qquad (3)$$

where $\Delta[Zn]_{exch}$ is the amount of metal in the exchangeably sorbed state that the soil absorbed from the equilibration solution or, conversely, gave into the equilibration solution in comparison to the initial content of exchangeably sorbed Zn in the soil; ΔAR is the ratio of the activities of the cations of the metal under study (a_{Zn}) and the cations of the macroelements Ca^{2+} and Mg^{2+} (a_{Ca+Mg}) in a quasi–equilibrium soil solution at different concentrations of Zn in this solution. In the graphical representation of the dependence $\Delta AR - \Delta[Zn]_{exch}$, the tangent of the slope angle of the linear part of the obtained dependence characterizes PBC_{Zn}.

Equation (3) follows from the law of acting masses for exchanging cations in the system cation exchange complex (CEC)—a quasi-equilibrium soil solution. Thus, the equivalent exchange of divalent zinc cations on negatively charged surfaces of soil particles for Ca^{2+} and Mg^{2+} cations (which in a sum of about 90% of the exchange positions in CEC [48]) can be expressed using the following exchange reaction equation:

$$[Ca + Mg]_{exch} + [Zn^{2+}] \leftrightarrow [Zn]_{exch} + [Ca^{2+} + Mg^{2+}] \qquad (4)$$

where $[Ca + Mg]_{exch}$ and $[Zn]_{exch}$ are the exchange cations in the composition of CEC, mmol kg^{-1} (determined using AAB-4.8 before the negative reaction to Ca^{2+} ions); $[Ca^{2+} + Mg^{2+}]$, and $[Zn^{2+}]$ are the same cations in the composition of the soil solution, mM. The proposal to collectively consider Ca^{2+} and Mg^{2+} ions in the soil–soil solution system as cations of the same type, denoting them with the symbol of the predominant Ca cation, was formulated by Beckett [49,50] with reference to the studies of other authors and their own previously obtained results.

Accordingly, the ratio of exchanging cations in the CEC phase and the soil solution can be expressed using Equation (5). The form of this equation shown below reflects the fact that in the case of an equivalent exchange—for example, $Zn/(Ca + Mg) - AR = AR_{Zn,(Ca+Mg)} = [Zn^{2+}]/[Ca^{2+} + Mg^{2+}]$:

$$\frac{[Zn]_{exch,i}}{[Ca + Mg]_{exch,i}} = K^S_{Zn,(Ca+Mg)} \frac{[Zn^{2+}]_i}{\left[Ca^{2+} + Mg^{2+}\right]_i} \tag{5}$$

where $i = 1$ or 2 is the index characterizing the state of the system, a_{Ca+Mg} is the total activity in a quasi-equilibrium soil solution of Ca^{2+} + Mg^{2+} ions, and a_{Zn} is the activity of Zn^{2+} ions. If we express the concentrations of exchange-bound Zn in the soil for states "1" and "2" through Equation (5), then the change in the amount of the corresponding metal form in the CEC can be described using the following equation:

$$[Zn]_{exch,1} - [Zn]_{exch,2} = K^S_{Zn,(Ca+Mg)} \times \left(AR_2 \times [Ca + Mg]_{exch,2} - AR_1 \times [Ca + Mg]_{exch,1}\right) \tag{6}$$

Or, due to the low loading of the CEC with Zn^{2+} ions, it can be expressed as: $[Ca + Mg]_{exch,1} \approx [Ca + Mg]_{exch,2} = [Ca + Mg]_{exch}$ obtain:

$$\Delta[Zn]_{exch} = K^S_{Zn,(Ca+Mg)} \times [Ca + Mg]_{exch} \times \Delta AR \tag{7}$$

Or, introducing the designation $PBC_{Zn} = K^S_{Zn,(Ca+Mg)} \times [Ca + Mg]_{exch}$, it can be expressed as:

$$\Delta[Zn]_{exch} = PBC_{Zn} \times \Delta AR \tag{8}$$

Considering Expressions (5), (7), and (8), the potential buffering capacity of the soil for the linear part of the ion exchange sorption isotherm of Zn can be expressed as the following expressions:

$$PBC_{Zn} = K^S_{Zn,(Ca+Mg)} \times [Ca + Mg]_{exch} \tag{9}$$

$$PBC_{Zn} = \frac{[Zn]_{exch}}{AR_{Zn,(Ca+Mg)}} \tag{10}$$

$$PBC_{Zn} = \frac{[Zn]_{exch}}{\left[Zn^{2+}\right]/\left[Ca^{2+} + Mga^{2+}\right]} \tag{11}$$

$$PBC_{Zn} = [Ca + Mg]_{exch} \times \frac{K_{d,\,exch}(Zn)}{K_{d,\,exch}(Ca + Mg)} \tag{12}$$

where $K_{d,exch}(Zn) = [Zn]_{exch}/[Zn]_{soil\ solution}$ and $K_{d,exch}(Ca + Mg) = [Ca + Mg]_{exch}/[Ca^{2+} + Mg^{2+}]_{soil\ solution}$, which represent the distribution coefficients of Zn^{2+} and (Ca^{2+} + Mg^{2+}), respectively, between the exchange form in the soil (mmol kg^{-1}) and quasi-equilibrium soil solution (mM).

It follows that the above conclusions would be correct if the load of CEC with Zn^{2+} ions is higher than 1%. In this case, highly selective CEC sorption sites with respect to zinc ions will be completely blocked, and low-selective sites will remain available for ion exchange sorption, [51]. At the same time, the CEC loading of the studied cation should not exceed 5–10% (according to F. Beckett [50]). If the above conditions are met, the following would be observed: $K^S_{Zn,(Ca+Mg)}$ = const, where $K^S_{Zn,(Ca+Mg)}$ is the selectivity coefficient of ion exchange $Zn^{2+}/(Ca^{2+} + Mg^{2+})$ [52,53]. The specified range was determined using the ion exchange equilibrium method and amounted to 0.5–19.6% of the total CEC of the studied soil. The average value of $K^S_{Zn,(Ca+Mg)}$ was equal to 13.5 $\pm$ 6.6. It should also be noted that the range of non-toxic concentrations of zinc in the soil fully fit into this range of CEC loadings, which corresponded to the indicative uptake of Zn by barley.

Considering that in the range of non-toxic concentrations of Zn in the soil and, accordingly, the soil solution, the values of $CR_{Zn,exch}$ = $[Zn]_{plant}/[Zn]_{exch}$ and CF_{Zn} = $[Zn]_{plant}/[Zn]_{soil\ solution}$ were found to be constant, so we substituted them in the $K_{d,exch}(Zn)$ parameter in Equation (12).

We also considered that for VPs (indicator part of barley plants):

- $CR_{(Ca+Mg),exch}$ = *const* (because of the constant concentration of Ca and Mg in the soils, which have been applied with the increasing doses of Zn, i.e., $[Ca + Mg]_{exch}$ = *const* = 23.7 $\pm$ 4.4 mmol kg^{-1}).

- $CF_{(Ca+Mg)}$ = *const* (because the concentrations of Ca^{2+} and Mg^{2+} in the soil solution with an increasing concentration of Zn^{2+} was not changed, and the $[Ca + Mg]_{plant}/[Ca^{2+} + Mg^{2+}]_{soil\ solution}$ ratio remained constant in the experiment with the water culture of barley. It was found to be equal to 266 $\pm$ 91 dm^3 kg^{-1}.

Based on these arguments, we express the parameters $K_{d,exch}(Zn)$ and $K_{d,exch}(Ca + Mg)$ through the parameters $CR_{Zn,exch}$, $CR_{(Ca+Mg),exch}$, CF_{Zn}, and $CF_{(Ca+Mg)}$ in Equation (12).

We obtained the final form of the equation for calculating the potential buffer capacity of soils with respect to technogenic Zn:

$$PBC(V)_{Zn} = [Ca + Mg]_{exch} \times \frac{CF_{Zn} \times CR_{(Ca+Mg),exch}}{CF_{(Ca+Mg)} \times CR_{Zn,exch}} \tag{13}$$

With the help of this equation while knowing the parameters CF_{Zn} and $CF_{(Ca+Mg)}$ for a specific agricultural crop (they can be obtained from experiments with water crops), it is easy to calculate the values of the potential buffer capacity of any soil with respect to zinc, for example, from agroecological examination materials in agricultural lands where amounts of Ca, Mg, and Zn in soil (exchangeable form) and conjugate plant samples have been determined.

Thus, considering the sum of green leaves and stems as indicator part of barley plants, we obtained the following results of experiments I and II (Table 2).

Table 2. The determined values of the parameters used to calculate $PBC(V)_{Zn}$ (according to experiments 1 and 2).

Parameter	Value
Experiment I	
$[Ca + Mg]_{VPs}$, mmol kg^{-1}	526 $\pm$ 9
$[Ca + Mg]_{exch}$, mmol kg^{-1}	23.7 $\pm$ 4.4
$CR_{(Ca+Mg),\ exch}$	22.2
$CR_{Zn,\ exch}$	5.68
Experiment II	
$[Ca + Mg]_{VPs}$, mmol kg^{-1}	613 $\pm$ 197
$[Ca^{2+} + Mg^{2+}]_{soil\ solution}$, mM	1.81 $\pm$ 0.26
$CF_{(Ca+Mg)}$, dm^3 kg^{-1}	339
CF_{Zn} dm^3 kg^{-1}	640

After substituting the values of the corresponding parameters into Equation (13), we obtained the value of $PBC(V)_{Zn}$ for the studied Albic Retisol (Loamic, Ochric) soil:

$$PBC(V)_{Zn} = [Ca + Mg]_{exch} \times (CF_{Zn} \times CR_{(Ca+Mg),\ exch})/(CR_{Zn,exch} \times CF_{(Ca+Mg)}) = 23.7 \times (640 \times 22.2)/(5.68 \times 339) = 175\ \text{mmol kg}^{-1} \tag{14}$$

Thus, using $CF_{Me(VPs)}$, (Me = Ca + Mg, Zn) with values established in experiments with aquatic crops of agricultural plants, it is possible to quantify the buffering capacity of soils based on data on the content of a "labile (accessible to plants)" forms of Zn in these soils and the metal concentrations in the vegetative mass of plants.

In parallel, the PBC_{Zn} of the studied Albic Retisol (Loamic, Ochric) soil was determined with the classical ion exchange equilibrium method under static conditions using a modification of the Beckett method [47–50]. In this method, the concentration of the macronutrient cation—Ca^{2+} in the balancing solution, with an increase in the amount of the dissolved metal under study (Zn in our case)—remained unchanged and equal to 2 mM (20 mM in the Beckett method), which is close to the concentration of the macronutrient cation in the free (gravity) soil solution of the humus horizons of most soils. The value of PBC_{Zn} was determined to be 150 mmol kg^{-1}. This value was comparable to the $PBC(V)_{Zn}$ value, obtained using test plants, despite the fundamental difference between the two methods (the relative difference was 15%).

2.4. Experiment III

During the vegetation experiment with a flow lysimeter, it was found that the decrease in the concentration of stable Zn and the volumetric activity density of ^{65}Zn in the soil solution before and after the vegetation vessels over time was satisfactorily described by a power function with a negative indicator of the type $[Zn]_{soil\ solution} = a \times t^b$, where a and b are parameters and t is time in days (Figure 6a,b).

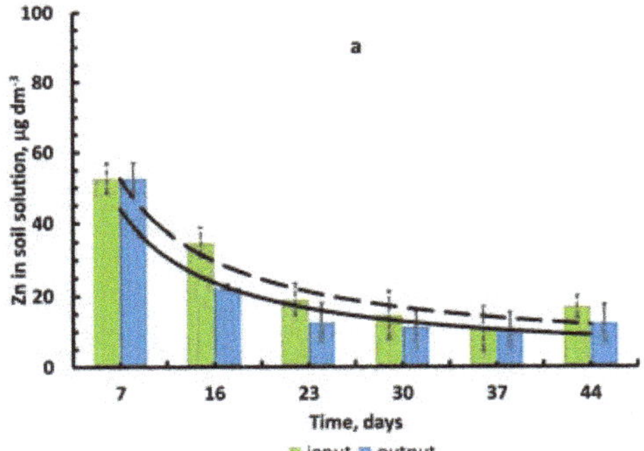

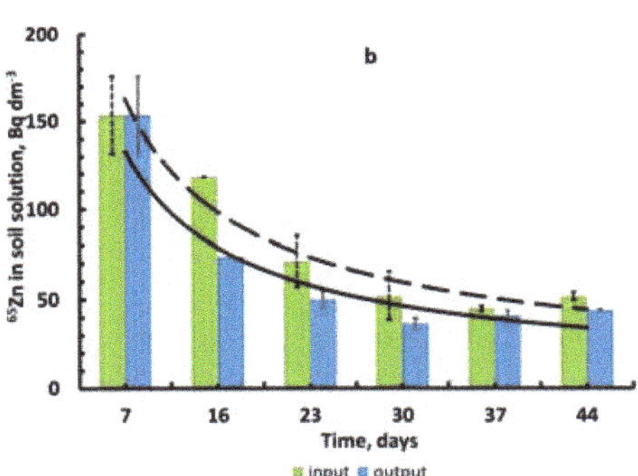

Figure 6. Experimental data for the soil solution in $[Zn]_{stable}$ in µg dm^{-3} (**a**); volumetric activity density (^{65}Zn) in Bq dm^{-3} (**b**) (mean value ± standard deviation; n = 3).

The values of parameters a and b and the determination coefficient (R^2) for Zn(^{65}Zn) in the soil solution at the output of the lysimeter (before the vegetation vessels) were: 263, −0.81 (0.85) and 666, −0.71 (0.90), respectively. In the soil solution after leaving the vegetation vessels and coarse filtration, they were: 262, −0.89 (0.90) and 609, −0.76 (0.90), respectively.

During the vegetative experiment, a significant decrease in the concentration of $[NO_3{}^-]$ (mg dm^{-3}) ions in the lysimetric solution was observed, while the average (at the input and output from the vegetative vessels) pH value increased from 5 to 7 (Figure 7a,b). The concentrations of $[K^+]$ and $[NH_4{}^+]$ ions in the solution also decreased during the vegetation experiment, respectively, from 40 ± 6 to 11 ± 0.5 mg dm^{-3} and from 6.0 ± 0.9 to 0.6 ± 0.1 mg dm^{-3}.

The values of the specific activity of ^{65}Zn/Zn in studied objects are important indicators for assessing the contribution of a particular form of zinc to the soil–soil solution–plant migration chain.

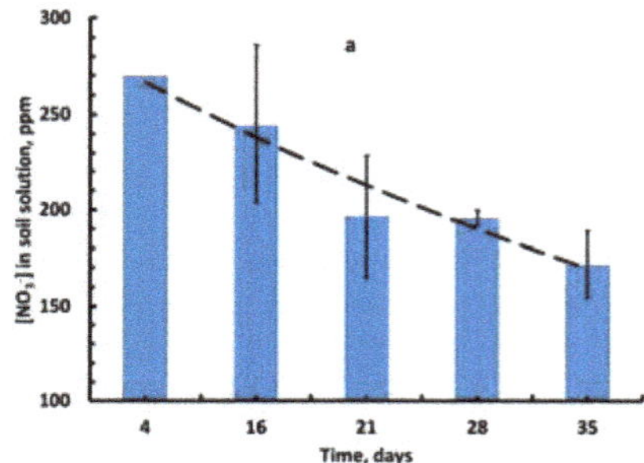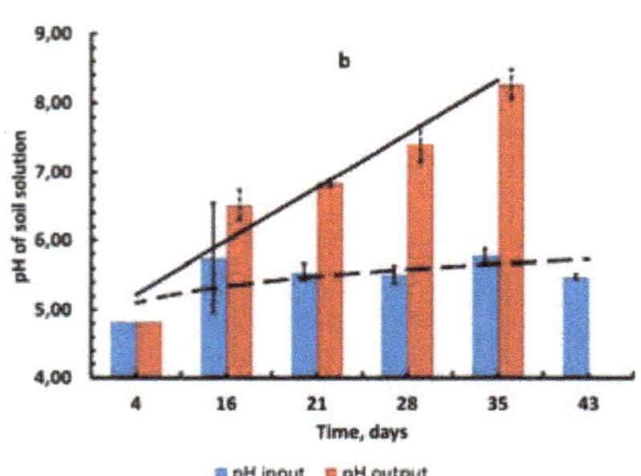

Figure 7. Concentration dynamics in soil solution (mean value ± standard deviation; $n = 3$): [NO$_3^-$] (**a**) and pH (**b**). Data on the 4th day were derived from single extractions.

Data on the dynamics of the specific activity of ^{65}Zn/Zn in the soil solution, vegetative parts (VPs), and roots of the test plant (barley) in terms of stable Zn contained in this solution (in Bq mg^{-1}) are shown in Figure 8a–c. The average values for the vegetation period of the corresponding parameters A$_{sp}$(^{65}Zn/Zn)$_{soil\ solution}$, A$_{sp}$(^{65}Zn/Zn)$_{VPs}$, A$_{sp}$(^{65}Zn/Zn)$_{roots}$ were, respectively, 3580 ± 390, 3210 ± 1250, and 3230 ± 780 Bq mg^{-1}. As a result of the alkalinization of the lysimetric solution, its extracting ability with respect to the "freshly applied" ^{65}Zn gradually decreased, although not as much as with respect to the less labile native zinc. This led to pronounced trends in increasing values of A$_{sp}$(^{65}Zn/Zn)$_{soil\ solution}$, A$_{sp}$(^{65}Zn/Zn)$_{VPs}$, and A$_{sp}$(^{65}Zn/Zn)$_{roots}$ during the growing season.

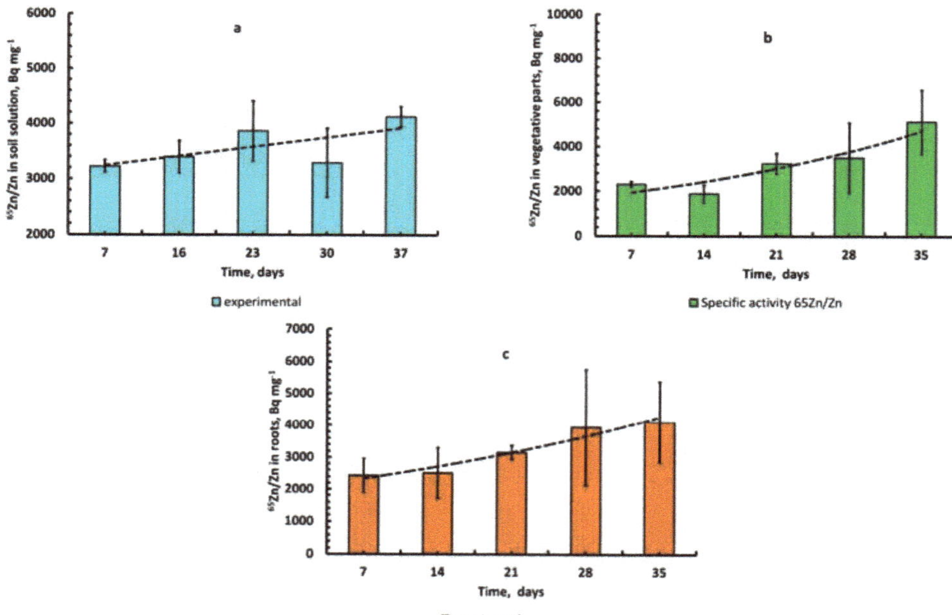

Figure 8. Experimental data for the specific activity: A$_{sp}$(^{65}Zn/Zn)$_{soil\ solution}$ (**a**), A$_{sp}$(^{65}Zn/Zn)$_{VPs}$ (**b**), and A$_{sp}$(^{65}Zn/Zn)$_{roots}$ (**c**), Bq×mg^{-1} (mean value ± standard deviation; $n = 3$).

The data obtained on the dynamics of the content of Zn(^{65}Zn) in the vegetative parts of barley (data for roots are not given) showed that the content of zinc in the plants generally increased during ontogenesis (Figure 9a,b).

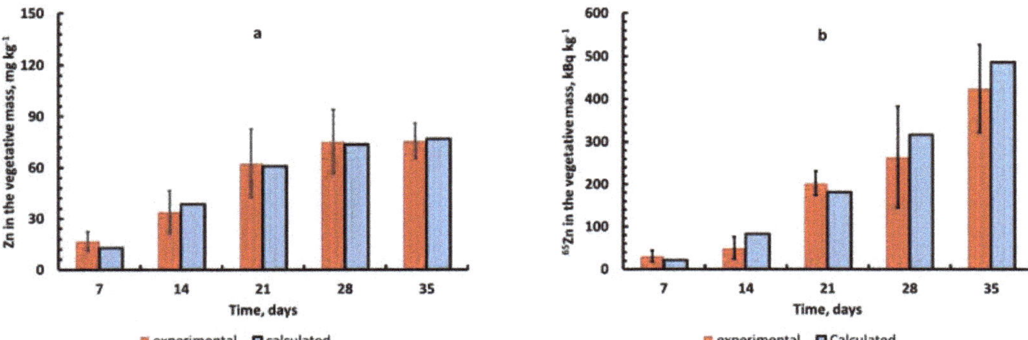

Figure 9. Dynamics of Zn concentration in mg kg^{-1} (**a**); mass activity concentration of ^{65}Zn (**b**) (mean values ± standard deviation; $n = 3$).

The values of the Zn(^{65}Zn) concentration factors grew with increasing plant age (Figure 10) due to both the accumulative effect when zinc is absorbed by plants and decreases in its content in the soil solution during the experiment. Due to the high variability of the data, it was not possible to identify a significant difference in the values of CF_{Zn} and CF_{Zn-65}.

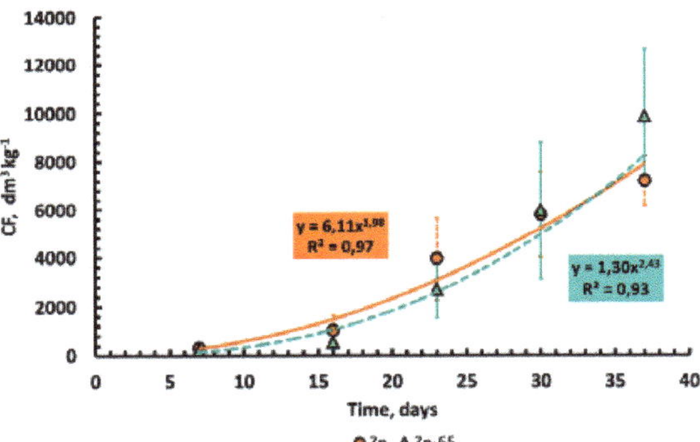

Figure 10. Dynamics of Zn(^{65}Zn) concentration factors: CF_{Zn}, and CF_{Zn-65}, dm^3 × kg^{-1} in the dry vegetative mass of barley (mean values ± standard deviation; $n = 3$).

In order to study the contribution of different soil forms to Zn content of the liquid phase of the soil in more detail, we used ^{65}Zn as a radioactive tracer when applying a sequential fractionation scheme of selected soil samples in accordance with the modified BCR method. The obtained results are presented in Figure 11a–c. A comparative analysis of the data showed the following ratios of different forms of stable Zn and radionuclide ^{65}Zn (data shown in parentheses) in soil in percent: I. 34.2 ± 3.4 (10.7 ± 0.4); II. 31.9 ± 3.7 (17.3 ± 0.1); III. 9.9 ± 2.6 (27.8 ± 2.3); and IV. 24.0 ± 7.7 (42.6 ± 3.9). According to Figure 11a,b, the relative contents of labile and conditionally labile forms of ^{65}Zn in the soil (Fractions I and II) significantly exceeded the content of the corresponding forms of the stable (natural) isotope Zn, respectively, by 3.2 and 1.8 times.

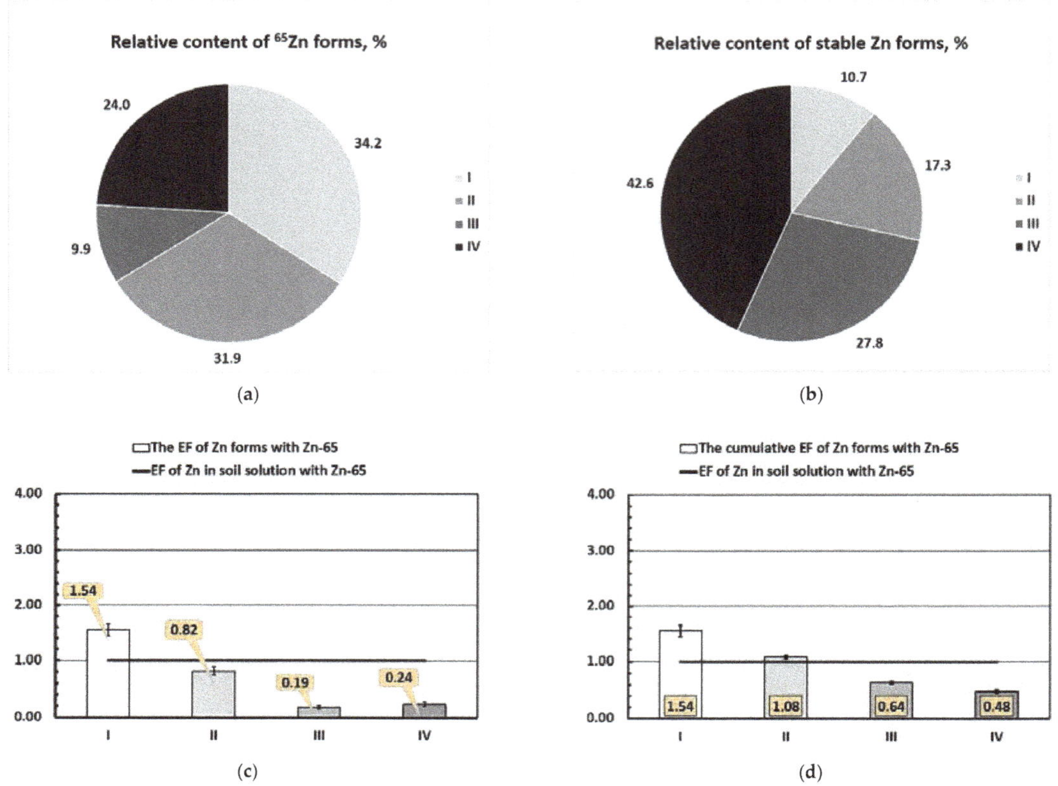

Figure 11. The relative content (%) of the forms of stable (natural) Zn (**a**) and ^{65}Zn (**b**) in Albic Retisol (Loamic, Ochric) soil, as determined by the modified BCR method [21]; the enrichment factors (*EF*) of Zn forms by ^{65}Zn in the soil (**c**); the cumulative EF of Zn forms by ^{65}Zn in the soil (**d**). Roman numerals I–IV indicate the forms of zinc in the soil, as given in Table 2.

At the same time, the values of the relative content of conditionally fixed and fixed forms of ^{65}Zn in the soil (fractions III–IV) were significantly lower than that of Zn by 2.8 and 3.0 times, respectively.

The *Enrichment Factor* values of the corresponding forms of natural Zn with the radioactive tracer ^{65}Zn in relation to the main component of our model system—the lysimetric soil solution at the time corresponding to the beginning of the growing experiment—were: 1.54 ± 0.11, 0.82 ± 0.07, 0.19 ± 0.03, and 0.24 ± 0.03 (Figure 11c). The value of $A_{sp}(^{65}Zn/Zn)_{soil\ solution}$ was equal to 3220 ± 110 Bq mg^{-1}. The ordinate 1.0 secant value was an EF $(^{65}Zn/Zn)_{soil\ solution}$.

The cumulative form of *EF* data obtained are represented as a curve (Figure 11d) on which each subsequent *EF* value is an integral value, where the ratio of the total specific activity of ^{65}Zn of this and previous fractions (Bq kg^{-1}) to the total mass fraction of the corresponding fractions of stable Zn (mg kg^{-1} of a soil) is normalized to the specific activity of ^{65}Zn/Zn in soil solution: $EF_{\Sigma Fr.\#}(^{65}Zn/Zn) = A_{sp}(^{65}Zn/Zn)_{\Sigma Fr.\#}/A_{sp}(^{65}Zn/Zn)_{soil\ solution}$. The values of *EF* for the sum of the relevant forms of natural Zn with tracer ^{65}Zn in relation to the soil solution were 1.54 ± 0.11 (Fraction I), 1.08 ± 0.03 (Σ of Fractions I–II), 0.64 ± 0.02 (Σ of Fractions I–III), and 0.48 ± 0.03 (Σ of Fractions I–IV). The secant value of the coordinate 1.00 was the EF $(^{65}Zn/Zn)_{soil\ solution}$.

2.5. Mobility of Native (Natural) Zn and Technogenic Zn in the Soil Solution–Plant System (According to Lysimetric Experience)

For stable Zn and radionuclide ^{65}Zn, it was found that their concentration and volumetric activity density in the soil solution decreased over time (Figure 6a,b), and at the input of vegetative vessels, the corresponding concentrations were higher than at the output from the vegetative vessels, which can be explained by the uptake of zinc by barley roots. The decrease in the concentration of stable Zn and the volumetric activity density of radionuclide ^{65}Zn in soil solutions over time were associated with both the depletion of the pool of water-soluble forms of metals in the soil as a result of continuous uptake by plant roots (not compensated for by desorption of Zn (^{65}Zn) from the cation exchange complex (CEC) into the soil solution) and the alkalinization of the soil solution by root secretions.

The decrease in the concentration of nitrate, as well as ammonium and potassium ions, in the soil solution (Figure 7a) can be described by exponential equations. It should be noted that the high concentration of nitrates led to an increase in the alkalinity of the quasi-equilibrium soil solution during the growing season (Figure 7b) due to the ability of barley to alkalize nutrient solutions in the light, releasing HCO_3^- or OH^- ions in the presence of a sufficient amount of nitrate ions and a low content of ammonium ions. This fact has been reported by many researchers [7,40,54]. It is caused by compliance with the principle of electroneutrality during the transmembrane transfer of anions and cations to the root symplast.

It was found that the content of Zn (^{65}Zn) in plants increased during ontogenesis (Figure 9a,b). This fact indicates that zinc is continuously accumulated in plants during the growth and development phases (before maturation) during ontogenesis, and this should be considered when comparing data on the metal contents in the vegetative mass of plants selected at different stages of development. We attribute a very significant increase in the values of CF_{Zn} and $CF_{Zn\text{-}65}$ during ontogenesis to both the accumulative effect of zinc plants uptake and a decrease in its content in the soil solution during the experiment (Figure 10).

2.6. Assessment of the Mobility of Natural Zn in the Soil–Soil Solution–Plant System Using the Radioisotope ^{65}Zn as a Tracer

We assumed that the isotope ^{65}Zn introduced into the soil as a radioactive label was not evenly distributed between the various forms of natural (stable) Zn, but it turned out to be primarily bound in the form of labile forms. In the future, its transformation into less labile forms should theoretically take place. This process is very long, so much so that it will not be possible to trace it for ^{65}Zn (during this time, the radionuclide will repeatedly decay). Nevertheless, it is possible to consider some theoretical aspects of the problem related to the direction and speed of transformation of ^{65}Zn forms in the soil.

It is known that in the case of the isotopic exchange of ^{65}Zn/Zn, we are dealing with an "ideal isotopic exchange" in which isotopic atoms that are identical in their physicochemical properties participate. In our case:

$$\text{Zn(soil form)} + {}^{65}\text{Zn}^{2+}\text{(soil solution)} \leftrightarrow {}^{65}\text{Zn(soil form)} + \text{Zn}^{2+}\text{ (soil solution)} \quad (15)$$

The processes of ideal isotope exchange are characterized by the absence of elemental (chemical) changes, as well as the immutability of the number of interacting particles and their concentrations. The reason for the spontaneous flow of ideal isotope exchange processes is only an increase in the entropy of the system, since the change in its enthalpy in this case will be zero ($\Delta H = 0$). From a physical point of view, the increase in the entropy of the system during isotope exchange corresponds to the transition of the system from a more ordered state (different amounts of the isotope ^{65}Zn are present in different forms) to a less ordered one (the isotope is evenly distributed between the forms involved in the isotope exchange process), which corresponds to the mixing of isotopes. When an equilibrium state occurs with an ideal isotope exchange, $\Delta G^0 = 0$. Accordingly, the equilibrium constant

$(K_p) = 1$ [29]. For the above equation of the ideal isotope exchange ^{65}Zn/Zn, this ratio looks like:

$$K_p = ([^{65}\text{Zn}]_{\text{soil}} \times [\text{Zn}]_{\text{soil solution}})/([\text{Zn}]_{\text{soil}} \times [^{65}\text{Zn}]_{\text{soil solution}}) = 1 \qquad (16)$$

This leads to the identity of the isotopic composition of the exchanging forms:

$$([^{65}\text{Zn}]/[\text{Zn}])_{\text{soil}} = ([^{65}\text{Zn}]/[\text{Zn}])_{\text{soil solution}} \qquad (17)$$

In the case of a non-equilibrium state (as in our case), we present the following inequality:

$$([^{65}\text{Zn}]/[\text{Zn}])_{\text{soil}} \neq ([^{65}\text{Zn}]/[\text{Zn}])_{\text{soil solution}} \qquad (18)$$

The value of $([^{65}\text{Zn}]/[\text{Zn}])_{\text{soil}}/([^{65}\text{Zn}]/[\text{Zn}])_{\text{soil solution}}$ is the same as $EF(^{65}\text{Zn}/\text{Zn})_{\text{soil}}$. The higher the value of EF $(^{65}\text{Zn}/\text{Zn})_{\text{soil}}$, the further away the soil is from the state of isotopic equilibrium between radioactive and naturally stable zinc isotopes present in the solid phase in all forms (labile, conditionally labile, and fixed) on the one hand and the water-soluble form on the other hand. Due to the insignificance of the values of $[^{65}\text{Zn}]_{\text{soil}}$ and $[^{65}\text{Zn}]_{\text{soil solution}}$, they are usually expressed by the values of mass and volumetric activity density, respectively, of the radionuclide: $A_m(^{65}\text{Zn})_{\text{soil}}$, $A_v(^{65}\text{Zn})_{\text{soil solution}}$.

The solution is the most important effective phase of soil [52] at the boundary of which with solid phases ion exchange processes occur (transformation function) and a certain quasi-equilibrium state is established, as described, for example, by Equation (15). However, due to its special properties, such as its high mobility (fluidity) and high rate of diffusion transfer of dissolved substances, it acts as a connecting link due to which ion-exchange reactions (for example, the zinc isotopes considered in this paper) indirectly occur between various forms localized in different parts of the solid phase of the soil (transport function). As a result, the total net process look such that the ^{65}Zn present in some forms in the solid phase will be predominantly desorbed into the soil solution, passing into other forms.

If $A_{sp}(^{65}\text{Zn}/\text{Zn})_{\text{Fr.\#}} > A_{sp}(^{65}\text{Zn}/\text{Zn})_{\text{soil solution}}$, i.e., the enrichment factor of any form of Zn in the soil with radionuclide ^{65}Zn relative to the soil solution is greater than 1, then the process of isotopic exchange of radionuclide between the corresponding form and the solution is shifted towards the latter. The place of $^{65}\text{Zn}^{2+}$ ions in the soil is occupied by ions of the natural stable isotopic carrier Zn^{2+} from soil solution. If $A_{sp}(^{65}\text{Zn}/\text{Zn})_{\text{Fr.\#}} < A_{sp}(^{65}\text{Zn}/\text{Zn})_{\text{soil solution}}$, the reverse process occurs. Thus, through the liquid phase, as already noted, there is an exchange of $\text{Zn}(^{65}\text{Zn})$ between competing binding sites in the solid phase of soils, forming with zinc ions the corresponding chemical forms. The concentration of [Zn] and the volumetric activity density of ^{65}Zn in a quasi-equilibrium soil (lysimetric) solution reflect the contribution of it different forms in the studied soil.

Values of the enrichment factor of the labile chemical Zn fraction (Fraction I) via the radioactive tracer ^{65}Zn (EF $(^{65}\text{Zn}/\text{Zn})_{\text{Fr.I}} = 1.54 > 1$, while EF $(^{65}\text{Zn}/\text{Zn})_{\text{Fr.II}} = 0.82$, EF $(^{65}\text{Zn}/\text{Zn})_{\text{Fr.III}} = 0.19$, and EF $(^{65}\text{Zn}/\text{Zn})_{\text{Fr.IV}} = 0.24$ were less then 1 (Figure 11c). This allowed us to assume that in the soil–soil solution system from chemical Fraction I, $^{65}\text{Zn}^{2+}$ ions were predominantly desorbed into the soil solution, and for fractions II–IV, in contrast, were sorbed from the solution, which gradually led to a decrease in $A_{sp}(^{65}\text{Zn}/\text{Zn})$ for the first fraction and an increase for the second group of fractions.

Based on the obtained results and the assumption of the increasing ability of extractants used in the sequential extraction of zinc ions [20], it was possible to calculate the pool of labile zinc in the unit of soil mass—the "E-value": $(E_{\text{Zn}}) = C(\text{Zn})_{\text{Fr.I}} = 3.95 \pm 0.16$ mg kg^{-1} (or $10.7 \pm 0.4\%$). The corresponding specific activity value was ^{65}Zn $(A_{sp}(^{65}\text{Zn}/\text{Zn})_{\text{Fr.I}})$ equal to 4660 ± 830 Bq $\times$ mg^{-1}.

Attention is drawn to the fact that the average value during the growing season $A_{sp}(^{65}\text{Zn}/\text{Zn})_{\text{soil solution}} = 3580 \pm 390$ Bq mg^{-1} $< A_{sp}(^{65}\text{Zn}/\text{Zn})_{\text{Fr.I}}$. This allowed us to assume the existence of a slow process of isotopic exchange between mobile and other (*"conditionally labile"*, *"conditionally fixed"*, and *"fixed"*) forms of zinc by means of the liquid phase of the studied soil–soil solution. It was possible to fix this moment thanks to a long process of preliminary equilibration of ^{65}Zn with the studied soil.

Later, during the vegetation experiment, as a result of the vital activity of plants, the soil solution was depleted with zinc and alkalized, which led to an increase in the value of $A_{sp}(^{65}Zn/Zn)_{\text{soil solution}}$ to values close to $A_{sp}(^{65}Zn/Zn)_{\text{Fr.I}}$ (Figure 9a), which actually reflected the ratio of ^{65}Zn/Zn in labile and *"conditionally labile"* forms in the soil and their contribution to the composition of a quasi-equilibrium soil solution.

For balance calculations and the study of the transformation of the forms of $Zn(^{65}Zn)$ in the soil, data on the total removal of both natural and radioactive zinc isotopes from the soil–soil solution system by vegetative parts and barley roots are of particular interest (recall that five plant selections were made in total). We found that the percentage of the total amount of $Zn(^{65}Zn)$ contained in the soil was insignificant and amounted to only 0.34 (0.70)%. Consequently, the removal of metal by plants had no noticeable effects on the ratio of the forms of zinc in the soil.

3. Materials and Methods

The behavior of Zn in a soil–plant system was studied in vegetation experiments with soil culture with increasing zinc concentration in the soil (greenhouse conditions), in vegetation experiments with water culture with increasing zinc concentration in a soil nutrient solution (greenhouse conditions), and in a lysimetric experiment with an aqueous culture with the sole application of radioactive ^{65}Zn to the soil (laboratory conditions).

Barley (*Hordeum vulgare* L.) of the Zazersky-85 variety and soddy-podzolic sandy loam cultivated soil—Albic Retisol (Loamic, Ochric) not containing free carbonates selected from the arable horizon of agricultural land near Obninsk (Kaluga region, Russia) were selected as our objects of research.

3.1. Experiment with the Soil Culture of Barley

During the research, most attention was paid to the root uptake and redistribution of Zn between different parts of barley plants. To do this, 5 kg of air-dry Albic Retisol (Loamic, Ochric) soil were placed into vessels. Nutrients were applied to the soil in the form of aqueous solutions of KH_2PO_4 and K_2SO_4 salts at rates of 100 mg kg^{-1} P and K, respectively, and Zn was applied in the form of nitrate solutions.

The amount of Zn applied to the soil was as follows: 0, 25, 50, 100, 175, 250 and 500 mg kg^{-1}. The amount of nitrogen introduced with $Zn(NO_3)_2$ was adjusted according to the variant with the sub-maximum dose of Zn using NH_4NO_3 (but no more than 1 g N/vessel). After applying salt solutions, the soils in the vessels were incubated for 30 days at a temperature of 20–23 °C. Twenty five barley seeds were sown in each vessel. The plants on the green mass were harvested at 7, 14, 21, 30, 45 and 70 days after sowing. The experiment was carried out in triplicate.

The experiment was carried out at a temperature of 20–29 °C, relative humidity of 55–75%, and soil mass moisture content of 60% of the full water capacity (FWC).

Exchangeable potassium, calcium, and magnesium were extracted from soils using different reagents: 1 M NH_4Cl (pH 6.5); 1 M CH_3COONH_4, pH 7.0 (AAB-7.0); and 1 M CH_3COONH_4, pH 4.8 (AAB-4.8) [55], respectively, before a negative reaction to Ca^{2+} ions. The physical and chemical parameters of the studied soil were determined with conventional methods [55,56]: pH_{KCl} (pH_{water}) was determined with a potentiometric method in a suspension of soil in 1 M solution of KCl (distilled water) with a ratio of solid and liquid phases (S:L) = 1:2.5 (1:25 for peat soil), the granulometric composition of soil was determined with the pipette method of N.A. Kachinsky [48,55], humus content was determined by means of bichromatic oxidation by the Tyurin method, hydrolytic acidity was determined by means of the Kappen method, the sum of exchange bases was determined by means of the Kappen–Gilkovitz method, and the content of labile forms of

P$_2$O$_5$ was determined by means of by the Kirsanov method in the modification of TSINAO (0.2 M HCl, S:L = 1:5).

The combination of easily and difficult-to-exchange forms of Zn(^{65}Zn) [18,34,57] was extracted using 1 M CH$_3$COONH$_4$ at pH 4.8 with the modified method proposed by N.G. Zyrin [32]—by successive, exhaustive extractions to a negative reaction to Ca^{2+}. The potential buffering capacity of soils with respect to Zn was determined by the Beckett method [47,49,50] using 0.01 M CaCl$_2$.

3.2. Experiment with Water Culture of Barley

Our experiments were carried out on constantly aerated and mixed nutrient solutions extracted from the studied soil at a narrow soil:solution ratio = 1:2, in which, in addition to the background concentration (0.07 µM), Zn was added in the form of nitrate at the following concentrations: 0, 0.63, 5.8, 29, 57, 114 and 430 µM. In the soil solutions, the amount of applied Zn(NO$_3$$^-$)$_2$ of N(NO$_3$$^-$) was adjusted according to the variant with a 114 µM dose of HM using Ca(NO$_3$)$_2$ ("puriss") at the rate of 0.2 mM N(NO$_3$$^-$). The contents of macroelement cations in nutrient solutions were (in mM): 1.60 ± 0.05 Ca^{2+} (considering the added in the form of Ca(NO$_3$)$_2$); 0.16 ± 0.02 Mg^{2+}, and 0.38 ± 0.02 K$^+$. The pH value of the nutrient solutions was adjusted to 5.8, corresponding to the acidity of a solution containing 114 µM Zn^{2+}, using solutions of CitH$_3$ and NaOH. The total content of citric acid anion in nutrient solutions was 0.05 mM. The plants were grown for 21 days, and we changed the solution to a fresh one daily.

3.3. Lysimetric Experiment with Water Culture of Barley

The vegetation experiment was carried out using a special stand, including a flow lysimetric installation of an original design (Figures 12 and A1 in Appendix A). In addition to the cyclic lysimetric installation, which provided the gravity runoff of soil moisture, the stand included:

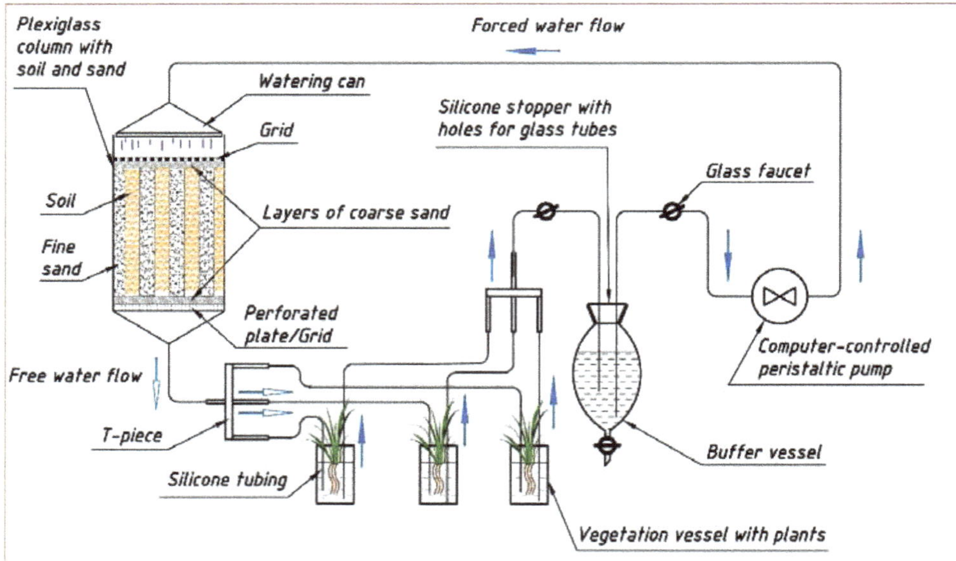

Figure 12. Structural and dynamic scheme of the stand used for studying the parameters of Zn(^{65}Zn) migration in the soil–soil solution–plant system (opened blue arrow is designated to soil solution before vegetation vessels (input), filled blue arrow—after them (output)).

- Flowing vegetative vessels with a soil solution in which sprouted plant seeds were placed on special stands in containers with a mesh bottom (nylon fabric) filled with coarse sand.
- A peristaltic pump.
- A system of tubes, taps, and adapter tees; a buffer tank.
- A lighting source for plants. From above, the flow lysimeter was non-hermetically covered with a plexiglass lid with a sprinkler device.

The preliminary preparation of the soil included the introduction of 200 kBq kg^{-1} of radionuclide ^{65}Zn ($T_{1/2}$ = 224 days) in the form of a working solution of ^{65}Zn(II) without an isotopic carrier such as stable Zn. The working solution was prepared from a sample solution of ^{65}Zn(II) in 0.1 M HCl ("CYCLOTRON Co., Ltd.", Obninsk, Russia), which contained 42.4 MBq ^{65}Zn at the time of certification. The mass activity density of the soil due to the presence of the radioisotope ^{65}Zn (Am^{65}Zn$_{soil}$) at the beginning of the vegetation experiment under consideration was 68,700 $\pm$ 2800 Bq kg^{-1}.

The resulting soil suspension after the application of the radionuclide was thoroughly mixed and incubated at room temperature for $\frac{1}{2}$ year, moistened twice a month, and dried in air (preventing complete drying). Then, the dried soil was ground and passed through a sieve with a diameter of 2 mm. The soil prepared in this way, containing ^{65}Zn, was placed in layers in a lysimetric installation with alternating layers of soil and washed quartz sand with a particle diameter of <1 mm as drainage (Figure 12). To prevent the colmatation of soil pores under the gravitational current of moisture, soil/sand layers were vertically oriented.

The total amount of deposited sand and soil was 5 kg each. The thickness of the layers was 2.5 cm. After that, the soil–soil solution system has been balanced in the lysimeter for 5 months by pouring deionized water onto the soil surface in the lysimeter and periodically returning the water flowing from the lysimeter back to the soil surface. Three weeks before the start of the vegetation experiment, a growing installation was assembled and water was added to the system up to a total volume of 5.5 dm^3 along with the nutrient solutions at the rates of 100 mg kg^{-1} N and K in the forms of NH$_4$NO$_3$ and K$_2$SO$_4$, respectively. Additional phosphorus was not applied, since the content of its labile forms in the soil was sufficient (Table 1). After assembly, the lysimetric installation and vegetative vessels were wrapped with a light-tight film. Then, the peristaltic pump was started and the system was left in operation for balancing for 3 weeks. Considering that the total volume of the liquid phase in the system was about 5.5. dm^3 and the rate of water supply by the peristaltic pump through the sprinkler device was 4 cm^3 per minute, the soil solution was subjected to complete regeneration, passing through the lysimeter, during the day.

The plants were grown in glass vegetation vessels with a volume of 2 dm^3 in triplicate. The volume of the soil lysimetric solution in each vessel was approximately 1.25 dm^3. The mixing of solutions was carried out by continuously bubbling air supplied to each vegetative vessel through thin silicone tubes using a low-power compressor. To compensate for evaporating moisture, the total volume of deionized water added daily to the soil surface in the lysimeter was 150–200 mL. In each vegetative vessel, a plastic separator was placed on stands with 6 large holes into which cartridges filled with large washed quartz sand (2–3 mm), closed from below with a nylon cloth, were inserted. Three barley seedlings sprouted within 3 days were planted in each cartridge. When barley roots appeared outside the cartridges, the latter were slightly raised above the water surface, ensuring that the roots were in the water. At the same time, the movement of water in the vegetative vessels caused by bubbling was sufficient to wet the substrate (sand) inside the cartridges. The duration of the vegetation experiment was 35 days. In total, 5 plant selections were made during the experiment each week (the 1st selection was double due to the small amount of plant material).

3.4. Determination of Zn(^{65}Zn) Forms (Chemical Fractions) in the Soil

Using AAB-4.8, "labile (available to plants)" forms of zinc were extracted from the soil using successive exhaustive extractions before a negative reaction to Ca^{2+} ions [32,35].

To determine the content of ^{65}Zn(Zn) associated with different organo–mineral fractions with the BCR method [21] (Table 3), the soil samples were preliminarily prepared.

Table 3. Sequental Extraction Procedure by BCR modified method [21].

#Form (Chemical Fraction of Zn(^{65}Zn)/Extraction with	Procedure
I. *Exchangeable and carbonate bound*/Acetic acid, 0.11 M	**I.** A sample of raw soil of known humidity (corresponding to 1 g of absolutely dry soil) without signs of gluing was placed in a 50 mL centrifuge tube. Then, 40 cm^3 of Solution A were added, the tube was closed with a lid, and the material was extracted by shaking for 16 h at 22.5 °C (or overnight) on a rotator. There was no delay between the addition of the extractant solution and the start of shaking. Then, the extract was separated from the solid precipitate by centrifugation at 3000 g for 20 min and the subsequent decantation of the supernatant into a volumetric glass flask (V = 100 mL) with a polished stopper. Next, 20 cm^3 of deionized water were added to the sediment, which was shaken for 15 min on a reciprocating shaker and centrifuged for 20 min at 3000 g, and the washing waters were separated by decantation and combined with the extract in a measuring flask. The solution in the flask was brought to the mark with deionized water, stirred, filtered through a 0.45 microns membrane filter, and analyzed for the content of Zn(^{65}Zn).
II. *Associated with reducible Fe–Mn oxides*/Hydroxylammonium Chloride (Hydroxylamine Hydrochloride), 0.5 M (pH 1.5, HNO$_3$, 2 M fixed vol.)	**II.** We added 40 cm^3 of freshly prepared Solution B to the remaining soil after stage (**I**) in a centrifuge tube (see above). The contents were mixed, achieving the complete dispersion of the residue by manual shaking. The centrifuge tube was closed with a lid, and the studied elements were extracted from the soil by mechanical shaking for 16 h at 22.5 °C (night). There was no delay between the addition of the extractant solution and the start of shaking. The procedure for separating the extract, washing the sediment, and preparing the analyte sample was performed in the same way as in step (**I**). It was necessary to carefully ensure that during the last operation we did not accidentally lose part of the solid residue.
III. *Associated with oxidizable organic matter and sulfides*/Solutions C and D. Solution C: Hydrogen peroxide, 300 mg g^{-1}, i.e., 8.8 M, stabilized HNO$_3$ to pH 2–3. Solution D: Ammonium acetate, 1.0 M, adjusted to pH 2.0 with HNO$_3$.	**III.** We carefully added 10 cm^3 of Solution C (in small aliquots to avoid losses due to a violent reaction) to the remainder of the soil in a centrifuge tube after stage (**II**). Then, we covered the tube with a lid (loosely) and kept it for 1 h at room temperature (shaking by hand periodically) to oxidize the organic components of the soil with hydrogen peroxide. Then, the oxidation was continued for another 1 h at 85 $\pm$ 2 °C in a water bath; during the first $\frac{1}{2}$ hour, centrifuge tubes with soil and extraction solution were periodically manually shaken. The volume of the contents in the test tube with the lid removed was evaporated to approximately V < 3 cm^3. Then, aliquots of Solution C with a volume of 10 cm^3 were repeatedly added to the contents of the centrifuge tube. We covered the tube with a lid (leaky) and again continued the oxidation of its contents for another 1 h at 85 $\pm$ 2 °C, periodically manually shaking the centrifuge tubes for the first $\frac{1}{2}$ hour. Then, we removed the lid and evaporated the liquid in the test tube to about V $\approx$ 1 cm^3, thus preventing the complete drying of the sample. We added 50 mL of Solution D to the cooled wet residue in the test tube and shook it for 16 h at a temperature of 22 $\pm$ 5 °C (or overnight). There was delay between the addition of the extractant solution and the start of shaking. The procedure for separating the extract, washing the sediment, and preparing the analysis sample was performed in the same way as in step (**I**).

Solution A. In a fume cupboard, we added 25 $\pm$ 0.2 cm^3 of glacial acetic acid to about 0.5 dm^3 of distilled water in a 1 dm^3 graduated polypropylene or polyethylene bottle and made up to 1 dm^3 with distilled water. We took 250 cm^3 of this solution (acetic acid, 0.43 M) and diluted it to 1 dm^3 with distilled water to obtain an acetic acid solution of 0.11 M.
Solution B. We dissolved 34.75 g of hydroxylammonium chloride in 400 cm^3 of distilled water. We transferred the solution to a 1 L volumetric flask, and added 25 cm^3 of 2 M HNO$_3$ (prepared by weighing from a suitable concentrated solution) by means of a volumetric pipette. We made up to 1 dm^3 with distilled water. We prepared this solution on the same day the extraction was carried out.
Solution C. 8.8 M water solution H$_2$O$_2$ (comprised 300 mg g^{-1} of hydrogen peroxide), was stabilized with HNO$_3$ to pH 2–3. It is recommended to use hydrogen peroxide acid-stabilized by the manufacturer to pH 2–3.
Solution D. We dissolved 77.08 g of ammonium acetate in 800 mL of distilled water and adjust the pH to 2.0 $\pm$ 0.1 with concentrated HNO$_3$ and made up to 1 L with distilled water.

Various forms of ^{65}Zn (Zn) were determined in the soil located in the lysimeter at moisture content (W = FMC (field moisture capacity)) just before the vegetation experiment. A mixed soil sample was obtained by combining micro-samples (m $\approx$ 5 g) extracted with the use of a special bore made of a chemically inert material (a total of 10 injections). After the careful homogenization of the combined sample, subsamples of raw soil material with known humidity underwent successive chemical extraction in accordance with the BCR fractionation scheme.

In order to facilitate the further analysis and discussion of the obtained data, we introduced the special classification of the forms of $Zn(^{65}Zn)$ in the soil: exchange and carbonate bonds denoted as *"labile"*; bonds associated with recoverable Fe–Mn oxides denoted as *"conditionally labile"*; bonds associated with oxidizable organic matter and sulfides denoted as *"conditionally fixed"*; and residue denoted as *"fixed"*.

3.5. Elemental Analysis and γ-Spectrometry of Samples

During the vegetation experiments, the following dynamics were determined:

- Concentrations of zinc and the mass activity density of ^{65}Zn in soil (including different forms) and plants: $[Zn]_{soil}$ and $[Zn]_{plant}$ (mg kg^{-1}); $A_m(^{65}Zn)_{soil}$ and $A_m(^{65}Zn)_{plant}$, (Bq kg^{-1}).

- Concentration of Zn and the volumetric activity density of ^{65}Zn in the phase of the soil solution: $[Zn]_{soil\ solution}$ (µg L^{-1}) and $A_v(^{65}Zn)_{soil\ solution}$ (Bq L^{-1}), respectively.

Values of mass and volumetric activity density of ^{65}Zn were calculated on the date of the beginning of the vegetation experiment.

Based on the above parameters, the values of the key parameter, conventionally called the "specific activity of $^{65}Zn/Zn$", were calculated in lysimetric waters, individual chemical fractions (forms) of zinc in the soil, and parts of plants. These were determined as the ratio of mass or volumetric activity density to the concentration of stable Zn contained in the corresponding objects of study (Bq mg^{-1}):

- In solution	$A_{sp}(^{65}Zn/Zn)_{solution} = A_v(^{65}Zn)/[Zn]_{solution}$
- In the soil as a whole, and in individual chemical fractions of the soil	$A_{sp}(^{65}Zn/Zn)_{Fr.\#} = A_m(^{65}Zn)_{Fr.\#}/[Zn]_{Fr.\#}$
- In vegetative parts (VPs) of plants	$A_{sp}(^{65}Zn/Zn)_{VP} = A_m(^{65}Zn)_{VP}/[Zn]_{VP}$
- In roots	$A_{sp}(^{65}Zn/Zn)_{roots} = A_m(^{65}Zn)_{roots}/[Zn]_{roots}$

Using the specific activity of $^{65}Zn/Zn$, the values of the parameter we called "Enrichment Factors—$EF(^{65}Zn/Zn)$ (or simply *EF*)—were calculated. These factors comprised the enrichment with the radioactive tracer ^{65}Zn of natural (stable) Zn contained in soil (including individual chemical fractions) and plants in relation to the main component of our model system–soil solution, e.g., $EF_{Fr.\#}(^{65}Zn/Zn) = A_{sp}(^{65}Zn/Zn)_{Fr.\#}/A_{sp}(^{65}Zn/Zn)_{soil\ solution}$.

The concentrations of Zn (^{65}Zn) in the roots and vegetative parts (VP) of barley plants were determined after the dry ashing of samples at 500 °C followed by the acid extraction of metal HNO$_3$; the total (gross) mass fraction of elements in the soil was determined by the Obukhov–Plekhanova method [58] after ashing at 500 °C and the subsequent decomposition of samples using heating with an HCl$_{conc.}$+HNO$_{3conc.}$+HF$_{conc.}$ mixture followed by successive treatments by HF$_{conc}$ (twice) and HNO$_{3conc}$ (triplicate).

Elemental analysis was carried out via the atomic absorption and optical emission methods with inductively coupled plasma using 140AA (Agilent) and Liberty II (Varian) spectrometers. The mass and volumetric activity densities of ^{65}Zn were determined on a gamma-spectrometric complex consisting of an ORTEC HPGe detector and analyzer (with a relative recording efficiency of 40%) and LSRM SpectraLine GP software.

3.6. Statistical Analysis

The accuracy of the approximating equations used to describe the obtained dependencies was estimated using the *t*-criterion and the coefficients of determination R^2. The statistical analysis of experimental data was carried out by standard methods using MS-Excel based on the theoretical aspects set out in the works [59–61].

4. Conclusions

For the studied soils, various types of plant responses to changes in the concentration of Zn were determined using barley as the test plant. Ranges of corresponding concentrations in soils and various parts of test plants were established.

It was shown that the edaphic factors determining the buffering capacity of soils play no less important roles in regulating the mobility of Zn in the soil–plant system than the biological factor (physiological characteristics of plants). For Zn, the concentration ratios (*CRs*) and concentration factors (*CFs*) in various parts of plants were found to be constant in the area of the indicative type of plant response to changes in the metal contents in soils (for example, we found values of CR_{labile} = 5.68 and *CF* = 640 L kg^{-1} for the vegetative mass of barley).

A method was proposed for determining the buffer capacity of soils with respect to HMs ($PBC(V)_{HM}$) using barley as the test plant (the tested part of the plant was leaves). With this method, we determined the value of $PBC(V)_{Zn}$ in the field of non-toxic zinc concentrations for the studied Albic Retisol (Loamic, Ochric)soil to be 175 mmol kg^{-1}. In parallel, the PBC_{Zn} of the studied soil was determined by the laboratory method of ion exchange equilibrium under static conditions. The PBC_{Zn} value was 150 mmol kg^{-1}. Thus, the values of the buffer capacity of soils in relation to a heavy metal (Zn), determined by two methods, turned out to be of the same order. The relative difference was 15%.

The considered methodological approach opened up opportunities for using data obtained during the agroecological monitoring of contaminated agricultural lands (such as the content of labile forms of HMs, the exchange forms of macroelements (Ca and Mg) in soils, and the concentrations of HMs, Ca, and Mg in plants) to calculate the values of $PBC(V)_{HM}$ of the studied soils. However, to do this, it is necessary to have a database of the concentration factors of these HMs and macroelements (such as Ca and Mg) for various crops that are intended to be used as test plants. It is desirable that the composition of nutrient solutions be as close as possible to real soil solutions, at least at the level of soil types.

With the help of an original lysimetric installation of cyclic action and the use of a radioactive tracer ^{65}Zn applied to the studied Albic Retisol (Loamic, Ochric)soil, various aspects of the process of zinc migration and transformation of its forms in the soil–soil solution–plant system were studied. This allowed us to obtain a number of important parameters characterizing the lability and bioavailability of zinc in quasi-equilibrium conditions. Thus, for the studied soil with a known content of natural (stable) Zn and additionally introduced radionuclide ^{65}Zn, the regularities of metal distribution between different soil forms were established by using sequential fractionation method.

It was established that the enrichment of labile chemical Zn fraction by the radioactive tracer ^{65}Zn (EF $(^{65}Zn/Zn)_{Fr.I}$ was 1.54 times higher than the same value for the soil solution at the time corresponding to the beginning of the vegetation experiment. At the same time, the ratio of the remaining ("conditionally labile", "conditionally fixed", and "fixed") forms of stable Zn in the soil with ^{65}Zn was 0.19–0.82 times lower than that of the soil solution. Based on the obtained value of EF $(^{65}Zn/Zn)_{Fr.I}$, we calculated the amount of the pool of labile "E-value" (E_{Zn}) compounds of native zinc in the unit of the mass of the studied soil as: $E_{Zn} \cong C(Zn)_{, Fr.I}$ = 3.95 ± 0.16 mg kg^{-1} (or 10.7 ± 0.4%). The corresponding specific activity value of ^{65}Zn ($A_{sp}(^{65}Zn/Zn)_{Fr.I}$) equaled 4660 ± 830 Bq mg^{-1}.

At the beginning of the vegetation experiment, we recorded a lower enrichment of native zinc contained in the quasi-equilibrium soil solution with ^{65}Zn (after balancing it with the soil for 1 year) compared to the stable zinc contained in Fraction I. This indicated the existence of a slow process of isotopic exchange between the mobile and other ("conditionally labile", "conditionally fixed", and "fixed") forms of zinc by means of the liquid phase of the studied soil–soil solution. Later, during the vegetation experiment, as a result of the vital activity of the plants, the soil solution was depleted with zinc and alkalinized, which led to an increase in the values of $A_{sp}(^{65}Zn/Zn)_{soil\ solution}$ to those close to $A_{sp}(^{65}Zn/Zn)_{Fr.I}$.

The dynamics of the transition to the soil solution phase and the parameters of the uptake of Zn(^{65}Zn) by test plants were also evaluated. According to the results of the experiment, the total removal of natural Zn and radionuclide ^{65}Zn by plants (roots and vegetative parts) was calculated as a percentage of the total amount of Zn(^{65}Zn) in the soil: 0.34 (0.70)%. These data suggest that the removal of zinc from soil by plants is insignificant and had no noticeable effects on the ratio of the forms of the metal in the soil.

Author Contributions: V.S.A. conceived and designed the experiments; V.S.A., L.N.A., and A.I.S. performed the experiments; V.S.A. analyzed the data; V.S.A., L.N.A., and A.I.S. contributed to manuscript preparation. All authors have read and agreed to the published version of the manuscript.

Funding: The study was carried out with the financial support of the Russian Foundation for Basic Research (RFBR Grant No. 19-29-05039).

Institutional Review Board Statement: Not applicable because no humans or animals were involved in our study.

Informed Consent Statement: Not applicable because no humans or animals were involved in our study.

Data Availability Statement: Data sharing is not applicable to this article.

Acknowledgments: The Russian Institute of Radiology and Agroecology (RIRAE) is highly acknowledged for providing access to the laboratory facilities needed for this study. We are grateful to Dikarev D.V., Frigidov R.A., Korneev Yu.N., Sarukhanov A.V., Tomson A.V., and Korovin S.V. for the help in conducting elemental and gamma spectrometric analyses.

Conflicts of Interest: The authors declare that they have no conflict of interest.

Appendix A

Figure A1. A stand for studying the parameters of ^{65}Zn(Zn) migration in the soil–soil solution–plant system: 1—the moment of equilibration of deionized water in the system with Albic Retisol (Loamic, Ochric) soil; 2—vegetative vessels with plants.

References

1. Pandey, R. Mineral Nutrition of Plants. In *Plant Biology and Biotechnology. Volume I: Plant Diversity, Organization, Function and Improvement*; Bahadur, B., Rajam, M.V., Sahijram, L., Krishnamurthy, K.V., Eds.; Springer: New Delhi, India, 2015; pp. 499–538.
2. Hacisalihoglu, G. Zinc (Zn): The Last Nutrient in the Alphabet and Shedding Light on Zn Efficiency for the Future of Crop Production under Suboptimal Zn. *Plants* **2020**, *9*, 1471. [CrossRef] [PubMed]
3. Grusak, M.A.; DellaPenna, D. Improving the Nutrient Composition of Plants to Enhance Human Nutrition and Health. *Annu. Rev. Plant Biol.* **1999**, *50*, 133–161. [CrossRef] [PubMed]
4. Khadhar, S.; Sdiri, A.; Chekirben, A.; Azouzi, R.; Charef, A. Integration of sequential extraction, chemical analysis and statistical tools for the availability risk assessment of heavy metals in sludge amended soils. *Environ. Pollut.* **2020**, *263*, 114543. [CrossRef]
5. Alloway, B.J. Soil factors associated with zinc deficiency in crops and humans. *Environ. Geochem. Health* **2009**, *31*, 537–548. [CrossRef] [PubMed]
6. Hacisalihoglu, G.; Kochian, L. How do some plants tolerate low levels of soil zinc? Mechanisms of zinc efficiency in crop plants. *New Phytol.* **2003**, *159*, 341–350. [CrossRef]
7. Barber, S.A. *Soil Nutrient Bioavailability: A Mechanistic Approach*, 2nd ed.; John Wiley & Sons: New York, NY, USA, 1995; p. 384.
8. Kabata-Pendias, A. *Trace Elements in Soils and Plants*; CRC Press: London, UK, 2011; p. 505.

9. Liu, Y.; Gao, T.; Xia, Y.; Wang, Z.; Liu, C.; Li, S.; Wu, Q.; Qi, M.; Lv, Y. Using Zn isotopes to trace Zn sources and migration pathways in paddy soils around mining area. *Environ. Pollut.* **2020**, *267*, 115616. [CrossRef] [PubMed]

10. Nevidomskaya, D.G.; Minkina, T.M.; Soldatov, A.V.; Bauer, T.V.; Shuvaeva, V.A.; Zubavichus, Y.V.; Trigub, A.L.; Mandzhieva, S.S.; Dorovatovskii, P.V.; Popov, Y.V. Speciation of Zn and Cu in Technosol and evaluation of a sequential extraction procedure using XAS, XRD and SEM–EDX analyses. *Environ. Geochem. Health* **2021**, *43*, 2301–2315. [CrossRef]

11. Nazif, W.; Marzouk, E.; Perveen, S.; Crout, N.; Young, S. Zinc solubility and fractionation in cultivated calcareous soils irrigated with wastewater. *Sci. Total Environ.* **2015**, *518–519*, 310–319. [CrossRef]

12. Voegelin, A.; Tokpa, G.; Jacquat, O.; Barmettler, K.; Kretzschmar, R. Zinc Fractionation in Contaminated Soils by Sequential and Single Extractions: Influence of Soil Properties and Zinc Content. *J. Environ. Qual.* **2008**, *37*, 1190–1200. [CrossRef]

13. Adriano, D.C. *Trace Elements in Terrestrial Environments*; Springer: New York, NY, USA; Berlin/Heidelberg, Germany, 2001; p. 868.

14. He, Q.; Ren, Y.; Mohamed, I.; Ali, M.; Hassan, W.; Zeng, F. Assessment of trace and heavy metal distribution by four sequential extraction procedures in a contaminated soil. *Soil Water Res.* **2013**, *8*, 71–76. [CrossRef]

15. Kersten, M.; Förstner, U. Chemical fractionation of heavy metals in anoxic estuarine and coastal sediments. *Water Sci Technol.* **1986**, *18*, 121–130. [CrossRef]

16. Ladonin, D.V. Fractional-Isotopic Composition of Lead Compounds in Soils of the Kologrivskii Forest Reserve. *Eurasian Soil Sci.* **2018**, *51*, 929–937. [CrossRef]

17. Ladonin, D.V. Heavy metal compounds in soils: Problems and methods of study. *Eurasian Soil Sci.* **2002**, *35*, 605–613.

18. Minkina, T.M.; Nevidomskaya, D.G.; Shuvaeva, V.A.; Soldatov, A.V.; Tsitsuashvili, V.S.; Zubavichus, Y.V.; Rajput, V.D.; Burachevskaya, M.V. Studying the transformation of Cu^{2+} ions in soils and mineral phases by the XRD, XANES, and sequential fractionation methods. *J. Geochem. Explor.* **2018**, *184*, 365–371. [CrossRef]

19. Minkina, T.M.; Mandzhieva, S.S.; Burachevskaya, M.V.; Bauer, T.; Sushkova, S.N. Method of determining loosely bound compounds of heavy metals in the soil. *MethodsX* **2018**, *5*, 217–226. [CrossRef] [PubMed]

20. Morabito, R. Extraction techniques in speciation analysis of environmental samples Fresenius. *J. Anal. Chem.* **1995**, *351*, 378–385.

21. Quevauviller, P. (Ed.) *Methodologies in Soil and Sediment Fractionation Studies. Single and Sequential Extraction Procedures*; Royal Society of Chemistry: Cambridge, UK, 2002; 180p.

22. Wali, A.; Colinet, G.; Ksibi, M. Speciation of Heavy Metals by Modified BCR Sequential Extraction in Soils Contaminated by Phosphogypsum in Sfax, Tunisia. *Environ. Res. Eng. Manag.* **2015**, *70*, 14–25. [CrossRef]

23. Young, S.; Zhang, H.; Tye, A.; Maxted, A.; Thums, C.; Thornton, I. Characterizing the availability of metals in contaminated soils. I. The solid phase: Sequential extraction and isotopic dilution. *Soil Use Manag.* **2005**, *21*, 450–458. [CrossRef]

24. Echevarria, G.; Morel, J.L.; Fardeau, J.C.; Leclerc-Cessac, E. Assessment of Phytoavailability of Nickel in Soils. *J. Environ. Qual.* **1998**, *27*, 1064–1070. [CrossRef]

25. Frossard, E.; Sinaj, S. The Isotope Exchange Kinetic Technique: A Method to Describe the Availability of Inorganic Nutrients. Applications to K, P, S and Zn. *Isot. Environ. Health Stud.* **1997**, *33*, 61–77. [CrossRef]

26. Garforth, J.M.; Bailey, E.H.; Tye, A.M.; Young, S.D.; Lofts, S. Using isotopic dilution to assess chemical extraction of labile Ni, Cu, Zn, Cd and Pb in soils. *Chemosphere* **2016**, *155*, 534–541. [CrossRef]

27. Huang, Z.-Y.; Chen, T.; Yu, J.; Zeng, X.-C.; Huang, Y.-F. Labile Cd and Pb in vegetable-growing soils estimated with isotope dilution and chemical extractants. *Geoderma* **2011**, *160*, 400–407. [CrossRef]

28. Tongtavee, N.; Shiowatana, J.; McLaren, R.G.; Gray, C.W. Assessment of lead availability in contaminated soil using isotope dilution techniques. *Sci. Total Environ.* **2005**, *348*, 244–256. [CrossRef] [PubMed]

29. Nefedov, V.D.; Texter, E.N.; Toropova, M.A. *Radiochemistry*; Higher School: Moscow, Russia, 1987; p. 271. (In Russian)

30. Mandzhieva, S.S.; Minkina, T.M.; Motuzova, G.V.; Golovatyi, S.E.; Miroshnichenko, N.N.; Fateev, A.I.; Lukashenko, N.K. Fractional and group composition of zinc and lead compounds as an indicator of the environmental status of soils. *Eurasian Soil Sci.* **2014**, *47*, 511–518. [CrossRef]

31. Motuzova, G.V. *Compounds of Microelements in Soils: Systemic Organization, Ecological Importance, Monitoring*; Editoral URSS: Moscow, Russia, 1999; p. 168. (In Russian)

32. Zyrin, N.G.; Sadovnikova, L.K. *Chemistry of Heavy Metals, Arsenic and Molybdenum in Soils*; MSU Publication: Moscow, Russia, 1985; p. 208. (In Russian)

33. Dobrovolsky, V.V. *Zinc and Cadmium in the Environment*; Nauka: Moscow, Russia, 1992; p. 200. (In Russian)

34. Anisimov, V.S.; Anisimova, L.N.; Frigidova, L.M.; Dikarev, D.V.; Frigidov, R.A.; Korneev, Y.N.; Sanzharov, A.I.; Arysheva, S.P. Evaluation of the Migration Capacity of Zn in the Soil–Plant System. *Eurasian Soil Sci.* **2018**, *51*, 407–417. [CrossRef]

35. Baker, A.J.M. Accumulators and excluders—Strategies in the response of plants to heavy metals. *J. Plant Nutr.* **1981**, *3*, 643–654. [CrossRef]

36. Koolman, J.; Roehm, K.-H. *Color Atlas of Biochemistry*, 2nd ed.; Georg Thieme Verlag: Stuttgart, Germany; New York, NY, USA, 2005; p. 467.

37. Hacisalihoglu, G.; Hart, J.J.; Kochian, L. High- and Low-Affinity Zinc Transport Systems and Their Possible Role in Zinc Efficiency in Bread Wheat. *Plant Physiol.* **2001**, *125*, 456–463. [CrossRef]

38. Hart, J.J.; Norvell, W.A.; Welch, R.M.; Sullivan, L.A.; Kochian, L. Characterization of Zinc Uptake, Binding, and Translocation in Intact Seedlings of Bread and Durum Wheat Cultivars. *Plant Physiol.* **1998**, *118*, 219–226. [CrossRef]

39. Pence, N.S.; Larsen, P.B.; Ebbs, S.; Letham, D.L.D.; Lasat, M.M.; Garvin, D.F.; Eide, D.; Kochian, L. The molecular physiology of heavy metal transport in the Zn/Cd hyperaccumulator Thlaspi caerulescens. *Proc. Natl. Acad. Sci. USA* **2000**, *97*, 4956–4960. [CrossRef] [PubMed]
40. Nye, P.H.; Tinker, P.B. *Solute Movement in the Soil-Root System*; Blackwell Scientific Publucations: Oxford, UK, 1977; p. 342.
41. Barbosa, B.C.F.; Silva, S.C.; de Oliveira, R.R.; Chalfun, A. Zinc supply impacts on the relative expression of a metallothionein-like gene in Coffea arabica plants. *Plant Soil* **2017**, *411*, 179–191. [CrossRef]
42. Cosio, C.; Martinoia, E.; Keller, C. Hyperaccumulation of Cadmium and Zinc in Thlaspi caerulescens and Arabidopsis halleri at the Leaf Cellular Level. *Plant Physiol.* **2004**, *134*, 716–725. [CrossRef].
43. Lin, Y.-F.; Aarts, M.G.M. The molecular mechanism of zinc and cadmium stress response in plants. *Cell. Mol. Life Sci.* **2012**, *69*, 3187–3206. [CrossRef]
44. Subhashini, V.; Swamy, A.V.V.S.; Krishna, R.H. Pot Experiment: To Study the Uptake of Zinc by Different Plant Species in Artificially Contaminated Soil. *World J. Environ. Eng.* **2013**, *1*, 27–33. [CrossRef]
45. Tiong, J.; McDonald, G.K.; Genc, Y.; Pedas, P.; Hayes, J.E.; Toubia, J.; Langridge, P.; Huang, C.Y. HvZIP 7 mediates zinc accumulation in barley (*Hordeum vulgare*) at moderately high zinc supply. *New Phytol.* **2014**, *201*, 131–143. [CrossRef]
46. Anisimov, V.S.; Sanzharova, N.I.; Anisimova, L.N.; Geras'kin, S.A.; Dikarev, D.V.; Frigidova, L.M.; Frigidov, R.A.; Belova, N.V. The assessment of the migration ability and phytotoxicity of Zn in the system soil-plant. *Agrochemistry* **2013**, *1*, 64–74. (In Russian)
47. Beckett, P.H.T. Studies on soil potassium II. The 'immediate' Q/I relations of labile potassium in the soil. *J. Soil Sci.* **1964**, *15*, 9–23. [CrossRef]
48. Sokolov, A.V. (Ed.) *Agrochemical Methods of Soil Research*; Nauka: Moscow, Russia, 1975; 656p. (In Russian)
49. Beckett, P. Potassium-Calcium Exchange Equilibria in Sois: Specific Adsorption Sites for Potassium. *Soil Sci.* **1964**, *97*, 376–383. [CrossRef]
50. Beckett, P.H.T.; Nafady, M.H.M. Potassium—Calcium exchange equilibria in soils: The location of non-specific (Gapon) and specific exchange sites. *J. Soil Sci.* **1967**, *18*, 263–281. [CrossRef]
51. Kruglov, S.V.; Anisimov, V.S.; Lavrent'eva, G.V.; Anisimova, L.N. Parameters of Selective Sorption of Co, Cu, Zn, and Cd by a Soddy-Podzolic Soil and a Chernozem. *Eurasian Soil Sci.* **2009**, *42*, 385–393. [CrossRef]
52. Pinsky, D.L. *Ion exchange Processes in Soils*; Pushchino, Russia, 1997; p. 166. (In Russian)
53. Sokolova, T.A.; Trofimov, S.Y. *Sorption Properties of Soils. Adsorption. Cation Exchange: A Textbook on Some Chapters of Soil Chemistry*; Grif and K: Tula, Russia, 2009; p. 172. (In Russian)
54. Blossfeld, S.; Perriguey, J.; Sterckeman, T.; Morel, J.-L.; Lösch, R. Rhizosphere pH dynamics in trace-metal-contaminated soils, monitored with planar pH optodes. *Plant Soil* **2010**, *330*, 173–184. [CrossRef]
55. Mineev, V.G. (Ed.) *Practical Course on Agrochemistry*; MSU Publication: Moscow, Russia, 2001; p. 689.
56. Arinushkina, E.V. *Chemical Analysis of Soils and Soils*; MSU Publication: Moscow, Russia, 1970; p. 480. (In Russian)
57. Sposito, G. *The Chemistry of Soils*, 2nd ed.; Oxford University Press Inc.: New York, NY, USA, 2008; p. 329.
58. Obukhov, A.I.; Plekhanova, I.O. *Atomic Absorption Analysis in Soil-Biological Investigations*; MSU Publication: Moscow, Russia, 1991; p. 184. (In Russian)
59. Dospekhov, B.A. *Methodology of Field Experience (with the Basics of Statistical Processing of Research Results)*; Agropromizdat: Moscow, Russia, 1985; p. 351. (In Russian)
60. Prichard, E.; Barwick, V. *Quality Assurance in Analytical Chemistry*; Wiley: Hoboken, NJ, USA, 2007; p. 293.
61. James, G.; Witten, D.; Hastie, T.; Tibshirani, R. *An Introduction to Statistical Learning*; Springer: New York, NY, USA, 2013; p. 426. [CrossRef]

plants

Article

Transcriptional Regulation of Genes Involved in Zinc Uptake, Sequestration and Redistribution Following Foliar Zinc Application to *Medicago sativa*

Alessio Cardini [1,†], Elisa Pellegrino [1,*,†], Philip J. White [2], Barbara Mazzolai [3], Marco C. Mascherpa [4] and Laura Ercoli [1]

[1]   Institute of Life Sciences, Scuola Superiore Sant'Anna, 56127 Pisa, Italy; alessio.cardini@santannapisa.it (A.C.); laura.ercoli@santannapisa.it (L.E.)
[2]   Department of Ecological Science, The James Hutton Institute, Invergowrie, Dundee DD2 5DA, UK; philip.white@hutton.ac.uk
[3]   Center for Micro-BioRobotics, Istituto Italiano di Tecnologia, Pontedera, 56025 Pisa, Italy; barbara.mazzolai@iit.it
[4]   Istituto di Chimica dei Composti Organo Metallici, National Research Council (CNR), 56124 Pisa, Italy; marcocarlo.mascherpa@pi.iccom.cnr.it
*   Correspondence: elisa.pellegrino@santannapisa.it
†   These authors contributed equally to this work.

Citation: Cardini, A.; Pellegrino, E.; White, P.J.; Mazzolai, B.; Mascherpa, M.C.; Ercoli, L. Transcriptional Regulation of Genes Involved in Zinc Uptake, Sequestration and Redistribution Following Foliar Zinc Application to *Medicago sativa*. *Plants* **2021**, *10*, 476. https://doi.org/10.3390/plants10030476

Academic Editor: Gokhan Hacisalihoglu

Received: 22 January 2021
Accepted: 25 February 2021
Published: 3 March 2021

Abstract: Zinc (Zn) is an essential micronutrient for plants and animals, and Zn deficiency is a widespread problem for agricultural production. Although many studies have been performed on biofortification of staple crops with Zn, few studies have focused on forages. Here, the molecular mechanisms of Zn transport in alfalfa (*Medicago sativa* L.) were investigated following foliar Zn applications. Zinc uptake and redistribution between shoot and root were determined following application of six Zn doses to leaves. Twelve putative genes encoding proteins involved in Zn transport (*MsZIP1-7*, *MsZIF1*, *MsMTP1*, *MsYSL1*, *MsHMA4*, and *MsNAS1*) were identified and changes in their expression following Zn application were quantified using newly designed RT-qPCR assays. These assays are the first designed specifically for alfalfa and resulted in being more efficient than the ones already available for *Medicago truncatula* (i.e., *MtZIP1-7* and *MtMTP1*). Shoot and root Zn concentration was increased following foliar Zn applications ≥ 0.1 mg plant^{-1}. Increased expression of *MsZIP2*, *MsHMA4*, and *MsNAS1* in shoots, and of *MsZIP2* and *MsHMA4* in roots was observed with the largest Zn dose (10 mg Zn plant^{-1}). By contrast, *MsZIP3* was downregulated in shoots at Zn doses ≥ 0.1 mg plant^{-1}. Three functional gene modules, involved in Zn uptake by cells, vacuolar Zn sequestration, and Zn redistribution within the plant, were identified. These results will inform genetic engineering strategies aimed at increasing the efficiency of crop Zn biofortification.

Keywords: ZIP transporters; nicotianamine; metal tolerance protein (MTP); yellow stripe-like protein (YSL); zinc-induced facilitators (ZIF); heavy metal transporters (HMA)

---

1. Introduction

A large proportion of the world's population suffers from Zn-related diseases (i.e., malabsorption syndrome, liver disease, chronic renal disease, sickle cell disease, and other chronic diseases), since they rely on cereal-based diets with low Zn content due to poor soil Zn availability [1–4]. Diversification of the human diet and biofortification of edible crops are therefore needed to alleviate Zn deficiency in humans. Similarly to humans, animals can suffer from Zn deficiencies that could be alleviated by biofortified feed or Zn supplementation, thus improving livestock health and quality of food products, which affect human health indirectly [5–8].

Zinc plays a major role as a co-factor of over 300 enzymes in plants and is an essential micronutrient [9]. Zinc is involved in various physiological functions, such as CO_2 fixation,

71

protein synthesis, free radical capture, regulation of growth and development, and disease resistance [9,10]. Many structural motifs in transcriptional regulatory proteins are stabilized by Zn, such as Zn finger domains [11]. Zinc deficiency reduces crop production, as does Zn excess [12]. Excessive Zn^{2+} can compete with other cations in binding to enzymes and for transport across membranes, thereby impairing cellular activities [12]. Thus, the uptake of Zn^{2+} by cells and its transport within the plant must be strictly regulated. Plant cells have evolved several homeostatic mechanisms for avoiding Zn^{2+} toxicity when exposed to large Zn availability in their environment. These include the reduction of Zn influx to cells, the stimulation of Zn efflux from the cytosol, the sequestration of Zn in vacuoles, and the chelation of Zn by Zn-binding ligands. In general, the concentration of Zn in plant tissues must be kept between 15 and 300 μg Zn g^{-1} dry matter (DM) to maintain cell structure and function [12,13]. Although tolerance to large tissue Zn concentrations varies among species [12,14], Zn concentrations above 400–500 μg g^{-1} DM often cause toxicity symptoms including impaired root and shoot growth, chlorosis and necrosis of leaves, reduced photosynthesis, nutrient imbalance, and ultimately loss of yield [9,12,15,16].

The process of producing crops with greater mineral concentrations in edible tissues is called biofortification and provides a solution to the problem of mineral deficiencies in human and animal nutrition [17]. There are various approaches to Zn biofortification of edible crops, including agronomic strategies and conventional or transgenic breeding strategies. Agronomic biofortification aims to increase Zn concentrations in edible tissues through the application of Zn fertilizers to the soil or to leaves. It is relatively inexpensive and efficient [18]. Foliar application of Zn is generally more effective than the application of Zn fertilizers to soil, since Zn uptake by plant roots is often limited by the low solubility of Zn salts, its binding to organic substrates, and its immobilization in the microbial biomass [19]. Both agronomic and genetic biofortification strategies have been studied extensively in cereal staple crops, such as rice, wheat, and maize, but less in legumes, such as beans, peas, or lentils [17,20,21]. An international program, the HarvestPlus Zinc Fertilizer Project, is exploring the potential of Zn fertilizers to enhance the yields and Zn concentrations in edible portions of staple crops in developing countries of Africa, Asia, and South America (www.harvestzinc.org (accessed on 2 March 2021)) [22], but this program does not include forage crops.

The natural direction of Zn flux in plants is from the soil via roots to the shoot and seeds [23]. Various transport proteins and ligands that are responsible for Zn^{2+} uptake by roots and its transport and sequestration within the plant have been characterized [12,24–26]. Among these, ZRT-IRT-like proteins (ZIPs) have been studied in several plants, including *Arabidopsis thaliana*, soybean (*Glycine max*), barley (*Hordeum vulgare*), barrel medic (*Medicago truncatula*), and rice (*Oryza sativa*) [27–31]. These proteins not only transport Zn^{2+} across membranes, but can also transport other transition metal cations, including Cd^{2+}, Fe^{3+}/Fe^{2+}, Mn^{2+}, Ni^{2+}, Co^{2+}, and Cu^{2+} [27,32,33]. Generally, the expression of *ZIP* genes is upregulated when plants become Zn-deficient [34–36], facilitating Zn influx to cells and movement of Zn between organs, and also when plants become Fe- or Mn-deficient [35,37–39]. Other proteins that transport Zn include the metal tolerance proteins (MTPs), which function as cation/proton antiporters and are thought to transport Zn into vacuoles [40], and the yellow stripe-like proteins (YSLs), which transport the Zn–nicotianamine complex (NA–Zn) and load Zn into the xylem and phloem [41]. The zinc-induced facilitators (ZIFs) and the heavy metal transporters (HMAs) are implicated in Zn influx to vacuoles and to the xylem, respectively [24]. Zinc is chelated by organic molecules, such as the carboxylic acid, citric acid, and nicotianamine (NA) in plants [42]. Nicotianamine is a non-proteinogenic amino acid with a high affinity for Fe, Cu, and Zn, and is involved in their homeostasis [43]. Nicotianamine mediates the intercellular and interorgan movement of Zn and was found to enable Zn hyperaccumulation in *Arabidopsis halleri* and *Noccaea caerulescens* [43,44]. In general, the functions of these transporters have been studied by expressing them in yeast, but to understand how the various Zn transport proteins and chelates act together to maintain appropriate cytosolic and tissue Zn concentrations, it is important to study the responses of an intact plant to fluctuations in Zn supply. Moreover, since there is a knowledge gap on the regulation of Zn transport following Zn foliar application, it is important also to elucidate plant transcriptional responses when Zn is not applied to roots.

Thus, in this study the transcriptional responses of genes encoding Zn transport-related processes facilitating Zn uptake by cells, vacuolar sequestration, and redistribution within the plant were studied following foliar Zn application to the most productive and widely cultivated forage legume, alfalfa (*Medicago sativa* L.). The study was designed to provide information on the molecular responses to Zn biofortification of forage crops [7,45]. A model for the roles of putative genes encoding proteins involved in Zn transport- related processes was built and used for the selection of genes (Figure 1).

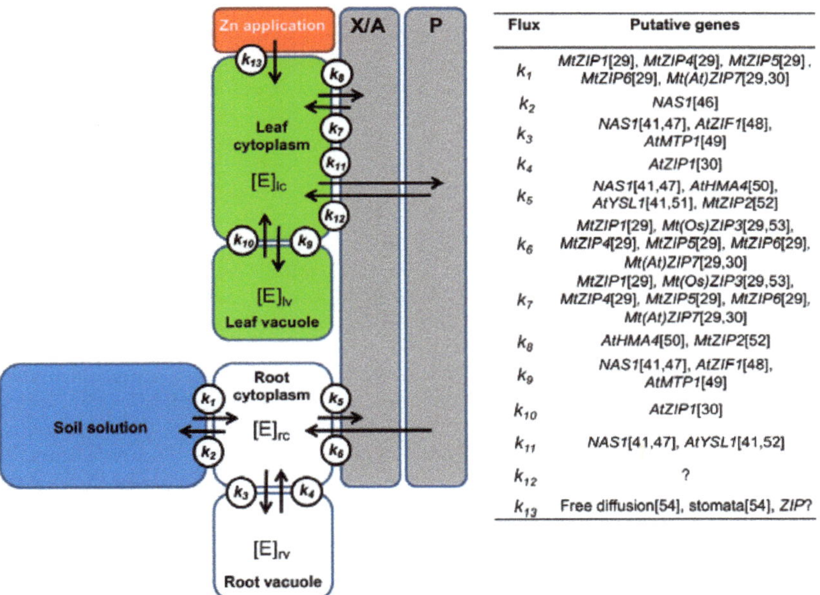

Figure 1. Suggested model for roles of putative genes encoding proteins involved in Zn transport- related processes. The sites of action in the plant (i.e., root cytoplasm, rc; root vacuole, rv; xylem and apoplast, X/A; phloem, P; leaf cytoplasm, lc; leaf vacuole, lv) and the element (E) fluxes (K_{1-13}) are reported. The concentration of the element is indicated in each site [E]. The scheme synthetizes information across studies in various plants. Gene abbreviations: *ZIP*, Zrt-/Irt-like protein; *NAS*, nicotianamine synthase; *ZIF*, zinc-induced facilitator; *MTP*, metal transporter protein; *HMA*, P1B-type heavy metal ATPase; *YSL*, yellow stripe like protein; *ZIP?* indicates a generic ZIP; free diffusion: diffusion through leaf epidermis; stomata: absorption through stomata. Plant abbreviations: Mt, *Medicago truncatula*; At, *Arabidopsis thaliana*; Os, *Oryza sativa*. References: [29,30,41,46–54].

The following hypotheses were tested: (i) foliar application of Zn determines Zn redistribution within the plant and is associated with changes in the expression of genes involved in Zn transport-related processes; (ii) genes encoding Zn transport-related processes are organized in functional modules that act in a concerted manner to redistribute Zn within the plant to maintain non-toxic cytosolic and tissue Zn concentrations. Genes encoding putative Zn transport-related processes were identified in alfalfa through phylogenetic comparisons and their likely roles are discussed. Changes in the expression of these genes following foliar Zn application were determined and the possible effects of these on the redistribution of Zn within cells and between tissues are also discussed. The knowledge gained from this study could help to optimize Zn biofortification strategies when using foliar Zn fertilizers and to provide strategies for breeding forage crops to addresses Zn deficiencies in livestock.

2. Results

2.1. Effects of Foliar Zn Application on Plant Zn Redistribution and Expression of Genes Involved in Zn Transport-Related Processes

The first aim of this study was to provide information on Zn redistribution within the plants and on the associated transcriptional responses of Zn transport-related genes following foliar Zn biofortification to alfalfa.

2.1.1. Shoot and Root Zn Concentration and Content

To provide novel information on Zn redistribution within alfalfa plants following foliar Zn application, we applied six Zn doses (0, 0.01, 0.1, 0.5, 1, 10 mg Zn plant^{-1}) and assessed Zn concentration and content in shoots and roots 5 days after application. The application of Zn to leaves did not modify shoot or root fresh and dry biomass (Table S1), and all *M. sativa* plants had a similar number of functional root nodules, irrespective of Zn treatments (data not shown). However, Zn concentrations in both shoots and roots were strongly affected by foliar Zn application ($F_{(5,17)}$ = 32.61, p < 0.001; $F_{(5, 17)}$ = 28.53, p < 0.001; respectively) (Figure 2a). A foliar Zn application of 0.01 mg Zn plant^{-1} produced a shoot Zn concentration similar to that of the control (no-Zn addition), but shoot Zn concentrations were increased progressively by larger doses (0.1 < 0.5/1 < 10 mg Zn plant^{-1}), from more than 3-fold to 35-fold more than that of the control (Figure 2). Foliar applications of 0.01, 0.1, and 0.5 mg Zn plant^{-1} did not produce root Zn concentrations greater than that of the control treatment, but foliar doses of 1 and 10 mg Zn plant^{-1} increased root Zn concentrations to 3-fold and 11-fold more than the control treatment, respectively. Shoot and root Zn contents were also strongly affected by foliar Zn application ($F_{(5,17)}$ = 53.73, p < 0.001; $F_{(5, 17)}$ = 32.45, p < 0.001; respectively) and their responses to increasing foliar Zn applications followed the corresponding Zn concentrations (Figure 2b). At all Zn dose plants did not show any visual symptom of Zn deficiency or toxicity. Moreover, plants grown for two months lacking Zn (i.e., 0 mg Zn plant^{-1}) had shoot and root Zn content of 4.5 and 2.7 µg plant^{-1}, respectively, probably relying on seed Zn content.

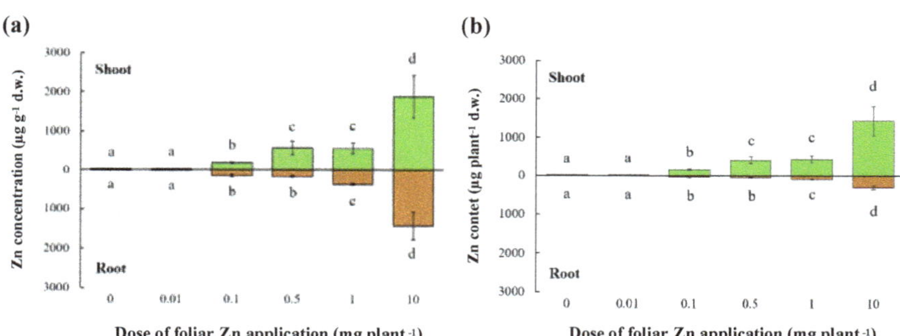

Figure 2. Zinc concentration (**a**) and content (**b**) in shoots and roots of alfalfa after the application of Zn to leaves. The Zn doses were 0, 0.01, 0.1, 0.5, 1, or 10 mg Zn plant^{-1}. Means ± standard error of three replicates are shown. Differences among the applied Zn doses were tested separately for shoot and root by one-way analysis of variance. Different letters denote significant differences in Zn concentrations in shoots and roots independently, according to Tukey-B honestly test (p < 0.05).

To summarize, the efficacy of Zn biofortification (i.e., shoot Zn concentrations in the range 15–400 µg Zn g^{-1} d.w.) was proved for the doses of 0.1 to 1 mg plant^{-1}, while the lowest dose (0.01 mg plant^{-1}) was ineffective, and the highest dose (10 mg plant^{-1}) produced toxic concentrations.

2.1.2. Phylogenetic Analysis

To infer the putative roles of the selected *M. sativa* Zn transport-related genes, we performed phylogenetic analyses. This was based on the assumption of a simple equivalence between a minimum similarity threshold in the phylogenetic comparisons and the function similarity between encoded proteins. Phylogenetic analysis of the coding sequences of the *ZIP* genes revealed several distinct clades (Figure S1). One clade contained sequences for *MsZIP2* and *MsZIP7*, which were similar to each other. In addition, the sequence of *MsZIP2* was closely related to those of *MtZIP2* and *GmZIP1-ZIP2*, and the sequence of *MsZIP7* was closely related to those of *MtZIP7* and *AtZIP11*. Another clade contained the sequences of *MsZIP1*, *MsZIP3*, *MsZIP5*, and *MsZIP6*. The sequence of *MsZIP1* clustered with that of *MtZIP1*. Sequences of *MsZIP3* and *MsZIP5* were similar to each other and clustered with the corresponding sequences for *M. truncatula* genes (Figure S1). Sequences for *MsZIP1*, *MsZIP3*, and *MsZIP5* were closely related to each other, whereas that of *MsZIP6* formed a separate cluster with the sequences of *MtZIP6* and *AtZIP12*. The sequence of *MsZIP4* was distant from the sequences of other *M. sativa* ZIPs and formed a cluster with the sequences of *MtZIP4* and *AtZIP4*.

Phylogenetic analyses of the coding sequences of the other genes related to Zn transport processes revealed that they were all similar to their *M. truncatula* counterparts. As regards *ZIF*, the sequence of *MsZIF1* clustered with the sequences of *MtZIF1* and *GmZIF1* (Figure S2a). As regards *MTP*, the sequence of *MsMTP1* formed a cluster with *MtMTP1* and *GmMTP1* and was also related to *AtMTP1* and *AtMTPA1* (Figure S2b). Similarly, the sequence of *MsYSL1* was most similar to those of *MtYSL1* and *GmYSL1* (Figure S3a) and the sequence of *MsHMA4* was most similar to those of *MtHMA4* and *GmHMA4* (Figure S3b). Finally, the sequence of *MsNAS1* was closely related to those of *MtNAS* and *GmNAS* (Figure S4).

To summarize, the genes selected for gene expression analysis were closely related to the homologous of *M. truncatula* and of other plant species. Thus, on the basis of the pattern of clustering and of the functions described in literature for the encoded proteins, we were able to infer the putative roles of the genes.

2.1.3. Gene Expression Analysis

To provide novel information on the transcriptional responses of genes encoding Zn transport-related processes following foliar Zn application, we analyzed the expression of *MsZIP1-7*, *MsMTP1*, *MsYSL1*, *MsHMA4*, and *MsNAS1* genes in shoots and roots, 5 days after Zn application of 0, 0.1, 1, and 10 mg Zn plant^{-1}. The Zn treatments were selected on the basis of the significance of the results on Zn redistribution in shoots and roots. The expression of *MsZIP3* was significantly downregulated only in shoots at foliar doses of 0.1, 1, and 10 mg Zn plant^{-1} ($F_{(3, 11)} = 28.46$, $p < 0.01$) (Figure 3). By contrast, the expression of *MsZIP2* was significantly upregulated in shoots and roots at the largest dose of 10 mg Zn plant^{-1} ($F_{(3, 11)} = 5.59$, $p < 0.05$; $F_{(3, 11)} = 9.26$, $p < 0.01$). The expression of *MsZIP1*, *MsZIP5*, and *MsZIP6* in shoots was not significantly affected by foliar Zn application, although a general trend towards downregulation with increasing foliar Zn doses was observed. The expression of *MsZIP4* and *MsZIP7* in shoots was unaffected by foliar Zn application.

In roots, all *ZIP* genes except *MsZIP2* were not significantly affected by foliar Zn application, although a general trend of *MsZIP1*, *MsZIP3*, *MsZIP5*, and *MsZIP7* towards upregulation with increasing foliar Zn doses was observed. Of the other genes related to Zn transport processes, the expression of *MsHMA4* was significantly upregulated in both shoots ($F_{(3, 11)} = 115.29$, $p < 0.01$) and roots ($F_{(3, 11)} = 14.23$, $p < 0.01$) following the application of 1 and 10 mg Zn plant^{-1} (shoots: +63% and +424%, respectively; roots: +86% and +66%, respectively; Figure 4).

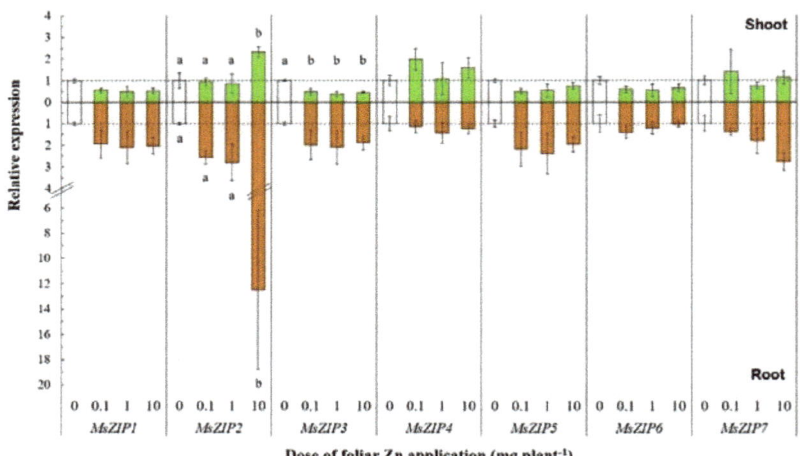

Figure 3. Relative expression of transmembrane Zn transporter genes after leaf Zn application to alfalfa. The Zn doses were 0, 0.1, 1, or 10 mg Zn plant^{-1}. Means $\pm$ standard error of three replicates are shown. The expression levels were calculated relative to reference genes (*MsACT-101* for shoot and *MsEF1-α* for root) and to the control (0 mg Zn plant^{-1}). The broken line denotes the threshold between up- and downregulation relative to the control. Differences in the expressions of each gene after different Zn doses were tested separately for shoot and root by one-way analysis of variance. Different letters denote significant differences among Zn doses, according to Tukey-B test ($p < 0.05$).

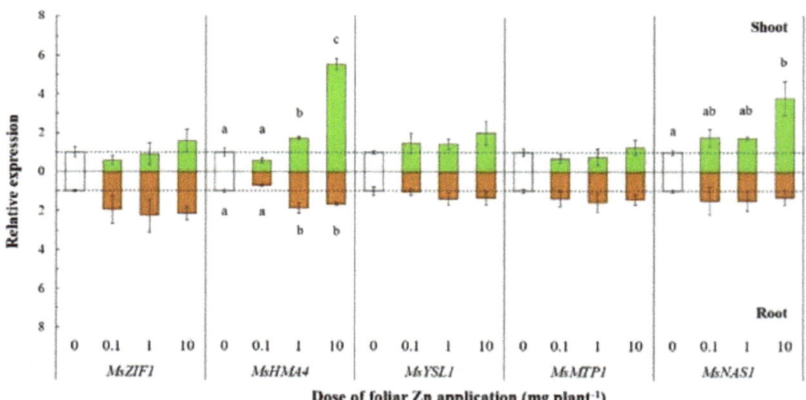

Figure 4. Relative expression of genes related to Zn transport processes after leaf Zn application to alfalfa. The Zn doses were 0, 0.1, 1, or 10 mg Zn plant^{-1}. Means $\pm$ standard error of three replicates are shown. The expression levels were calculated relative to reference genes (*MsACT-101* for shoot and *MsEF1-α* for root) and to the control (0 mg Zn plant^{-1}). The broken line denotes the threshold between up- and downregulation relative to the control. Differences in the expression of each gene at the different Zn doses were tested separately for shoot and root by one-way analysis of variance. Different letters denote significant differences among Zn doses, according to Tukey-B test ($p < 0.05$).

In shoots, the expression of *MsHMA4* was about threefold higher following a dose of 10 mg Zn plant^{-1} than following a dose of 1 mg Zn plant^{-1}, whereas the expression of MsHMA4 in roots was similar when 1 or 10 mg Zn plant^{-1} was applied.

The expression of *MsNAS1* was also significantly upregulated ($F_{(3,\,11)} = 6.46$, $p < 0.05$) at the largest foliar Zn dose (10 mg plant^{-1}), whereas its expression in roots was unaltered following foliar Zn application (Figure 4). In shoots, *MsYSL1* and *MsZIF1* were not significantly affected by foliar Zn application, although there was a trend towards upregulation of the expression with increasing Zn doses, while the expression of *MsMTP1* remained unaltered following the application of Zn (Figure 4). Finally, in roots, *MsMTP1* and *MsZIF1* were not significantly affected by foliar Zn application, although there was a trend towards upregulation of the expression with increasing Zn doses, whereas the expression of *MsYSL1* remained unchanged (Figure 4).

The permutation analysis of variance (PERMANOVA) showed that the expression of *ZIP* genes was significantly affected by foliar Zn application dose and differed between shoots and roots, which explained 29% and 23% of the total variance, respectively (Table 1). The expression of other genes related to Zn transport processes that were studied (*MsZIF1*, *MsNAS1*, *MsHMA4*, *MsYSL1*, and *MsMTP1*) were also affected by foliar Zn application dose and the organ examined. Zinc application dose explained 17% of the total variance, while plant organ explained 19%. PERMANOVA on all studied genes highlighted a significant effect of Zn application dose, plant organ, and their interaction on gene expression, explaining 68% of the total variance.

Table 1. Permutation analyses of variance (PERMANOVAs) on the effect of application of three doses of zinc (Zn) (0.1, 1, and 10 mg Zn plant^{-1}) and plant compartment (shoot and root) on the expression of seven *MsZIP* genes and separately on the expression of other five genes (*MsZIF1*, *MsNAS1*, *MsHMA4*, *MsYSL1*, and *MsMTP1*). A PERMANOVA was also performed on the response of all the genes. The analysis of homogeneity of multivariate dispersion (PERMDISP) was also performed. The studied plant was alfalfa (*Medicago sativa* L.). The analysis also included no-Zn addition control.

Response Variables	Explanatory Variables	Zn Application (Zn)	Plant Compartment (Comp)	Zn x Comp	Residual
ZIP genes	Pseudo F	5.56	8.16	1.76	
	P(perm)	**0.002**	**0.001**	0.082	
	Explained variance (%)	29.1	22.9	9.7	38.3
	PERMDISP P(perm)	0.412	0.852		
Other genes	Pseudo F	3.06	5.59	1.76	
	P(perm)	**0.007**	**0.015**	0.1	
	Explained variance (%)	17.3	19.35	12.78	
	PERMDISP P(perm)	0.412	0.852		
All genes	Pseudo F	4.27	10.49	3.41	
	P(perm)	**0.001**	**0.001**	**0.003**	
	Explained variance (%)	17.3	25.2	25.6	31.9
	PERMDISP P(perm)	0.152	**0.030**		

To summarize, among the 12 studied genes, only the expression of *MsZIP2*, *MsZIP3*, *MsHMA4*, and *MsNAS1* changed after foliar Zn application. *MsZIP2* and *MsHMA4* were upregulated in shoots and roots, whereas *MsZIP3* was downregulated and *MsNAS1* upregulated only in shoots.

2.2. Functional Modules of Genes Encoding Zn Transport-Related Processes

The second aim of this study was to provide novel information on how Zn transport-related genes are organized in functional modules in alfalfa. Using correlation analysis to reveal functional modules of genes whose expression is co-regulated in plants, we observed three functional modules for co-expression (Figure 5; $r > 0.6$). The first functional module of genes consisted of *MsZIP1*, *MsZIP5*, and *MsZIP6* in shoots and of *MsZIP1*, *MsZIP3*, *MsZIP4*, *MsZIP5*, and *MsZIP6* in roots. The second functional module of genes consisted of *MsMTP1* and *MsZIF1* in both shoots and roots. The third functional module of genes consisted of *MsHMA4*, *MsYSL1*, and *MsNAS1* in both shoots and roots. Moreover, while the expression pattern of ZIPs in shoots did not diverge from the one in roots, the expression pattern of the other genes involved in Zn transport-related processes strongly diverged (Figure 5).

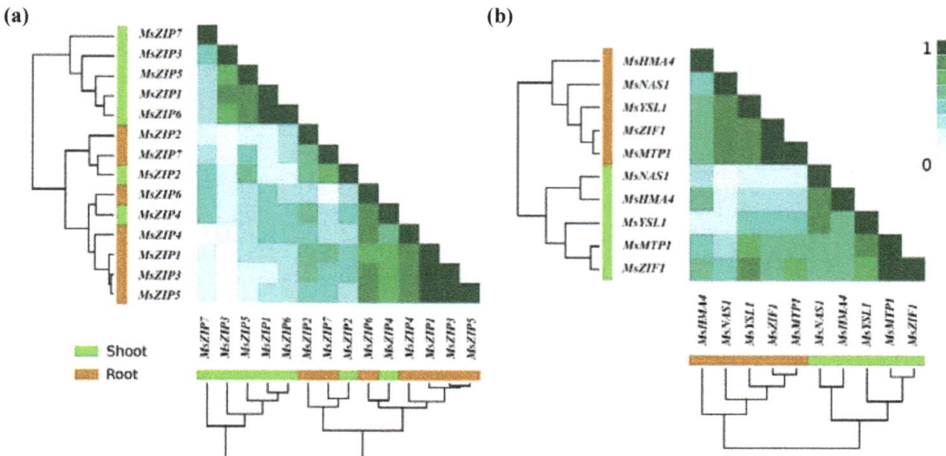

Figure 5. Heatmaps reporting correlations between expression of genes related to Zn transport after foliar Zn application. Gene expression is calculated as the difference between 0.1, 1, or 10 mg Zn plant^{-1} and a control of 0 mg Zn plant^{-1}. The similarity in the degree of correlation in fold-change of gene expression to Zn application relative to the control was based on the average linkage clustering of the Pearson correlations (*r*). In the clustering trees, the genes are indicated in brown for roots and in green for shoots, while the ranks of correlations of the heatmap are indicated by color intensity (*r* 0 to 1: from low to strong intensity of green). Seven genes encoding transmembrane Zn transporter (*MsZIP1-7*) (**a**); four genes encoding cellular Zn transporters (including vacuolar transporters) (*MsZIF1*, *MsHMA4*, *MsYSL1*, and *MsMTP1*) and a gene encoding a nicotianamine synthase (*MsNAS1*) (**b**).

The identification of these modules may allow for the definition of how the genes act in a concerted manner to redistribute Zn within the plant, maintaining non-toxic cytosolic and tissue Zn concentrations.

3. Discussion

In this work, for the first time, Zn biofortification was applied to the most productive and widely cultivated forage legume, alfalfa. Specific qPCR assays were designed for this crop and were successfully validated to study the gene expression response to foliar Zn application. We firstly characterized the expression of Zn transport-related genes after foliar Zn application to alfalfa and provide new molecular insights by identifying three functional gene modules involved in Zn influx to cells, Zn sequestration in the vacuole, and Zn redistribution within the plant.

3.1. Zn Redistribution within the Plant after Foliar Zn Application

The critical leaf concentration for Zn deficiency approximates 15–20 µg Zn g^{-1} dry weight and the critical leaf concentration for Zn toxicity approximates 400–500 µg Zn g^{-1} [12,13]. Before foliar Zn application, the alfalfa plants used in the experiments reported here were probably Zn-deficient, since their shoot Zn concentrations were below the critical leaf concentration for Zn deficiency (Figure 2). After the application of the lowest foliar Zn dose (0.01 mg plant^{-1}), plants probably remained Zn-deficient (7.6 µg Zn g^{-1} dry weight), but all other foliar Zn doses increased Zn concentrations in shoots above the critical concentration for Zn deficiency (Figure 2). Plants treated with 0.1 mg Zn plant^{-1} probably had an optimal Zn status for plant growth, whereas plants treated with 0.5 and 1 mg Zn plant^{-1} had shoot Zn concentrations close to the toxicity threshold. When a foliar dose of 10 mg Zn plant^{-1} was applied, shoot Zn concentrations greatly exceeding the threshold for Zn toxicity (Figure 2). Plants often exhibit characteristic visual symptoms of Zn deficiency and Zn toxicity when these occur [12,13], but 5 days after foliar Zn application, no visual symptoms of Zn deficiency or toxicity, nor differences in plant biomass, were observed among plants receiving contrasting foliar Zn doses (data not shown). Foliar Zn doses larger than 0.1 mg Zn plant^{-1} resulted in incremental increases in the Zn concentration and content of roots (Figure 2), despite Zn having limited mobility in the phloem [20,55]. This observation suggests that roots can act as a sink for Zn applied to leaves, thereby mitigating excessive Zn accumulation in shoot tissues.

In previous work, foliar application of Zn was shown to increase Zn concentration in phloem-fed tissues, such as fruits, seed, and tubers [56–59]. The shoot to root Zn concentration ratio shifted from values below one in conditions of Zn deficiency (0.4) to values greater than one in Zn-replete or Zn-intoxicated plants (1.3–3.2) (Figure 2). When the plants are Zn-deficient, the recirculation of Zn between organs via the xylem and phloem is required to meet minimal growth demands and the application of foliar Zn to Zn-deficient plants must be effectively redistributed within the plant [42,60], whereas when excessive foliar Zn is applied, Zn must be chelated in the cytoplasm, sequestered in the vacuole, and redistributed via the phloem or xylem to other organs to avoid toxicity [12].

3.2. Phylogenetic and Gene Expression Analysis

Despite several genes encoding Zn transporters having been identified in plants, and the encoded proteins characterized, the mechanisms of Zn uptake and transport in alfalfa are still largely unknown. However, the recently sequenced alfalfa genome has allowed for the discovery of genes involved in Zn uptake and distribution within this species [61].

The influx and efflux of Zn across the plasma membrane of plant cells must be tightly controlled to allow optimal cell functioning and hence to ensure normal plant growth and development [42]. The expression of only two of the seven *ZIP* genes studied, *MsZIP2* and *MsZIP3*, showed statistically significant responses to foliar Zn application (Figure 3). The expression of *MsZIP2* was significantly upregulated in both shoots and roots in response to the largest dose of foliar Zn applied (10 mg Zn plant^{-1}). It is likely that this dose is toxic to both shoot and root cells. The relative induction in the expression of *MsZIP2* was greatest in roots. The phylogenetic analysis of *ZIP* transporters revealed that *MsZIP2* is closely related to *MtZIP2* and *AtZIP2* (Figure S1). The *AtZIP2* protein was previous found in the same clade with HsZIP2 [62]. Thus, *MsZIP2* is probably located in the plasma membrane performing similar functions to *MtZIP2*, *AtZIP2*, and *HsZIP2*. The authors of [52] reported that *M. truncatula* plants grown with adequate soil Zn availability expressed *MtZIP2* in roots and stems, but not in leaves. The expression of *MtZIP2* in roots increased with increasing Zn fertilizer applications to soil, with the greatest expression being found at toxic Zn doses [52]. Similarly, the authors of [30] found that the expression of *AtZIP2* was ≈10-fold higher in roots than shoots in Zn-replete *Arabidopsis thaliana* plants and that Zn deficiency reduced the expression of *AtZIP2* in both roots and shoots. The localization of *ZIP2* at the plasma membrane was observed in both *M. truncatula* [49] and *A. thaliana* [30]. The expression of *AtZIP2* was localized to the stele of the root [30], supporting a role of *AtZIP2* in long distance transport of Zn between roots and shoots. It is possible that the increased expression of *MsZIP2* observed in our study when plants experience Zn toxicity might be a detoxification strategy, either through storing excess Zn in xylem parenchyma cells or recirculating Zn in the xylem.

The expression of *MsZIP3* was significantly downregulated in shoots following the foliar application of Zn (Figure 3). The *ZIP3* transporter is thought to mediate Zn influx to the cell from the apoplast [42]. Therefore, the downregulation of *MsZIP3* in shoots of plants receiving more Zn is consistent with the ability of plant cells to control their Zn uptake to affect cytoplasmic Zn homeostasis. Reduced expression of *MsZIP3* in plants with a greater Zn supply is also in agreement with previous studies of *M. truncatula* and *A. thaliana* [27,29], despite the higher phylogenetic similarity of *MsZIP3* to *MtZIP3* than to *AtZIP3* (Figure S1). However, although *AtZIP3* could restore growth to a Zn uptake-defective yeast [30], *MtZIP3* was not found to be able to restore the growth of a Zn uptake-defective yeast in Zn-limited media, although it did restore the growth of a Fe uptake-defective yeast in

Fe-limited media [29]. Thus, the MsZIP3 transporter could have a higher affinity for Fe than Zn. In *O. sativa*, the *ZIP3* gene is expressed in the xylem parenchyma and transfer cells and might be responsible for unloading transition metal cations from the xylem to the parenchyma in plants receiving an excessive Zn supply [53]. The role of OsZIP3 in unloading Zn from the vascular tissues suggests that the reduced expression of *MsZIP3* in shoots of *M. sativa* receiving an excessive foliar Zn dose might be a detoxification strategy to reduce Zn uptake by shoot cells.

The observation that foliar Zn applications had no effect on the expression of *ZIP* genes, except *MsZIP2* and *MsZIP3* (Figure 3), might be explained by the roles of ZIP proteins in the transport of other transition metals. For example, evidence of Cu and Mn transport by *ZIP4* were provided through yeast complementation studies [29,63]. Moreover, applying the same technique, a role of *ZIP6* was highlighted in the transport of Fe by [29], whereas the authors of [63] did not find any involvement of *ZIP6* in the transport of Cu, Zn, or Fe. Although the changes in the expression of *MsZIP1*, *MsZIP5*, and *MsZIP6* following foliar Zn application were not statistically significant, changes in their expression in shoots were positively correlated with changes in the expression of *MsZIP3*, showing a general trend for them to be downregulated following foliar Zn application and suggesting that these four ZIPs might act as a functional module in the shoot (Figure 5). By contrast, the expression of *MsZIP1*, *MsZIP3*, *MSZIP4*, and *MsZIP5* were positively correlated in roots, suggesting that these genes behave as a functional module in roots.

The expression of *MsHMA4*, which is implicated in Zn redistribution within the plant [50–64], was increased in both shoots and roots of plants whose shoot Zn concentration suggested they were close to, or experiencing, Zn toxicity (Figure 4). The significant upregulation of *MsHMA4* following foliar application of ≥ 1 mg Zn plant^{-1} might be related to the removal of excess Zn from both shoots and roots. This interpretation is consistent with the role of HMA4 in *A. thaliana* and in the metal hyperaccumulators *Arabidopsis halleri* and *Noccea caerulescens* [12,50,65–67], in which greater expression of *HMA4* results in greater Zn flux to the xylem and Zn translocation to transpiring leaves. However, the phylogenetic similarity of *MsHMA4* to *MtHMA4* and, particularly, to *AtHMA5* (Figure S3b) suggests a role in Cu transport [68–70]. The implication of the latter observation is unclear.

Since Zn^{2+} concentrations are low in the alkaline phloem sap, the transport of most Zn in the phloem is as Zn ligand complexes, such as zinc–nicotianamine (NA–Zn) [71]. Nicotianamine is the main Zn chelate in phloem transport and is also important for Zn sequestration in vacuoles [43], and tolerance of excessive Zn uptake [46]. Nicotianamine concentrations generally correlate with those of *NAS* transcripts, and for this reason *NAS* expression can be used as a proxy for NA content [48,72]. Accordingly, in the work reported here the increased expression of *MsNAS1* in shoots following the application of ≥ 1 mg Zn plant^{-1} (Figure 4) probably reflects the role of NA in Zn detoxification through its sequestration within vacuoles and its redistribution from shoot to root after excessive foliar Zn applications. This observation is consistent with the work of [71], who found that the expression of *NAS2* in wheat increased following foliar Zn application, despite the high phylogenetic distance of the *NAS* genes in *M. sativa* and wheat (Figure S4). Moreover, other authors reported that *NAS* expression is constitutively high in plants that hyperaccumulate Zn [12,72–74].

Homologs of *MsMTP1* and *MsZIF1* were previously found to encode transporters loading Zn and NA into the vacuoles of *Thlaspi geosingense* and *A. thaliana* cells, respectively [48,75]. Unexpectedly, the expression of these genes was unaffected by foliar Zn application (Figure 4). This observation suggests that the proteins encoded by these genes might not contribute to Zn detoxification in *M. sativa*. Nevertheless, only *MsZIF1* of all the genes studied here showed a trend towards increased expression in roots with increasing foliar Zn dose, which might indicate a role in detoxification of excess Zn in roots through its sequestration with NA in the vacuole.

In *A. thaliana*, *AtYSL1* has a role in the long-distance transport of the NA–Zn complex and in loading Zn into seeds [41,76]. For this reason and according to the similarities in the phylogenetic tree between *MsYSL1/MtYSL1* and *AtYSL1* (Figure S3a), an increase in the expression of *MtYSL1* was expected to occur in parallel with the increased expression of *MsNAS1* in shoots. However, the expression of *MsYSL1* did not show any significant change in shoots or roots in response to foliar Zn application, although there was a trend towards greater *MsYSL1* expression in shoots with increasing foliar Zn doses (Figure 4).

3.3. Functional Modules of Genes Encoding Zn Transport-Related Processes

The responses of gene expression to foliar Zn applications suggest three functional modules that effect cytoplasmic Zn homeostasis through Zn transport-related processes in *M. sativa*: genes involved in Zn influx to cells (shoots: *MsZIP1*, *MsZIP5*, and *MsZIP6*; roots: *MsZIP1*, *MsZIP3*, *MSZIP4*, *MsZIP5*, and *MsZIP6*), genes involved in Zn sequestration in the vacuole (shoots and roots: *MsMTP1* and *MsZIF1*), and genes involved in Zn redistribution within the plant (shoots and roots: *MsHMA4*, *MsYSL1*, and *MsNAS1*) (Figure 5). In a previous work that jointly analyzed the structures and phylogenetic trees of 21 *ZIP* genes in *Populus trichocarpa* in response to metal stress, four classes of genes were identified [77]. Among them, class I and class II were identified as involved in the transportation and absorption of metal ions (i.e., Zn, iron, copper, and manganese) during nutritional surpluses, while class III and class IV were identified as induced for metal ion transport during stress. Similarly to our results, in *PtrZIP1*, *PtrZIP4*, *PtrZIP5*, and *PtrZIP6* belonged to the same class (i.e., class I), but the pattern of gene expression under Zn deficiency and Zn application differed. Accordingly, a joint sequence and expression analysis of ZIP transporter genes revealed coexpression networks in iron acquisition strategies in land plants as well as in green algae [78].

The high correlation found in the present work between the expression of *MsMTP1* and *MsZIF1* in both shoots and roots (Figure 5) is also supported by previous works reporting a synergistic action of these gene for the sequestration of Zn in the vacuole [48,75,79]. Finally, the high correlation in the expression of *MsHMA4*, *MsYSL1*, and *MsNAS1* found in shoots and roots in response to foliar Zn applications (Figure 5) supports the expectation that these genes are components of a functional module affecting the long-distance transport of Zn in the plant, as it was previously highlighted in *A. thaliana* by [80].

However, more effort should be made in further studies to verify the localization of those proteins within cell and tissue of *M. sativa* as well as of other known orthologs involved in Zn transport. Moreover, additional time course studies should be performed to account for time-dependent responses.

4. Materials and Methods

4.1. Plant Growth and Experimental Design

Surface-sterilized seeds of alfalfa (*M. sativa* L.) were germinated on moist sterilized silica sand (1–4 mm size) in a climatic chamber at 24/21 °C day/night temperature, 16/18 h light/dark cycle, and 200 µmol photons m^{-2} s^{-1}. After 2 weeks of growth, 3 seedlings were transplanted to 1500 mL volume pots and filled with sterilized silica sand (number of pots 18), and *Sinorhizobium meliloti* was supplied as a filtrate to all plants to ensure that the plants produced nodules in all treatments. A Hoagland nutrient solution lacking Zn [81] was used to fertilize the plants, with 10 mL solution being applied every week. After 2 months of growth, when plants were in the vegetative growth stage, plants were treated with 1 of 6 doses of Zn (0, 0.01, 0.1, 0.5, 1, and 10 mg Zn $plant^{-1}$) (3 replicates per dose). Six $ZnSO_4 \cdot 7H_2O$ solutions of 0, 0.05, 0.5, 2.5, 5, and 50 g Zn L^{-1} were prepared to supply these doses. A drop of Tween 20 detergent was added to the 6 solutions to break the surface tension of the leaves and enhance Zn uptake. Zinc was applied to the middle leaf laminae of the 3 plants in each pot as twenty 10 µL droplets. The experiment was arranged in a fully randomized design, with 3 replicates for each Zn dose. The shoots and roots of the plants were harvested separately 5 days after Zn application. At harvest, 1 mM $CaCl_2$ solution and water were used to remove any residual Zn from the leaf surface [82]. Shoot and root fresh weight was measured, whereas shoot and root dry weight was determined on subsamples after oven drying at 70 °C to constant weight.

4.2. Measurement of Zn Concentration

Approximately 100 mg of shoot or root dry biomass was carefully weighed and mineralized in a microwave medium pressure digestor (Milestone Start D, FKV Srl, Torre Boldone, Italy) with 7 mL of 69% HNO_3 and 2 mL of 30% H_2O_2 (ultrapure grade). Zinc concentration in the resulting solutions was determined by inductively coupled plasma optical emission spectroscopy (ICP-OES) using an Optima 8000 spectrometer (Perkin Elmer, Waltham, MA, USA), following the procedure of [83].

4.3. Gene Selection and Design and Validation of New RT-qPCR Assay

Seven genes encoding putative ZRT-IRT-like proteins (ZIP) were selected for investigation (i.e., *ZIP1-7*) (Figure 1). The selection was based on information gathered by [29,52] on the expression of genes encoding Zn transporters in the model legume *M. truncatula* and on the structure of the neighbor-joining (NJ) tree built using available ZIP sequences of several plant species. Five more genes, whose products are involved in Zn transport-related processes [24], were also chosen for investigation on the basis of information gathered by other authors and on sequence similarity with other plant species. The *NAS1* gene encoding nicotianamine synthase (NAS) was chosen because this enzyme synthesizes nicotianamine (NA), which is involved in long-distance Zn transport [41] (Figure 1). The *HMA4* gene, which encodes a transmembrane P-type ATPase heavy metal transporter, was chosen because this transporter loads Zn into the xylem in roots for its transport to shoots [51] (Figure 1). The *MTP1* gene, which encodes a transporter of the CDF family, was selected because this transporter is implicated in the sequestration of excess Zn in the vacuole [49,75] (Figure 1). The *ZIF1* gene, which encodes the Zn-induced facilitator 1 transporter, was chosen because it transports NA into the vacuole to chelate vacuolar Zn [48,84] (Figure 1). The *YSL1* gene, which encodes a transporter of NA–Zn complexes, was chosen because it is implicated in Zn loading and transport of Zn in the phloem [60] (Figure 1). To standardize the expression of genes encoding Zn transport-related processes, we selected 2 reference genes: actin (*ACT-101*) and elongation factor 1-α (*EF1-α*) [85].

Using the draft genome sequence of alfalfa in the Alfalfa Gene Index and Expression Atlas Database (AEGD) [61] (http://plantgrn.noble.org/AGED/index.jsp (accessed on 2 March 2021)), we retrieved homologous gene sequences of *M. sativa* by Basic Local Alignment Search Tool (BLAST) similarity searches using the gene sequences of *M. truncatula*. The chosen genes for *M. sativa* were named *MsZIP1-7* for the seven *ZIP* genes and *MsNAS1*, *MsHMA4*, *MsZIF1*, *MsYSL1*, and *MsMTP1* for the other selected genes. The two reference genes were named *MsACT-101* and *MsEF1-α*. The gene sequences and their annotations have been deposited in the National Center for Biotechnology Information (NCBI) under the submission # 2338923.

Forward and reverse new PCR primers for the 12 Zn transport-related genes and the 2 reference genes suitable for SYBR Green II RT-qPCR assays (Biorad, Hercules, CA, USA) were designed (Table 2). The Primer-BLAST online tool in the National Center for Biotechnology Information (NCBI; https://www.ncbi.nlm.nih.gov/tools/primer-blast/ (accessed on 2 March 2021)) was used to design primers. The newly designed RT-qPCR assays are suitable for both *M. sativa* and *M. truncatula*. These assays are the first designed specifically for alfalfa and resulted to be more efficient than the ones already available for *M. truncatula* (i.e., *MtZIP1-7* and *MtMTP1*; [29,52,86]). The length of the fragment, the Sanger sequences of the PCR amplicons (Table 1), and the single melting temperature peaks confirmed the specificity of the new RT-qPCR assays (Figure S5). Sanger sequencing was performed on PCR amplicons of 3 complementary DNA (cDNA) samples (Material and Methods S1). Examples of electropherograms of the sequences are reported in Figures S6 and S7. The sequences of the obtained PCR amplicons have been deposited in the NCBI under the submission # 2338930. Amplification efficiencies (E) in the range of

96.1–111.0% were evidence of accurate quantification, while the coefficients of correlation (R^2 > 0.998) indicated a high precision of measurements across concentration ranges of at least 3–4 orders of magnitude (Table 2 and Figure S8). The concentration ranges over which the relationship between the relative fluorescence and the logarithm of the concentration was linear, and the precision of quantification (standard curves) as reflected in the coefficient of correlation (R^2), was determined using 3 independent 10-fold serial dilutions of a cDNA sample of *M. sativa*. The accuracy of quantification was determined by the efficiency (E) of each qPCR amplification using the equation $E = [10^{-1/S} - 1] \times 100$, where S is the slope of the standard curve. The evaluation of the reference genes based on the cycle threshold (Ct) values made us choose the actin gene (*MsACT-101*) for quantifying relative gene expression in the shoots and the elongation factor 1-α (*MsEF1-α*) gene for quantifying relative gene expression in roots (Figure S9a,b). This choice was based on the observations that there was no statistical difference in the expression of the reference genes in tissues following foliar Zn applications and that *MsACT-101* and *MsEF1-α* showed the smallest overall variation in the shoot and root, respectively (Figure S9c,d).

4.4. RNA Extraction and Gene Expression Analysis

Total RNA was extracted from 50 mg subsamples of fresh shoot and root tissue using the RNeasy Mini Kit (Qiagen, Hilden, Germany). The extractions were performed from tissues of plants treated with the foliar Zn doses that produced a significant increase in Zn concentration in shoots (0.1, 1, and 10 mg Zn plant^{-1}) and the control plants to which no foliar Zn had been applied (24 RNA extractions). Any DNA in the RNA extracts was removed by a DNase treatment (Promega, Madison, WI, USA). The purity of the RNA extracts was verified by spectroscopic light absorbance measurements at 230, 260, and 280 nm using the NanoDrop 2000 (Thermo Scientific, Worcester, MA, USA) [87]. The integrity and approximate concentration of the extracted RNA was determined by electrophoresis of the RNA extracts in a 1% agarose gel containing Sybr Safe (Invitrogen, Carlsbad, CA, USA). One microgram of total RNA was reverse transcribed to complementary DNA (cDNA) using the iScript cDNA Synthesis Kit (Biorad, Hercules, CA, USA) in a 20 µL reaction volume. The RT-qPCRs for gene expression analysis were run as 3 technical replicates with a final reaction volume of 20 µL, containing 10 µL of SYBR Green Supermix (Biorad, Hercules, CA, USA), 5 µL of 100-fold diluted cDNA, and 0.4 µM final concentrations of the gene-specific PCR primers on a CFX Connect Real-Time System thermal cycler (Biorad, Hercules, California). The qPCR conditions were 95 °C for 3′, followed by 40 cycles of 95 °C for 5′, and 60 °C for 30″. A dissociation curve of each reaction was performed (65° C to 95° C, 0.5° C increment every 5″) to check that PCR amplified only one product. The most suitable reference gene for relative gene expression analysis was determined by comparing the expression levels of the reference genes *MsACT-101* and *MsEF1-α* across all cDNA samples. Relative gene expression was calculated using the double standardization (ΔΔCq) method that requires a reference gene and a control treatment [88].

Table 2. Gene name and forward and reverse sequences of 14 newly designed primer pairs for the quantification of the expression of genes of alfalfa (*Medicago sativa*) encoding proteins involved in cellular zinc (Zn) influx and efflux and Zn chelation. Two reference genes (i.e., *MsACT-101* and *MsEF1-α*) were also designed. The length of the amplicons, the primer amplification efficiency (%), and R^2 of the standard curve are indicated. The reference sequences are indicated by the accession number of the *Medicago truncatula* sequences and by the contig number of the *M. sativa* sequences. The primers were designed online Primer-Basic Local Alignment Search Tool (BLAST). See Figure 1 for the full names of the genes. Reference sequence accession number from NCBI—GenBank Accession number—https://www.ncbi.nlm.nih.gov/genbank/ (accessed on 2 March 2021). Reference sequence contig number from AGED—The Alfalfa Gene Index and Expression Atlas Database—http://plantgrn.noble.org/AGED/ (accessed on 2 March 2021).

Gene *	Reference Sequence Accession Number and Contig Number	Forward Primer (5′-3′)	Reverse Primer (5′-3′)	Amplicon Size (bp)	Efficiency (%)	R^2
MsZIP1	AY339054 [†]/19855 [‡]	ATGATTAAAGCCTTCGCGGC	TCTGCTGGAACTTGTTTAGAAGG	233	99.8	0.999
MsZIP2	AY007281/82450	AGCCCAATTGGCGTAGGAAT	ACAGCAACACCAAAAAGCACA	215	99.3	0.999
MsZIP3	AY339055/33860	TGGTGTGATTTTGGCAACCG	TGACGGACCCGAAGAAACAG	325	104.9	0.999
MsZIP4	XM_003603101/92651	GGAGGGTGCATTTCTCAAGC	AGCAATGCCTGTTCCAATGC	108	97.1	0.999
MsZIP5	XM_013605712/66451	TGAAGGCATGGGACTTGGAA	CCAGCTGAAGCTGCATTGAA	192	99.3	0.998
MsZIP6	AY339058/9668	CTTGGCGACACGTTCAATCC	CCACAAGTCCCGAAAAGGGA	188	106.0	0.998
MsZIP7	AY339059/62098	GGCTTGTGCTGGTTATTTGAT	TTTCCATGCGTCTGCTTTTGT	310	96.1	0.999
MsZIF1	XM_003601836/59165	TGCCTGCATTTGGTTACCG	CTGCAGCTTCCACATTGTCAG	77	105.9	0.999
MsHMA4	XM_003626900/19210	TGCTCAACTTGCCAAAGCAC	GGAATGAACCATCCCAGCCA	111	108.9	0.999
MsYSL1	XM_024781439/4892	CAAGAAGCAAGTGCATGGGT	TCCACAGTCTTCTTTGCCTGAG	94	111.0	0.999
MsMTP1	FJ389717/67347	TGCAGCATTTGCCATCTCCT	TGCATAGAAACCAAAGCACCA	114	104.5	0.999
MsNAS1	XM_003594705/61146	GCTAGCTTGGCTGAAGATTGG	AGATACAAAGCACTCGGAGACA	87	100.5	0.999
MsACT-101	XM_003593074/89028	TCTCTGTATGCCAGTGGACG	TCTGTTAAATCACGCCCAGCA	140	102.4	0.999
MsEF1-α	XM_003618727/56897	CCACAGACAAGCCCCTCAG	TCACAACCATACCGGGCTTC	114	100.2	0.999

* See Figure 1 for the full names of the genes; [†] NCBI—GenBank Accession number—https://www.ncbi.nlm.nih.gov/genbank/ (accessed on 2 March 2021); [‡] AGED—The Alfalfa Gene Index and Expression Atlas Database—http://plantgrn.noble.org/AGED/ (accessed on 2 March 2021).

4.5. Bioinformatic and Statistical Analyses

A BLAST search was performed in the Alfalfa Gene Index and Expression Atlas database using the ZIP1-7, ZIF1, MTP1, YSL1, HMA4, and NAS coding sequences from *M. truncatula*. This allowed for the identification of gene sequences encoding potential metal transporters and chelators in the whole *M. sativa* genome. The sequences obtained were aligned with the corresponding sequences from *M. truncatula*, and the length of the *M. sativa* genes was determined after removing the external unaligned nucleotides. The *M. sativa* and *M. truncatula* ZIP gene sequences were also aligned with those of other plant species (*A. thaliana, G. max, H. vulgare, O. sativa, Triticum aestivum,* and *Zea mays*) obtained from a search of GenBank. Similarly, the *M. sativa* and *M. truncatula* gene sequences of ZIF1, MTP1, YSL1, HMA4, and NAS were aligned with their corresponding sequences of other plant species (*A. thaliana, G. max, H. vulgare, O. sativa, T. aestivum,* and *Z. mays*) obtained from a search of GenBank. Sequence alignments were performed using the algorithm ClustalW in MEGA X [89]. Phylogenetic comparisons were performed to infer the putative roles of the selected *M. sativa* Zn transport-related gens. This was based on the assumption of a simple equivalence between a minimum similarity threshold in the phylogenetic comparisons and the function similarity between encoded proteins. For some proteins belonging to the same family, this assumption can hold true, since they have been shown to have very tightly correlating functions, such as those considered in this study. Thus, functions are indicated with high probability by annotations based on similarities. The phylogenetic trees were inferred by neighbor-joining (NJ) analysis [90] in MEGA X, and the evolutionary distances were calculated using the p-distance method [91]. Branch support bootstrap values were derived from 500 bootstrap replicates. The phylograms were drawn by MEGA X and edited using Adobe Illustrator CC 2017.

The effect of the application of the foliar Zn on tissue Zn concentration and on the expression of the selected genes was analyzed in shoots and roots separately by one-way analysis of variance (ANOVA), followed by a Tukey-B test in the case of significance of the response to foliar Zn application. When required, gene expression data were log-transformed to meet the ANOVA assumptions. The data displayed graphically are the means and associated standard errors of the untransformed raw data. All statistical analyses were performed using the software package SPSS version 21.0 (SPSS Inc., Chicago, IL, USA). Permutational analysis of variance (PERMANOVA) [92] was used to test the effect of foliar Zn application and plant organ (shoot and root) on the expression of the 7 ZIP genes and of the other 5 genes encoding Zn transport-related processes separately. In addition, the PERMANOVA was performed on the expression of all the genes together. The response data matrices were standardized by sample and total, and then Euclidean distances were calculated among samples. *P*-values were calculated using the Monte Carlo test [93]. Since PERMANOVA is sensitive to differences in multivariate location and dispersion, analysis of homogeneity of multivariate dispersion (PERMDISP [94]) was performed to check the homogeneity of dispersion among groups. The analyses were performed using PRIMER 7 and PERMANOVA+ software [95]. Finally, heatmaps were constructed to illustrate correlations in expression among ZIPs and among other genes encoding Zn transport-related processes using the R package ggplot2 [96], using the average linkage clustering of the Pearson correlations calculated from relative gene expression following foliar Zn application.

5. Conclusions

This is the first study to characterize the expression of genes related to Zn transport processes following foliar Zn application to a forage legume, providing new molecular insights to the responses of Zn transport-related processes to foliar Zn applications. A significant increase in the expression of *MsZIP2* as foliar Zn doses increased suggests the detoxification of excess Zn through the accumulation of Zn in xylem parenchyma cells. A decrease in the expression of *MsZIP3* as foliar Zn doses increased suggests a reduction in the Zn influx capacity of shoot cells to reduce Zn uptake. An increase in the expression of *MsHMA4* in roots and shoots as foliar Zn doses increased suggests an increase in the transport of Zn in the xylem when plants are subject to Zn toxicity, while an increase in the expression of *MsNAS1* in the shoot suggests the chelation of excess Zn in the shoot, enabling Zn sequestration in vacuoles or the redistribution of Zn to roots via the phloem. The elucidation of three functional modules of genes involved in (a) Zn influx to cells, (b) sequestration of Zn in the vacuole, and (c) redistribution of Zn within the plant are fundamental to understanding the molecular mechanisms of cytoplasmic Zn homeostasis and might inform the selection of appropriate genotypes enabling greater Zn accumulation in edible portions or increased tolerance of Zn in the environment.

Supplementary Materials: The following are available online at https://www.mdpi.com/2223-7747/10/3/476/s1: Figure S1: Neighbor-joining (NJ) phylogenetic tree of ZIP gene sequences of alfalfa and other plant species. Figure S2: NJ phylogenetic trees of ZIF and MTP genes sequences ((a) and (b), respectively) of alfalfa and other plant species. Figure S3: NJ phylogenetic trees of YSL and HMA gene sequences ((a) and (b), respectively) of alfalfa and other plant species. Figure S4: NJ phylogenetic trees of NAS gene sequences of alfalfa and other plant species. Figure S5: Melting curve analysis of PCR products. Figure S6: Examples of electropherograms of PCR products of ZIP genes (*MsZIP1-7*). Figure S7: Examples of electropherograms of PCR products of the genes *MsZIF1*, *MsHMA4*, *MsYSL1*, *MsNAS1*, and *MsMTP1* and of the reference genes *MsACT-101* and *MsEF1-α*. Figure S8: Standard curves for the 14 newly designed primer pairs. Figure S9: Cycle threshold value of the reference genes. Table S1: Fresh and dry weight of shoots and roots of alfalfa after leaf Zn application. Material and methods S1: PCR details.

Author Contributions: Conceptualization, E.P. and L.E.; methodology, A.C., E.P., and L.E.; validation A.C. and B.M.; formal analysis, A.C. and M.C.M.; investigation, A.C., E.P., and L.E.; data curation, A.C. and E.P.; Writing—Original draft preparation, A.C., E.P., and L.E.; Writing—Review and editing, A.C., E.P., P.J.W., and L.E.; funding acquisition, L.E. All authors have read and agreed to the published version of the manuscript.

Funding: This research received no external funding.

Data Availability Statement: The gene sequences and their annotations have been deposited in the NCBI under the submission # 2338923. The sequences of the obtained PCR amplicons have been deposited in the NCBI under the submission # 2338930.

Acknowledgments: We acknowledge Hannes A. Gamper for the technical support in real-time RT-PCR.

Conflicts of Interest: The authors declare no conflict of interest.

References

1. Prasad, A.S. Discovery of human zinc deficiency: Its impact on human health and disease. *Adv. Nutr.* **2013**, *4*, 176–190. [CrossRef]
2. Cakmak, I.; McLaughlin, M.J.; White, P. Zinc for better crop production and human health. *Plant Soil* **2017**, *411*, 1–4. [CrossRef]
3. Koren, O.; Tako, E. Chronic dietary zinc deficiency alters gut microbiota composition and function. *Multidiscip. Digital Publ. Inst. Proc.* **2020**, *61*, 16. [CrossRef]
4. Desta, M.K.; Broadley, M.R.; McGrath, S.P.; Hernandez-Allica, J.; Hassall, K.L.; Gameda, S.; Amede, T.; Haefele, S.M. Plant vailable zinc is influenced by landscape position in the Amhara region, Ethiopia. *Plants* **2021**, *10*, 254. [CrossRef]
5. McDonald, P.; Edwards, R.A.; Greenhalgh, J.F.D.; Morgan, C.A. *Animal Nutrition*; Pearson Education Limited: Harlow, UK, 2002.
6. Ciccolini, V.; Pellegrino, E.; Coccina, A.; Fiaschi, A.I.; Cerretani, D.; Sgherri, C.; Quartacci, M.F.; Ercoli, L. Biofortification with iron and zinc improves nutritional and nutraceutical properties of common wheat flour and bread. *J. Agric. Food Chem.* **2017**, *65*, 5443–5452. [CrossRef] [PubMed]
7. Capstaff, N.M.; Miller, A.J. Improving the yield and nutritional quality of forage crops. *Front. Plant Sci.* **2018**, *9*, 1–18. [CrossRef] [PubMed]
8. Huma, Z.E.; Khan, Z.I.; Noorka, I.R.; Ahmad, K.; Bayat, A.R.; Wajid, K. Bioaccumulation of zinc and copper in tissues of chicken fed corn grain irrigated with different water regimes. *Int. J. Environ. Res.* **2019**, *13*, 689–703. [CrossRef]
9. Broadley, M.R.; White, P.J.; Hammond, J.P.; Zelko, I.; Lux, A. Zinc in plants. *New Phytol.* **2007**, *173*, 677–702. [CrossRef]
10. Sasaki, H.; Hirose, T.; Watanabe, Y.; Ohsugi, R. Carbonic anhydrase activity and CO_2-transfer resistance in Zn-deficient rice leaves. *Plant Physiol.* **1998**, *118*, 929–934. [CrossRef]
11. Albert, I.L.; Nadassy, K.; Wodak, S.J. Analysis of zinc binding sites in protein crystal structures. *Protein Sci.* **1998**, *7*, 1700–1716. [CrossRef]
12. White, P.J.; Pongrac, P. Heavy-metal toxicity in plants. In *Plant Stress Physiology*; CABI: Wallingford, UK, 2017; pp. 301–331.

13. Broadley, M.R.; Brown, P.; Cakmak, I.; Rengel, Z.; Zhao, F. Function of nutrients: Micronutrients. In *Marschner's Mineral Nutrition of Higher Plants*; Academic Press: London, UK, 2012; pp. 191–248.

14. Alloway, B.J. *Zinc in Soils and Crop Nutrition*; IZA and IFA: Brussels, Belgium; Paris, France, 2008.

15. Chaney, R.L. Zinc phytotoxicity. In *Zinc in Soils and Plants*; Springer: Dordrecht, The Netherlands, 1993; pp. 135–150.

16. Di Baccio, D.; Tognetti, R.; Minnocci, A.; Sebastiani, L. Responses of the *Populus x euramericana* clone I-214 to excess zinc: Carbon assimilation, structural modifications, metal distribution and cellular localization. *Environ. Exp. Bot.* **2009**, *67*, 153–163. [CrossRef]

17. White, P.J.; Broadley, M.R. Biofortifying crops with essential mineral elements. *Trends Plant Sci.* **2005**, *10*, 586–593. [CrossRef]

18. Saltzman, A.; Birol, E.; Bouis, H.E.; Boy, E.; De Moura, F.F.; Islam, Y.; Pfeiffer, W.H. Biofortification: Progress toward a more nourishing future. *Glob. Food Secur.* **2013**, *2*, 9–17. [CrossRef]

19. Gregory, P.J.; Wahbi, A.; Adu-Gyamfi, J.; Heiling, M.; Gruber, R.; Joy, E.J.M.; Broadley, M.R. Approaches to reduce zinc and iron deficits in food systems. *Glob. Food Secur.* **2017**, *15*, 1–10. [CrossRef]

20. White, P.J.; Broadley, M.R. Physiological limits to zinc biofortification of edible crops. *Front. Plant Sci.* **2011**, *2*, 80. [CrossRef]

21. Rawat, N.; Neelam, K.; Tiwari, V.K.; Dhaliwal, H.S. Biofortification of cereals to overcome hidden hunger. *Plant Breeding* **2013**, *132*, 437–445. [CrossRef]

22. Cakmak, I. HarvestPlus zinc fertilizer project: HarvestZinc. *Better Crops* **2012**, *96*, 17–19.

23. White, P.J.; Broadley, M.R. Biofortification of crops with seven mineral elements often lacking in human diets–iron, zinc, copper, calcium, magnesium, selenium and iodine. *New Phytol.* **2009**, *182*, 49–84. [CrossRef] [PubMed]

24. Olsen, L.I.; Palmgren, M.G. Many rivers to cross: The journey of zinc from soil to seed. *Front. Plant Sci.* **2014**, *5*, 30. [CrossRef]

25. Caldelas, C.; Weiss, D.J. Zinc homeostasis and isotopic fractionation in plants: A review. *Plant Soil* **2017**, *411*, 17–46. [CrossRef]

26. Hacisalihoglu, G. Zinc (Zn): The last nutrient in the alphabet and shedding light on Zn efficiency for the future of crop production under suboptimal zn. *Plants* **2020**, *9*, 1471. [CrossRef] [PubMed]

27. Grotz, N.; Fox, T.; Connolly, E.; Park, W.; Guerinot, M.L.; Eide, D. Identification of a family of zinc transporter genes from Arabidopsis that respond to zinc deficiency. *Proc. Natl. Acad. Sci. USA* **1998**, *95*, 7220–7224. [CrossRef] [PubMed]

28. Zhao, H.; Eide, D. The yeast ZRT1 gene encodes the zinc transporter protein of a high-affinity uptake system induced by zinc limitation. *Proc. Natl. Acad. Sci. USA* **1996**, *93*, 2454–2458. [CrossRef] [PubMed]

29. López-Millán, A.F.; Ellis, D.R.; Grusak, M.A. Identification and characterization of several new members of the ZIP family of metal ion transporters in *Medicago truncatula*. *Plant Mol. Biol.* **2004**, *54*, 583–596. [CrossRef]

30. Milner, M.J.; Seamon, J.; Craft, E.; Kochian, L.V. Transport properties of members of the ZIP family in plants and their role in Zn and Mn homeostasis. *J. Exp. Bot.* **2013**, *64*, 369–381. [CrossRef]

31. Tiong, J.; McDonald, G.; Genc, Y.; Shirley, N.; Langridge, P.; Huang, C.Y. Increased expression of six ZIP family genes by zinc (Zn) deficiency is associated with enhanced uptake and root-to-shoot translocation of Zn in barley (*Hordeum vulgare*). *New Phytol.* **2015**, *207*, 1097–1109. [CrossRef] [PubMed]

32. Mäser, P.; Thomine, S.; Schroeder, J.I.; Ward, J.M.; Hirsch, K.; Sze, H.; Talke, I.N.; Amtmann, A.; Maathuis, F.J.M.; Sanders, D.; et al. Phylogenetic relationship within cation transporter families of Arabidopsis. *Plant Physiol.* **2001**, *126*, 1646–1667. [CrossRef]

33. Eckhardt, U.; Marques, A.M.; Buckhout, T.J. Two iron-regulated cation transporters from tomato complement metal uptake-deficient yeast mutants. *Plant Mol. Biol.* **2001**, *45*, 437–448. [CrossRef]

34. Ramesh, S.A.; Shin, R.; Eide, D.J.; Schachtman, D.P. Differential metal selectivity and gene expression of two zinc transporters from rice. *Plant Physiol.* **2003**, *133*, 126–134. [CrossRef]

35. Ishimaru, Y.; Suzuki, M.; Tsukamoto, T.; Suzuki, K.; Nakazono, M.; Kobayashi, T.; Wada, Y.; Watanabe, S.; Matsuhashi, S.; Nakanishi, H.; et al. Rice plants take up iron as an Fe3þ-phytosiderophore and as Fe2þ. *Plant J.* **2006**, *45*, 335–346. [CrossRef]

36. Eide, D.; Broderius, M.; Fett, J.; Guerinot, M.L. A novel iron-regulated metal transporter from plants identified by functional expression in yeast. *Proc. Natl. Acad. Sci. USA* **1996**, *93*, 5624–5628. [CrossRef] [PubMed]

37. Bughio, N.; Yamaguchi, H.; Nishizawa, N.K.; Nakanishi, H.; Mori, S. Cloning an iron-regulated metal transporter from rice. *J. Exp. Bot.* **2002**, *53*, 1677–1682. [CrossRef]

38. Vert, G.; Grotz, N.; Dédaldéchamp, F.; Gaymard, F.; Guerinot, M.L.; Briat, J.F.; Curie, C. IRT1, an Arabidopsis transporter essential for iron uptake from the soil and for plant growth. *Plant Cell* **2002**, *14*, 1223–1233. [CrossRef]

39. Pedas, P.; Ytting, C.K.; Fuglsang, A.T.; Jahn, T.P.; Schjoerring, J.K.; Husted, S. Manganese efficiency in barley: Identification and characterization of the metal ion transporter HvIRT1. *Plant Physiol.* **2008**, *148*, 455–466. [CrossRef] [PubMed]

40. Kolaj-Robin, O.; Russell, D.; Hayes, K.A.; Pembroke, J.T.; Soulimane, T. Cation diffusion facilitator family: Structure and function. *FEBS Lett.* **2015**, *589*, 1283–1295. [CrossRef]

41. Curie, C.; Cassin, G.; Couch, D.; Divol, F.; Higuchi, K.; Le Jean, M.; Misson, J.; Shikora, A.; Czernic, P.; Mari, S. Metal movement within the plant: Contribution of nicotianamine and yellow stripe 1-like transporters. *Ann. Bot.* **2009**, *103*, 1–11. [CrossRef] [PubMed]

42. Sinclair, S.A.; Krämer, U. The zinc homeostasis network of land plants. *BBA Mol. Cell. Res.* **2012**, *1823*, 1553–1567. [CrossRef]

43. Deinlein, U.; Weber, M.; Schmidt, H.; Rensch, S.; Trampczynska, A.; Hansen, T.H.; Husted, S.; Schjoerring, J.K.; Talke, I.N.; Krämer, U.; et al. Elevated nicotianamine levels in *Arabidopsis halleri* roots play a key role in zinc hyperaccumulation. *Plant Cell* **2012**, *24*, 708–723. [CrossRef] [PubMed]

44. Foroughi, S.; Baker, A.J.M.; Roessner, U.; Johnson, A.A.T.; Bacic, A.; Callahan, D.L. Hyperaccumulation of zinc by *Noccaea caerulescens* results in a cascade of stress responses and changes in the elemental profile. *Metallomics* **2014**, *6*, 1671–1682. [CrossRef] [PubMed]

45. Foyer, C.H.; Lam, H.-M.; Nguyen, H.T.; Siddique, K.H.M.; Varshney, R.K.; Colmer, T.D.; Cowling, W.; Bramley, H.; Mori, T.A.; Hodgson, J.M.; et al. Neglecting legumes has compromised human health and sustainable food production. *Nat. Plants* **2016**, *2*, 16112. [CrossRef]

46. Aarts, M.G. Nicotianamine secretion for zinc excess tolerance. *Plant Physiol.* **2014**, *166*, 751–752. [CrossRef] [PubMed]

47. Clemens, S.; Deinlein, U.; Ahmadi, H.; Höreth, S.; Uraguchi, S. Nicotianamine is a major player in plant Zn homeostasis. *Biometals* **2013**, *26*, 623–632. [CrossRef] [PubMed]

48. Haydon, M.J.; Kawachi, M.; Wirtz, M.; Hillmer, S.; Hell, R.; Krämer, U. Vacuolar nicotianamine has critical and distinct roles under iron deficiency and for zinc sequestration in *Arabidopsis*. *Plant Cell* **2012**, *24*, 724. [CrossRef] [PubMed]

49. Desbrosses-Fonrouge, A.G.; Voigt, K.; Schröder, A.; Arrivault, S.; Thomine, S.; Krämer, U. Arabidopsis thaliana MTP1 is a Zn transporter in the vacuolar membrane which mediates Zn detoxification and drives leaf Zn accumulation. *FEBS Lett.* **2005**, *579*, 4165–4174. [CrossRef] [PubMed]

50. Hussain, D.; Haydon, M.J.; Wang, Y.; Wong, E.; Sherson, S.M.; Young, J.; Camakaris, J.; Harper, J.F.; Cobbett, C.S. P-type ATPase heavy metal transporters with roles in essential zinc homeostasis in Arabidopsis. *Plant Cell* **2004**, *16*, 1327–1339. [CrossRef]

51. Palmer, C.M.; Guerinot, M.L. Facing the challenges of Cu, Fe and Zn homeostasis in plants. *Nat. Chem. Biol.* **2009**, *5*, 333–340. [CrossRef] [PubMed]

52. Burleigh, S.H.; Kristensen, B.K.; Bechmann, I.E. A plasma membrane zinc transporter from *Medicago truncatula* is up-regulated in roots by Zn fertilization, yet down-regulated by arbuscular mycorrhizal colonization. *Plant Molec. Biol.* **2003**, *52*, 1077–1088. [CrossRef] [PubMed]

53. Sasaki, A.; Yamaji, N.; Mitani-Ueno, N.; Kashino, M.; Ma, J.F. A node-localized transporter OsZIP3 is responsible for the preferential distribution of Zn to developing tissues in rice. *Plant J.* **2015**, *84*, 374–384. [CrossRef]

54. Fageria, N.K.; Filho, M.B.; Moreira, A.; Guimarães, C.M. Foliar fertilization of crop plants. *J. Plant Nutr.* **2009**, *32*, 1044–1064. [CrossRef]

55. White, P.J. Long-distance transport in the xylem and phloem. In *Marschner's Mineral Nutrition of Higher Plants*; Academic Press: London, UK, 2012; pp. 49–70.

56. Cakmak, I.; Torun, A.; Millet, E.; Feldman, M.; Fahima, T.; Korol, A.; Nevo, E.; Braun, H.J.; Özkan, H. Triticum dicoccoides: An important genetic resource for increasing zinc and iron concentration in modern cultivated wheat. *J. Soil Sci. Plant Nut.* **2004**, *50*, 1047–1054. [CrossRef]

57. Cakmak, I. Enrichment of cereal grains with zinc: Agronomic or genetic biofortification? *Plant Soil* **2008**, *302*, 1–17. [CrossRef]

58. Cakmak, I.; Pfeiffer, W.H.; McClafferty, B. Biofortification of durum wheat with zinc and iron. *Cereal Chem.* **2010**, *87*, 10–20. [CrossRef]

59. White, P.J.; Thompson, J.A.; Wright, G.; Rasmussen, S.K. Biofortifying Scottish potatoes with zinc. *Plant Soil* **2017**, *411*, 151–165. [CrossRef]

60. Erenoglu, E.B.; Kutman, U.B.; Ceylan, Y.; Yildiz, B.; Cakmak, I. Improved nitrogen nutrition enhances root uptake, root-to-shoot translocation and remobilization of zinc (^{65}Zn) in wheat. *New Phytol.* **2011**, *189*, 438–448. [CrossRef] [PubMed]

61. O'Rourke, J.A.; Fu, F.; Bucciarelli, B.; Yang, S.S.; Samac, D.A.; Lamb, J.F.; Li, J.; Dai, X.; Zhao, P.X.; Vance, C.P. The *Medicago sativa* gene index 1.2: A web-accessible gene expression atlas for investigating expression differences between *Medicago sativa* subspecies. *BMC Genom.* **2015**, *16*, 502. [CrossRef]

62. Hanikenne, M.; Krämer, U.; Demoulin, V.; Baurain, D. A comparative inventory of metal transporters in the green alga *Chlamydomonas reinhardtii* and the red alga *Cyanidioschizon merolae*. *Plant Physiol.* **2005**, *137*, 428–446. [CrossRef]

63. Wintz, H.; Fox, T.; Wu, Y.Y.; Feng, V.; Chen, W.; Chang, H.S.; Zhu, T.; Vulpe, C. Expression profiles of *Arabidopsis thaliana* in mineral deficiencies reveal novel transporters involved in metal homeostasis. *J. Biol. Chem.* **2003**, *278*, 47644–47653. [CrossRef]

64. Sinclair, S.A.; Senger, T.; Talke, I.N.; Cobbett, C.S.; Haydon, M.J.; Kraemer, U. Systemic upregulation of MTP2-and HMA2-mediated Zn partitioning to the shoot supplements local Zn deficiency responses. *Plant Cell* **2018**, *30*, 2463–2479. [CrossRef] [PubMed]

65. Baker, A.J.; Whiting, S.N. In search of the Holy Grail—A further step in understanding metal hyperaccumulation? *New Phytol.* **2002**, *155*. [CrossRef]

66. Hanikenne, M.; Talke, I.N.; Haydon, M.J.; Lanz, C.; Nolte, A.; Motte, P.; Kroymann, J.; Weigel, D.; Krämer, U. Evolution of metal hyperaccumulation required cis-regulatory changes and triplication of HMA4. *Nature* **2008**, *453*, 391–395. [CrossRef] [PubMed]

67. Ó Lochlainn, S.; Bowen, H.C.; Fray, R.G.; Hammond, J.P.; King, G.J.; White, P.J.; Broadley, M.R. Tandem quadruplication of HMA4 in the zinc (Zn) and cadmium (Cd) hyperaccumulator *Noccaea caerulescens*. *PLoS ONE* **2011**, *6*, e17814. [CrossRef]

68. Andrés-Colás, N.; Sancenón, V.; Rodríguez-Navarro, S.; Mayo, S.; Thiele, D.J.; Ecker, J.R.; Puig, S.; Peñarrubia, L. The Arabidopsis heavy metal P-type ATPase HMA5 interacts with metallochaperones and functions in copper detoxification of roots. *Plant J.* **2006**, *45*, 225–236. [CrossRef] [PubMed]

69. Sankaran, R.P.; Huguet, T.; Grusak, M.A. Identification of QTL affecting seed mineral concentrations and content in the model legume Medicago truncatula. *Theor. Appl. Genet.* **2009**, *119*, 241–253. [CrossRef] [PubMed]

70. Hermand, V.; Julio, E.; de Borne, F.D.; Punshon, T.; Ricachenevsky, F.K.; Bellec, A.; Gosti, F.; Berthomieu, P. Inactivation of two newly identified tobacco heavy metal ATPases leads to reduced Zn and Cd accumulation in shoots and reduced pollen germination. *Metallomics* **2014**, *6*, 1427–1440. [CrossRef] [PubMed]

71. Deshpande, P.; Dapkekar, A.; Oak, M.D.; Paknikar, K.M.; Rajwade, J.M. Zinc complexed chitosan/TPP nanoparticles: A promising micronutrient nanocarrier suited for foliar application. *Carbohyd. Polym.* **2017**, *165*, 394–401. [CrossRef] [PubMed]

72. Talke, I.N.; Hanikenne, M.; Krämer, U. Zinc-dependent global transcriptional control, transcriptional deregulation, and higher gene copy number for genes in metal homeostasis of the hyperaccumulator *Arabidopsis halleri*. *Plant Physiol.* **2006**, *142*, 148–167. [CrossRef] [PubMed]

73. Becher, M.; Talke, I.N.; Krall, L.; Krämer, U. Cross-species microarray transcript profiling reveals high constitutive expression of metal homeostasis genes in shoots of the zinc hyperaccumulator *Arabidopsis halleri*. *Plant J.* **2004**, *37*, 251–268. [CrossRef]

74. Weber, M.; Harada, E.; Vess, C.; Roepenack-Lahaye, E.V.; Clemens, S. Comparative microarray analysis of Arabidopsis thaliana and Arabidopsis halleri roots identifies nicotianamine synthase, a ZIP transporter and other genes as potential metal hyperaccumulation factors. *Plant J.* **2004**, *37*, 269–281. [CrossRef] [PubMed]

75. Gustin, J.L.; Loureiro, M.E.; Kim, D.; Na, G.; Tikhonova, M.; Salt, D.E. MTP1-dependent Zn sequestration into shoot vacuoles suggests dual roles in Zn tolerance and accumulation in Zn-hyperaccumulating plants. *Plant J.* **2009**, *57*, 1116–1127. [CrossRef]

76. Jean, M.L.; Schikora, A.; Mari, S.; Briat, J.F.; Curie, C. A loss-of-function mutation in AtYSL1 reveals its role in iron and nicotianamine seed loading. *Plant J.* **2005**, *44*, 769–782. [CrossRef] [PubMed]

77. Zhang, H.; Zhao, S.; Li, D.; Xu, X.; Li, C. Genome-wide analysis of the ZRT, IRT-Like protein (ZIP) family and their responses to metal stress in *Populus trichocarpa*. *Plant Mol. Biol. Rep.* **2017**, *35*, 534–549. [CrossRef]

78. Ivanov, R.; Bauer, P. Sequence and coexpression analysis of iron-regulated ZIP transporter genes reveals crossing points between iron acquisition strategies in green algae and land plants. *Plant Soil* **2017**, *418*, 61–73. [CrossRef]

79. Sharma, S.S.; Dietz, K.J.; Mimura, T. Vacuolar compartmentalization as indispensable component of heavy metal detoxification in plants. *Plant Cell Environ.* **2016**, *39*, 1112–1126. [CrossRef]

80. Pita-Barbosa, A.; Ricachenevsky, F.K.; Wilson, M.; Dottorini, T.; Salt, D.E. Transcriptional plasticity buffers genetic variation in zinc homeostasis. *Sci. Rep.* **2019**, *9*, 19482. [CrossRef] [PubMed]

81. Li, S.; Zhou, X.; Huang, Y.; Zhu, L.; Zhang, S.; Zhao, Y.; Chen, R. Identification and characterization of the zinc-regulated transporters, iron-regulated transporter-like protein (ZIP) gene family in maize. *BMC Plant Biol.* **2013**, *13*, 114. [CrossRef] [PubMed]

82. Yilmaz, O.; Kazar, G.A.; Cakmak, I.; Ozturk, L. Differences in grain zinc are not correlated with root uptake and grain translocation of zinc in wild emmer and durum wheat genotypes. *Plant Soil* **2017**, *411*, 69–79. [CrossRef]

83. Nölte, J. *ICP Emission Spectrometry: A Practical Guide*; Wiley-VCH: Weinheim, Germany, 2003; Volume 1.

84. Haydon, M.J.; Cobbett, C.S. A novel major facilitator superfamily protein at the tonoplast influences zinc tolerance and accumulation in *Arabidopsis*. *Plant Physiol.* **2007**, *143*, 1705–1719. [CrossRef] [PubMed]

85. Nicot, N.; Hausman, J.F.; Hoffmann, L.; Evers, D. Housekeeping gene selection for real-time RT-PCR normalization in potato during biotic and abiotic stress. *J. Exp. Bot.* **2005**, *56*, 2907–2914. [CrossRef] [PubMed]

86. Chen, M.; Shen, X.; Li, D.; Ma, L.; Dong, J.; Wang, T. Identification and characterization of MtMTP1, a Zn transporter of CDF family, in the *Medicago truncatula*. *Plant Physiol. Biochem.* **2009**, *47*, 1089–1094. [CrossRef] [PubMed]

87. Desjardins, P.; Conklin, D. NanoDrop microvolume quantitation of nucleic acids. *JOVE* **2010**, *5*, e2565. [CrossRef] [PubMed]

88. Livak, K.J.; Schmittgen, T.D. Analysis of relative gene expression data using real-time quantitative PCR and the $2^{-\Delta\Delta CT}$ method. *Methods* **2001**, *25*, 402–408. [CrossRef] [PubMed]

89. Kumar, S.; Stecher, G.; Li, M.; Knyaz, C.; Tamura, K. MEGA X: Molecular evolutionary genetics analysis across computing platforms. *Mol. Biol. Evol.* **2018**, *35*, 1547–1549. [CrossRef] [PubMed]

90. Saitou, N.; Nei, M. The neighbor-joining method: A new method for reconstructing phylogenetic trees. *Mol. Biol. Evol.* **1987**, *4*, 406–425. [CrossRef] [PubMed]

91. Nei, M.; Kumar, S. *Molecular Evolution and Phylogenetics*; Oxford University Press: Oxford, UK, 2000.

92. Anderson, M.J. A new method for non-parametric multivariate analysis of variance. *Austral. Ecol.* **2001**, *26*, 32–46. [CrossRef]

93. Anderson, M.; Braak, C.T. Permutation tests for multi-factorial analysis of variance. *J. Stat. Comput. Sim.* **2003**, *73*, 85–113. [CrossRef]

94. Anderson, M.J.; Ellingsen, K.E.; McArdle, B.H. Multivariate dispersion as a measure of beta diversity. *Ecol. Lett.* **2006**, *9*, 683–693. [CrossRef] [PubMed]

95. Clarke, K.R.; Gorley, R.N. *Getting Started with PRIMER v7*; Plymouth Marine Laboratory: Plymouth, UK, 2015.

96. Wickham, H. ggplot2. *WIREs Comput. Stat.* **2011**, *3*, 180–185. [CrossRef]

Article

Plant Available Zinc Is Influenced by Landscape Position in the Amhara Region, Ethiopia

Mesfin K. Desta [1,2,*], Martin R. Broadley [2], Steve P. McGrath [1], Javier Hernandez-Allica [1], Kirsty L. Hassall [1], Samuel Gameda [3], Tilahun Amede [4] and Stephan M. Haefele [1]

[1] Sustainable Agriculture Sciences Department, Rothamsted Research, West Common, Harpenden, Hertfordshire AL5 2JQ, UK; steve.mcgrath@rothamsted.ac.uk (S.P.M.); javier.hernandez@rothamsted.ac.uk (J.H.-A.); kirsty.hassall@rothamsted.ac.uk (K.L.H.); stephan.haefele@rothamsted.ac.uk (S.M.H.)

[2] Future Food Beacon of Excellence and School of Biosciences, University of Nottingham, Nottingham LE12 5RD, UK; Martin.Broadley@nottingham.ac.uk

[3] International Maize and Wheat Improvement Center (CIMMYT), ILRI Campus P.O. Box 5689, Addis Ababa, Ethiopia; S.Gameda@cgiar.org

[4] International Crops Research Institute for the Semi-Arid Tropics (ICRISAT), ILRI Campus P.O. Box 5689, Addis Ababa, Ethiopia; T.Amede@cgiar.org

* Correspondence: Mesfin.Desta@Nottingham.ac.uk or Mesfin.Kebede-Desta@rothamsted.ac.uk

Abstract: Zinc (Zn) is an important element determining the grain quality of staple food crops and deficient in many Ethiopian soils. However, farming systems are highly variable in Ethiopia due to different soil types and landscape cropping positions. Zinc availability and uptake by plants from soil and fertilizer sources are governed by the retention and release potential of the soil, usually termed as adsorption and desorption, respectively. The aim of this study was to characterize the amount of plant available Zn at different landscape positions. During the 2018/19 cropping season, adsorption-desorption studies were carried out on soil samples collected from on-farm trials conducted at Aba Gerima, Debre Mewi and Markuma in the Amhara Region. In all locations and landscape positions, adsorption and desorption increased with increasing Zn additions. The amount of adsorption and desorption was highly associated with the soil pH, the soil organic carbon concentration and cation exchange capacity, and these factors are linked to landscape positions. The Freundlich isotherm fitted very well to Zn adsorption (r^2 0.87–0.99) and desorption (r^2 0.92–0.99), while the Langmuir isotherm only fitted to Zn desorption (r^2 0.70–0.93). Multiple regression models developed by determining the most influential soil parameters for Zn availability could be used to inform Zn fertilizer management strategies for different locations and landscape positions in this region, and thereby improve plant Zn use efficiency.

Keywords: adsorption; desorption; landscape position; isotherm; plant available Zn

Citation: Desta, M.K; Broadley, M.R; McGrath, S.P; Hernandez-Allica, J.; Hassall, K.L; Gameda, S.; Amede, T.; Haefele, S.M. Plant Available Zinc Is Influenced by Landscape Position in the Amhara Region, Ethiopia. *Plants* **2021**, *10*, 254. https://doi.org/10.3390/plants10020254

Academic Editor: Gokhan Hacisalihoglu

Received: 25 December 2020

Accepted: 25 January 2021

Published: 28 January 2021

Publisher's Note: MDPI stays neutral with regard to jurisdictional claims in published maps and institutional affiliations.

1. Introduction

Zinc (Zn) is a trace metal essential to all forms of life because of its fundamental role in gene expression, cell development and replication [1]. In plants, it plays a key role in various enzymatic reactions such as the synthesis of auxin, metabolic processes, and oxidation reduction reactions. It also participates in chlorophyll formation and is essential for many enzymes which are vital for nitrogen metabolism, energy transfer and protein synthesis [2]. Zn has been classed as a catalytic, structural, and regulatory ion [3]. It also has a critical effect on cellular homeostasis. Deficiencies of Zn in people are also widespread due to a lack of dietary intake, which is of public health importance [4–6].

Zn deficiencies are common on many cultivated soils in Ethiopia. Soil types, texture, pH, soil organic carbon (SOC), available phosphorus (P), total and available copper (Cu) and iron (Fe), exchangeable cations and cation exchange capacity (CEC) are the main contributors to the extent of Zn deficiency [7,8]. Zinc deficiency has been reported on several soil types in Ethiopia, for example, on Nitisols [9], Nitisols, Vertisols, Fluvisols and Cambisols [10], Vertisols [11], and in a review on Vertisols, Cambisols, Fluvisols, Nitisols, Andisols and Alfisols [12]. In addition, Zn deficiencies were also linked to Cambisols, Luvisols and Regosols of the Tigray region [8], and to salt affected soils of Eastern Ethiopia [13]. These deficiencies along with the potentially low Zn concentration in the crops grown on these soils may cause serious impacts on human health [5,6].

Adsorption and desorption of nutrient ions are the primary processes that affect transport of nutrients and contaminants in soils [14]. Adsorption refers to the quantity of a nutrient that is retained on soil exchange surfaces while desorption is the release from these surfaces; both occur in a system in the state of equilibrium. These are usually described through isotherms, showing the amount of adsorbed/desorbed nutrient in the solid phase (soil colloids) as a function of the concentration of that nutrient in the liquid phase (soil solution), determined at equilibrium conditions and a constant temperature. Although various isotherms have been developed, the two most commonly used isotherms are the Langmuir and Freundlich isotherms. The relationships between adsorption-desorption characteristics and soil properties have been extensively studied on metals such as Zn, Cu and others. Amongst soil properties, pH, clay content, cation exchange capacity (CEC), SOC and hydrous oxides exert the most significant influence on the adsorption-desorption reactions of Zn in soils and, thus, regulate the amount of Zn dissolved in soil solution [7,8,15–18].

Generally, the solubility of Zn in the soil decreases 100-fold for each unit increase in soil pH [19]. This is due to the greater adsorptive capacity of the soil solid surfaces resulting from increased pH-dependent negative charges, the formation of hydrolyzed forms of Zn, chemisorption on calcite and co-precipitation as Fe oxides [7]. Similarly, [7] reported that high pH and electrical conductivity (EC) are responsible for low availability of Zn in soils. For example, Zn concentration of teff (*Eragrostis tef*, (Zucc.) Trotter]) and wheat (*Triticum asetivum*, L.) leaves were significantly and positively correlated with soil Zn and soil organic carbon, respectively while negatively correlated with pH and CEC of soils in the Tigray Region [8].

Ethiopian farming systems and landscape positions are highly variable and hence nutrient mobility in the soil and their effect on plant Zn uptake and grain quality are also likely to vary. Although the application of Zn as a fertilizer proved to enhance the productivity and quality of staple food crops to some extent in Ethiopia [8], this is not always the case. Therefore, it is important to devise a mechanism for stratified nutrient management options for these systems. Improving the grain Zn content of staple food crops can only be achieved through a better understanding of Zn dynamics in these soils. One way to do this is through adsorption-desorption studies.

The aim of this study was, therefore, to better understand the influence of different landscape positions (upslope, midslope, and footslope) and the associated soil properties on the amount of Zn adsorbed and desorbed in typical soils of Ethiopia. The fitness of the most common adsorption-desorption isotherms for these soils was tested, to identify the dominant soil characteristics driving these processes. Multiple regression models were used, which can be used to inform the amount of adsorbed and desorbed Zn and which in turn could be used to help devise stratified Zn fertilizer recommendations and improve crop Zn use efficiency for these systems and landscape positions.

2. Materials and Methods

2.1. Site Description

The soils used in this study were collected from on-farm trials during the 2018/19 cropping season in the Amhara Region of Ethiopia. Experimental sites were at locations in three districts of the Amhara Region (Bahir Dar Zuriya, Enarj Enawega and Bure Districts), named Aba Gerima, Debre Mewi, and Markuma, respectively (Figure 1). The climate in the region is subtropical with an average annual rainfall of 1022 mm at Aba Gerima, 1240 mm at Debre Mewi and 1450 mm at Markuma and annual minimum and maximum temperatures of 12 and 30 °C respectively [20]. The experimental locations are characterized by hilly landscapes on a plateau at about 1800 to 2200 m ASL. Experimental fields were chosen based on landscape position which in this region has strong effects on soil characteristics (Table 1). Landscape position determines erosion/accumulation of soil particles, causes a notable shift of clay and organic matter concentrations, and of soil colour. The soils at Aba Gerima are highly degraded on the upslope with a clear clay movement to the footslope (Table 2). Few landscape position effects were observed at Markuma which has relatively gentle slopes. The most dominant soil types for all locations were Nitisols but Vertisols were observed in the footslope of Debre Mewi.

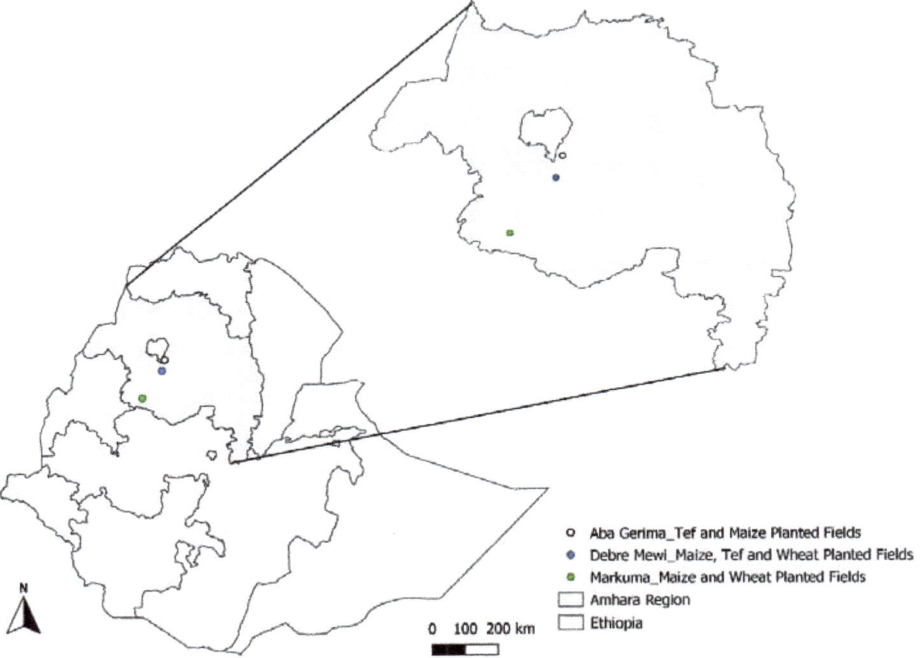

Figure 1. Location of the study sites in the Amhara region, Ethiopia.

The dominant crops grown were: Aba Gerima—tef, maize and finger millet; Debre Mewi—tef, wheat and maize, and Markuma—maize and wheat. At Aba Gerima and going down the slope, there was a shift of crops from the other cereals to finger millet which was sown in high planting density on the footslope. Likewise at Debre Mewi, but the crops are limited to tef and wheat on the footslope with predominantly maize on the upslope, whereas the Markuma sites consistently grow maize and wheat across all landscape positions.

Table 1. Soil pH, Total Nitrogen, Soil Organic C, Olsen P, and total Ca, Cu, Fe, Mg, Mn, Mo, P, S and Zn. Shown are average values from 5 fields for sites in the same landscape position.

Location_Crop *	Landscape Position	pH	Total N (%)	SOC (%)	Olsen P (mg kg^{-1})	Total Concentrations (mg kg^{-1})									
						Ca	Cu	Fe	K	Mg	Mn	Mo	P	S	Zn
Aba Gerima_T	Upslope	6.0	0.11	1.34	4.2	4470	106	122,623	1514	9469	1815	0.03	289	140	94
	Midslope	5.8	0.08	0.94	3.2	3603	140	125,029	1509	6358	2140	0.02	156	88	114
	Footslope	4.9	0.13	1.41	5.4	1185	78	120,950	2131	3081	1675	0.08	504	180	96
Aba Gerima_M	Upslope	5.2	0.14	1.53	4.8	1769	74	116,409	2220	3276	1513	0.10	605	213	98
	Midslope	5.6	0.12	1.38	3.6	3952	64	115,441	1446	7905	1600	0.03	402	171	97
	Footslope	5.5	0.11	1.31	4.9	3564	69	117,034	1307	7327	1808	0.03	395	144	103
Debre Mewi_MTW	Upslope	5.1	0.17	1.90	5.0	2178	97	112,587	3012	3158	2328	0.20	552	239	101
	Midslope	5.6	0.12	1.37	3.2	3607	64	106,311	2708	4331	1918	0.09	355	148	91
	Footslope	6.2	0.12	1.51	3.3	6190	60	106,482	3084	5334	1991	0.08	233	133	99
Markuma_MW	Upslope	4.8	0.18	2.44	3.9	805	65	102,562	2770	2281	1890	0.48	532	254	55
	Midslope	4.9	0.17	2.27	2.5	909	66	103,972	2967	2287	1800	0.33	499	253	54
	Footslope	4.9	0.15	2.09	2.0	764	70	105,920	2905	2318	1761	0.27	460	231	59
LSD		0.37	0.04	0.47	1.6	1378	36	9987	597	1842	418	0.11	162	61	14
DF		46	46	46	46	46	46	46	46	46	46	46	46	46	46

* Crops grown at the sites were tef (T), maize (M) and wheat (W). LSD is the average Least Significant Differences while DF is degree of freedom.

Table 2. Ammonium-Oxalate extractable (Al, Fe, Mn and P), exchangeable cations (Ca, K, Mg and Na), effective cation exchange capacity (eCEC), base saturation and soil texture. Shown are average values from 5 fields per each site in the same landscape position.

Location_Crop *	Landscape Position	AmmOx (mg kg^{-1})				Exchangeable Cations (cMolc kg^{-1})				eCEC (cMolc kg^{-1})	Base Saturation	Soil Texture (%)			Texture Class, USDA
		Al	Fe	Mn	P	Ca	K	Mg	Na		%	Sand	Silt	Clay	
Aba Gerima_T	Upslope	6882	14,172	1218	232	16.7	0.3	10.0	0.10	29.8	91	34	28	38	CL
	Midslope	4358	11,411	1514	138	15.2	0.1	9.1	0.10	27.8	89	36	28	36	CL
	Footslope	5031	12,348	1127	168	6.3	0.1	3.0	0.04	12.4	76	25	25	50	C
Aba Gerima_M	Upslope	4636	13,103	1036	180	8.6	0.3	3.2	0.05	13.6	89	25	28	47	C
	Midslope	5473	12,957	1085	170	15.4	0.3	10.2	0.06	28.3	91	29	28	43	C
	Footslope	5139	14,245	1288	176	13.8	0.1	9.0	0.07	25.9	89	31	32	37	CL
Debre Mewi_MTW	Upslope	3940	10,347	1894	153	9.9	0.3	4.4	0.05	17.0	86	23	27	50	C
	Midslope	3706	9012	1537	94	16.2	0.3	6.7	0.07	25.8	90	21	22	57	C
	Footslope	3351	9759	1657	82	25.7	0.4	7.8	0.06	37.3	91	13	17	70	C
Markuma_MW	Upslope	4767	12,527	1396	162	7.7	0.1	2.8	0.03	12.7	84	29	32	39	CL
	Midslope	4057	10,492	1284	136	7.9	0.2	2.9	0.02	13.0	85	28	30	42	C
	Footslope	4422	10,602	1242	142	7.2	0.2	2.8	0.02	12.5	82	27	29	44	C
LSD		784	2781	392	83	5	0.2	2.6	0.02	7.5	6	5	5	8	
DF		46	46	46	46	46	46	46	46	46	46	46	46	46	

* Crops grown at the sites were tef (T), maize (M) and wheat (W); CL = Clay Loam, C = Clay; LSD is the average Least Significant Differences while DF is degree of freedom; AmmOx-Ammonium-Oxalate extractable Aluminium (Al), Iron (Fe), Manganese (Mn) and Phosphorus (P).

2.2. Soil Sample Collection, Preparation and Standard Analysis

Geo-referenced representative soil samples were collected from 60 on-farm experiments in the 2018/19 cropping season. The soil samples were top-soils (0–20 cm depth), combined from 5 sub-samples in each plot, and were collected just before the cropping season. The individual experimental fields were chosen based on landscape position (upslope, midslope, and footslope) and crop grown, i.e., teff (*Eragrostis tef*, (Zucc.) Trotter]), maize (*Zea maize*, L.), wheat (*Triticum asetivum*, L.) or finger millet (*Eleusine coracana* (L.) Gaertn.). These soil samples were used for adsorption-desorption studies. Approximately 100 g of each sample were air dried at ambient temperature (25 °C), ground with mortar and pestle and passed through a 2-mm-sieve.

All samples were subjected to wet chemistry analysis following standard procedures. The soil pH was measured in deionized water, a soil: water ratio of 1:2.5 (10 g of soils with 25 mL of water) and with a temperature compensated two combination pH electrode. Total carbon and nitrogen concentrations were determined by dry combustion [21] using a Leco TruMac CN Combustion Analyser (Leco Corporation, St Joseph, Michigan), and because the pH of all soils was below 6.5, total carbon was assumed to be equivalent to SOC. Available phosphorus (Olsen's P) was extracted by the sodium bicarbonate method [22]. Phosphorus in the bicarbonate solution was determined using the phospho-molybdenum blue method on the Skalar SANPLUS System (continuous colorimetric flow analysis; Skalar Analytical BV). Total elemental concentrations were measured after an aqua regia extraction [23], followed by inductively coupled plasma optical emission spectrometry (ICP-OES; model, Perkin Elmer Life and Analytical, Shelton, USA). Acid oxalate extractable Fe, Al, Mn and P were determined following extraction with a mixed solution of ammonium oxalate and oxalic acid at a soil: solution ratio of 1:100 [24]. Samples were shaken in the dark (4 h, 20 °C) using a reciprocal shaker, filtered, then acidified and analyzed by ICP-OES. The eCEC determination started with a one-step centrifuge extraction with a 0.0166 M cobalt (III) hexamine chloride solution (Cohex) $[Co[NH_3]_6]Cl_3$. All exchangeable cations are in the extract while the decrease in Co concentration is a measure of the eCEC, and concentrations were measured by ICP-OES analysis [25]. Soil texture was analyzed using a Laser Scattering Particle Size Distribution Analyser (LA-960, Horbia Ltd., Kyoto, Japan).

2.3. Adsorption and Desorption Isotherms

For the adsorption experiments, 0.50 g of soil was equilibrated with 10 mL 0.01 M $CaCl_2$ solution containing varying concentrations of $ZnSO_4 \times 7H_2O$ (0, 2, 5, 10, 15 and 30 mg Zn L^{-1}) and shaken end-over-end for 24 h at room temperature. $CaCl_2$ was used as the aqueous solvent phase to improve centrifugation and minimize cation exchange [14]. Controls were prepared with only Zn in 0.01 M $CaCl_2$ solution (no soil added), for calibration and checking the stability of the test substance in $CaCl_2$ solution. A 24-h shaking period was sufficient for complete equilibration of the Zn solutions and the soil in Zn solutions ranging between 1 to 160 mg L^{-1} [26]. The soil and stock solution mixtures were then centrifuged at 3600 rpm for half an hour and the clear supernatant solution was decanted and analyzed for the Zn concentration. This value was set as the Zn equilibrium concentration (Ce), and the difference between the initial stock solution (Co) and the equilibrium solution concentration (Ce) is the adsorbed Zn. To derive desorption isotherms, the original samples were re-suspended with 10 mL of fresh 0.01 M $CaCl_2$ stock solution and shaken for 24 h. Again, the mixture was centrifuged, and the supernatant solutions was analyzed for the desorbed Zn concentration (Cde). The amount of Zn adsorbed at equilibrium Qe (mg kg^{-1}) was calculated from the following equation [27]:

$$Qe = (Co - Ce)\frac{V}{W} \tag{1}$$

where Co and Ce (mg L^{-1}) are the initial and equilibrium concentrations of Zn in the solution, respectively; Qe (mg kg^{-1}) is the amount of adsorbate per unit mass of soil. V is the volume of the solution added (L), and W is the weight of the adsorbent (soil) used (kg).

The percentage of Zn adsorbed or desorbed by the soil was determined from the difference between the initial and equilibrium concentrations for adsorbed Zn and the ratio between desorbed to initial Zn for desorbed Zn [27]:

$$\% \text{ Adsorption} = \frac{(Co - Ce)}{Co} * 100\% \tag{2}$$

$$\% \text{ Desorption} = \frac{Cde}{Co - Ce} * 100\% \tag{3}$$

where Co, Ce and Cde (mg L^{-1}) are the initial, equilibrium and desorbed Zn concentrations in the soil solution, respectively.

2.4. Zn Analysis in the Soil Solutions

Portable X-ray fluorescence (pXRF, Tracer 5i, Bruker) was used to measure the amount of Zn in the adsorption and desorption extracts. For this, the pXRF was set to spectrometer mode, selecting the precious metals calibration, configuring the settings to voltage 40 KV and current 40 µA, and 90 s scanning time with Ti/AL filters. First, the equipment was calibrated with the Ag-925 (sterling silver metal for calibrating the Tracer 5i) and the average of fifteen readings was within the range set by the laboratory (8.010–8.323 for Cu and 91.677–91.990 for Ag). Regression analysis between the concentrations of the standard stock solutions (0, 2, 5, 10, 15 and 30 mg L^{-1}) and the pXRF readings in pulses gave an r^2 of 0.99.

2.5. Langmuir and Freundlich Isotherm Models

The adsorption and desorption data were fitted to the two most commonly used isotherms in soils. The linear form of the Langmuir isotherm [28] is represented by the following equation:

Langmuir adsorption

$$\frac{Ce}{Qe} = \frac{1}{Kb} + \frac{Ce}{b} \tag{4}$$

Langmuir desorption

$$\frac{Ce}{Qde} = \frac{1}{Kb} + \frac{Ce}{b} \tag{5}$$

where Ce (mg L^{-1}) is the equilibrium concentration, Qe and Qde (mg kg^{-1}) are the amount of adsorbate adsorbed and desorbed per unit mass of adsorbent, and b and K are the Langmuir constants related to adsorption capacity and rate of adsorption and desorption, respectively. The essential characteristics of Langmuir can be expressed by a dimensionless constant called separation factor or equilibrium parameters, R$_L$, defined as:

$$RL = \frac{1}{1 + KCo} \tag{6}$$

where K is the Langmuir constant and Co (mg L^{-1}) is the initial Zn concentration. The value of R$_L$ indicates the type of isotherm to be either unfavorable (>1), linear (R$_L$ = 1), favorable (0 < R$_L$ < 1) or reversible (R$_L$ = 0).

The linear form of the Freundlich equation [29] is:

Freundlich adsorption

$$\log^{Qe} = \log^{Kf} + \frac{1}{n} \log^{Ce} \tag{7}$$

Freundlich desorption

$$\log^{Qde} = \log^{Kf} + \frac{1}{n} \log^{Ce} \tag{8}$$

where Qe and Qde (mg kg^{-1}) are the amount of adsorbed and desorbed at equilibrium and Ce (mg L^{-1}) is the equilibrium concentration; Kf and n are Freundlich constants, where n gives an indication of how favorable the adsorption process is; Kf is the adsorption capacity of the adsorbent.

2.6. Statistics and Modelling

Multiple linear regression models were developed for adsorption and desorption trends by including those independent variables pH, SOC, eCEC and clay that are known to significantly affect these processes; by including all soil parameters and eliminating those which were not significant through backward elimination; forward selection, forcing the model to have pH, eCEC and SOC in and backwards selection, but first removing high Variance Inflation Factor (VIFs), respectively. The aim was to determine the most explanatory factors affecting Zn adsorption/desorption, and to search for new important factors. Models were fitted in the R statistical environment v. 3.6.2.

3. Results

3.1. General Soil Physico-Chemical Properties

Generally, the study sites were characterized by increasing pH and decreasing soil organic carbon and total soil N towards lower landscape positions except in the field planted with teff at Aba Gerima (Table 1). No consistent trend with landscape positions could be detected for Olsen P or any of the total elements determined. These soils are classified as strongly to slightly acidic at Aba Gerima teff planted fields and Debre Mewi, strongly to moderately acidic at Aba Gerima maize planted fields while Markuma is characterized by strongly acidic soils [30]. Soil organic carbon (SOC in %) contents of these soils are classified by the same author as low (0.5–1.5%) except at Markuma, which has medium (1.5–3.0%) SOC concentrations. Total nitrogen concentrations (%) are rated as low to moderate for all except for moderate values at Markuma [30]. Available Olsen P concentrations (mg kg^{-1}) are generally classified as low [31] and total P (mg kg^{-1}) ranges from low to medium [32]. The total concentration of all the secondary macronutrients (Mg and S) fall into medium classes whilst calcium was found to be low at Markuma, medium at Aba Gerima and high at Debre Mewi [32]. With the exception of total Fe and Mn concentrations, which are very high, all the other micronutrients determined (Cu, Mo and Zn) fall in the medium class [32].

Table 2 shows the ammonium oxalate extracts, exchangeable cations, eCEC and soil texture for each site and landscape positions. Generally, ammonium oxalate extractable elements were highly variable and no consistent trend with landscape position could be detected for Al, Fe, Mn and P. Exchangeable cations decreased in the sequence Ca > Mg > K > Na and can be characterized as high for Ca, very low to medium for K, medium to high for Mg and very low for Na [32,33]. The eCEC varied and can be considered medium [32,33], but was generally low at Markuma. Exchangeable cations and eCEC indicate a base saturation between 76% and 91% which corresponds well with the soil pH values in Table 1. The soil texture at all sites ranges from clay-loam to clay, with clay contents between 38% to 70%, and sand contents between 13% to 36%. Again, none of these soil characteristics indicated any clear trend corresponding with the landscape position except texture which usually showed increasingly finer texture (more clay) from the top to the bottom (except for the Aba Gerima fields planted with maize).

3.2. Effect of Stock Solution on Equilibrium, Adsorbed and Desorbed Zn

Regardless of the rate of adsorption, the amount of Zn adsorbed (mg kg^{-1}) on the soil particles increased with increasing added Zn concentrations for all sites and landscape positions (Figure 2). The variation in the ranges of Zn adsorption at the different sites could be due to differences in soil characteristics such as pH, clay and soil organic carbon content or CEC. The subsequent Zn desorption also followed similar trends at all sites; the desorbed amount increased with increasing concentration of the previously used adsorption solution, but relatively smaller amounts of Zn desorbed than adsorbed (Figure 2).

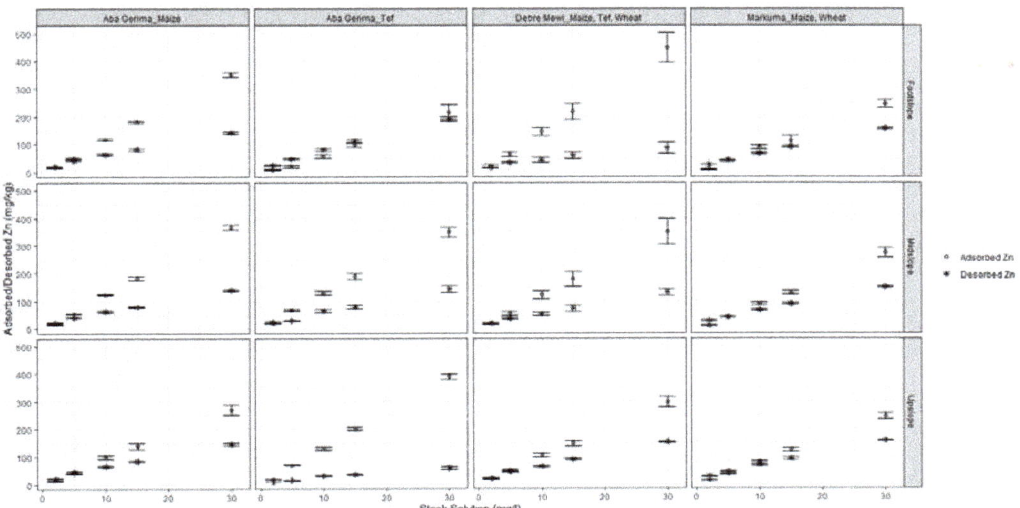

Figure 2. Stock solution concentration on adsorbed and desorbed Zn along the landscape positions.

Generally, the percentage of adsorbed Zn increased with increasing initial Zn concentrations at all sites (Figure 3). However, the adsorption rate reached a maximum with stock solutions of 5 mg Zn L^{-1} at for Aba Gerima fields planted with tef while it needed 10 mg Zn L^{-1} for this at Aba Gerima fields planted with maize and at Debre Mewi, except for the footslope at all sites which did not reach a plateau. However in Markuma, the percentage and rate of Zn adsorption kept increasing with increasing initial Zn concentrations in all landscape positions (Figure 3) because most of the soil parameters were similar across the topo sequence (Tables 1 and 2).

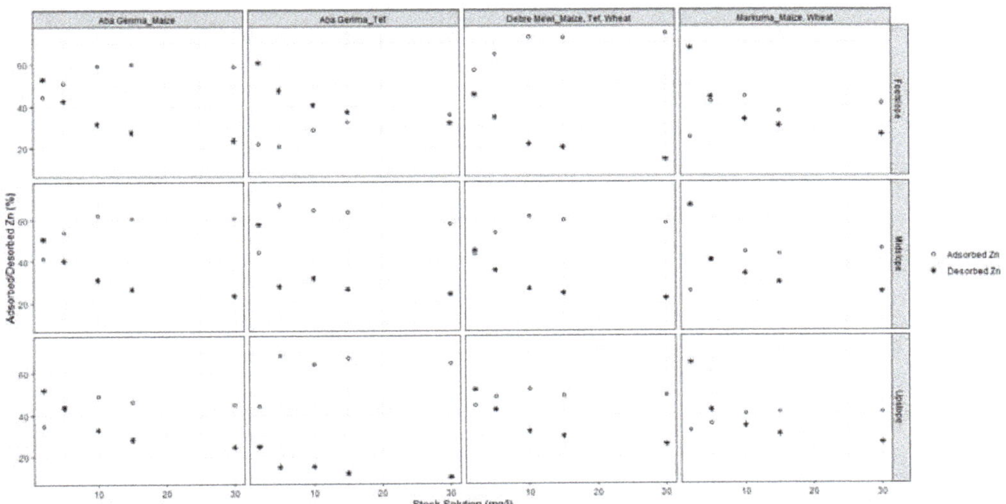

Figure 3. Percentage of adsorbed and desorbed Zn along the landscape positions.

The separation factor R_L ranges between 0 and 1 (Table 3) for all soils which indicates that the situation is favorable for reversible adsorption and desorption processes; a $R_L > 1$

means that ad/des is not strong whereas values close to 0 indicate non-reversible processes. The separation factors generally decrease with increasing initial Zn concentrations for all locations and landscape positions. However, at the highest concentration of the stock solution (30 mg L^{-1}) the RL factor approaches 0.25 which might indicate a declining rate of desorption for all landscape positions (Figure 3).

Table 3. Summary of the RL factor for adsorption and desorption.

Landscape Position	Co (mg L^{-1})	Aba Gerima_Tef		Aba Gerima_Maize		Debre Mewi_Maize, Tef, Wheat		Markuma_Maize, Wheat	
		RL_Ad	RL_De	RL_Ad	RL_De	RL_Ad	RL_De	RL_Ad	RL_De
Upslope	2	0.92	0.84	0.87	0.89	0.90	0.90	0.84	0.89
	5	0.83	0.67	0.73	0.77	0.78	0.78	0.67	0.77
	10	0.71	0.50	0.57	0.63	0.64	0.64	0.51	0.62
	15	0.62	0.40	0.47	0.53	0.54	0.55	0.41	0.52
	30	0.45	0.25	0.31	0.36	0.37	0.38	0.26	0.35
Midslope	2	0.93	0.91	0.90	0.90	0.91	0.89	0.82	0.89
	5	0.84	0.80	0.79	0.79	0.80	0.77	0.65	0.77
	10	0.72	0.67	0.65	0.65	0.66	0.62	0.48	0.62
	15	0.63	0.58	0.55	0.56	0.57	0.52	0.38	0.52
	30	0.46	0.41	0.38	0.39	0.40	0.35	0.24	0.35
Footslope	2	0.73	0.88	0.90	0.91	0.95	0.93	0.83	0.90
	5	0.52	0.74	0.79	0.80	0.89	0.84	0.66	0.77
	10	0.35	0.58	0.66	0.66	0.80	0.73	0.49	0.63
	15	0.26	0.48	0.56	0.57	0.73	0.64	0.39	0.53
	30	0.15	0.32	0.39	0.40	0.57	0.44	0.24	0.36
LSD		0.10	0.11	0.10	0.11	0.09	0.11	0.12	0.11
DF		72	72	67	67	67	67	72	72

RL-Ad and RL_De refers to separation factor for adsorption and desorption, respectively; LSD is the average Least Significant Differences while DF is degree of freedom.

In contrast to adsorption, the percentage of desorbed Zn decreases with increasing initial Zn regardless of the location and landscape position (Figure 3). Separation factors followed similar trends to that of adsorption and decreased with increasing initial Zn concentrations.

3.3. Comparing the Adsorption-Desorption Results with the Langmuir and Freundlich Isotherms

The Langmuir and Freundlich isotherm coefficients for adsorption and desorption, and the respective functions are presented in Tables 4 and 5 and Figures 4–6, respectively. Both Langmuir and Freundlich isotherms, assume linearity in their equations for their respective variables. Accordingly, the Langmuir isotherm assumes linearity when equilibrium Zn vs. the ratio between equilibrium to adsorbed Zn is plotted while the Freundlich isotherm assumes the same for the log equilibrium vs. log adsorbed. The same assumptions are valid for desorption.

Freundlich isotherms were found to fit well for the observed adsorption (Figure 4) and desorption process (Figure 6) for all locations and landscape positions, and similar results were found in [8]. In contrast, Langmuir isotherms described only the desorption processes well (Table 4, Figure 5). These results were confirmed by good relationships (r^2) between log equilibrium vs. log of adsorbed, log equilibrium vs. log of desorbed and equilibrium vs. ratio of equilibrium to desorbed, respectively (Table 5). Unlike the Freundlich isotherm, which had regression coefficients ≥ 0.87 for adsorption and desorption across all sites and landscape positions, the Langmuir isotherm achieved variable regressions of between 0.70–0.93 for desorption and between 0.12 and 0.53 for adsorption (Figure 5, Table 4).

Table 4. Langmuir coefficients for adsorption and desorption isotherms.

| Location_Crop * | Landscape Position | Langmuir Isotherm Coefficients | | | | | |
| | | Adsorption | | | Desorption | | |
		b	K	r^2	b	K	r^2
Aba Gerima_T	Upslope	−646	−0.04	0.16	152	0.09	0.75
	Midslope	−890	−0.02	0.17	365	0.07	0.72
	Footslope	−452	−0.03	0.41	753	0.02	0.55
Aba Gerima_M	Upslope	3020	−0.02	0.18	305	0.06	0.93
	Midslope	−420	−0.05	0.35	377	0.05	0.80
	Footslope	−671	−0.03	0.36	334	0.06	0.85
Debre Mewi_MTW	Upslope	−27554	−0.01	0.12	329	0.06	0.86
	Midslope	−1061	−0.05	0.20	45	0.04	0.72
	Footslope	−654	−0.18	0.53	165	0.21	0.70
Markuma_MW	Upslope	−877	−0.02	0.30	313	0.05	0.80
	Midslope	−324	−0.04	0.44	333	0.05	0.81
	Footslope	−211	−0.02	0.21	289	0.06	0.83
LSD		4716	0.02		90	0.02	
DF		285	285		285	285	

* Crops grown at the sites were tef (T), maize (M) and wheat (W).

Table 5. Freundlich coefficients for adsorption and desorption isotherms.

| Location_Crop * | Landscape Position | Freundlich Isotherm Coefficients | | | | | |
| | | Adsorption | | | Desorption | | |
		1/n	Kf	r^2	1/n	Kf	r^2
Aba Gerima_T	Upslope	1.24	1.39	0.90	0.78	0.99	0.96
	Midslope	0.91	1.43	0.88	1.33	1.33	0.95
	Footslope	0.79	0.63	0.91	1.21	1.21	0.99
Aba Gerima_M	Upslope	1.17	1.06	0.97	0.75	1.27	0.99
	Midslope	0.76	1.24	0.96	1.21	1.29	0.98
	Footslope	0.80	1.23	0.97	1.28	1.31	0.98
Debre Mewi_MTW	Upslope	0.93	1.23	0.99	1.34	1.29	0.99
	Midslope	0.80	1.27	0.95	1.20	1.24	0.98
	Footslope	0.64	1.68	0.94	1.40	1.36	0.92
Markuma_MW	Upslope	0.87	0.95	0.97	1.44	1.30	0.98
	Midslope	0.75	0.86	0.96	1.38	1.28	0.98
	Footslope	0.86	0.92	0.87	1.46	1.32	0.98
LSD		0.09	0.10		0.03	0.03	
DF		285	285		285	285	

* Crops grown at the sites were tef (T), maize (M) and wheat (W).

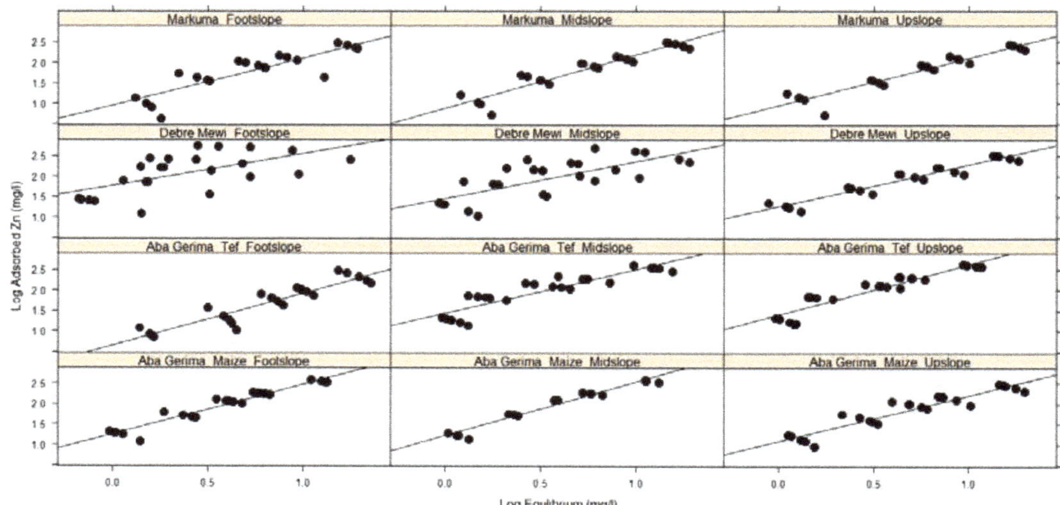

Figure 4. Adsorption data fitted to the Freundlich isotherm.

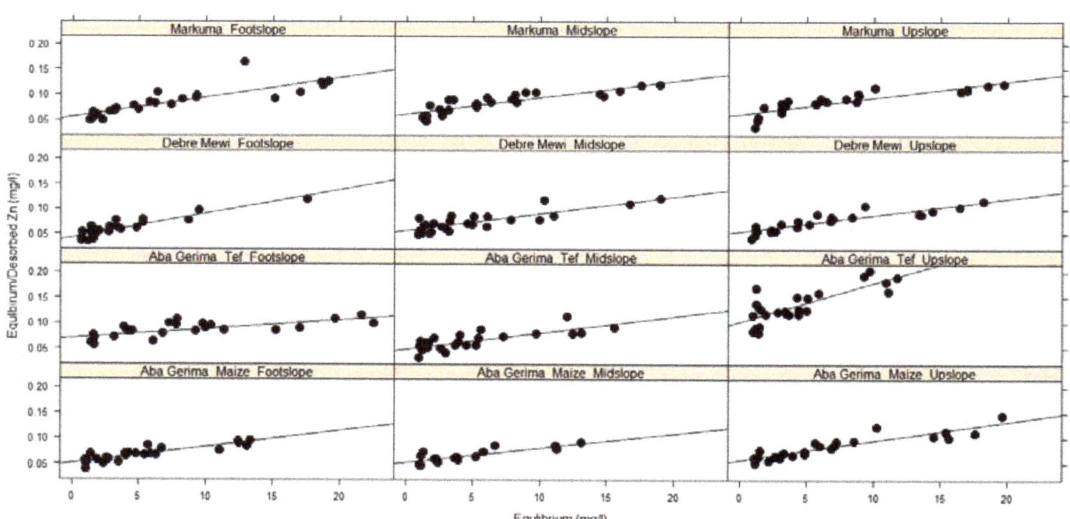

Figure 5. Desorption data fitted to the Langmuir isotherm.

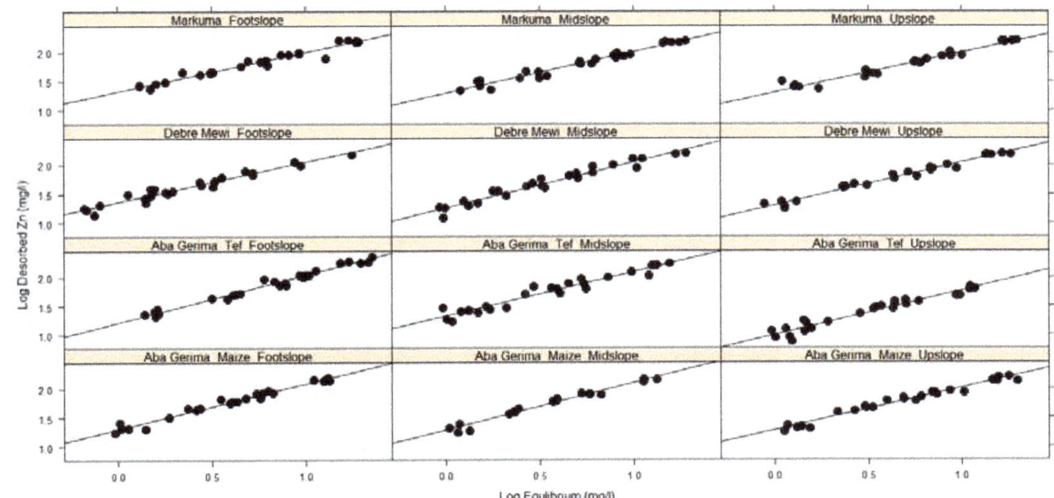

Figure 6. Desorption data fitted to the Freundlich isotherm.

3.4. Empirical Models with Soil Parameters

Soil factors play different roles with different magnitudes in any dynamic soil nutrient processes. Accordingly, not all soil factors are equally important and do affect the adsorption and desorption processes, and only some will have significant affects [8]. Therefore, we decided to develop functional models with the most relevant soil factors, both from the literature and the current experiments, in order to describe the observed adsorption and desorption processes. Nevertheless, an attempt was made to develop different models including other known factors affecting Zn adsorption/desorption as well as to search for new important factors. However, all models including more factors than pH, SOC and eCEC provided no substantial improvement and tended to overfit the functions for adsorption and desorption. The selected models below conform to statistical assumptions of normality and residual plots and were fitted in the R statistical environment v. 3.6.1. The resulting multiple regression model for adsorption and desorption were:

Multiple regression model for adsorption

$$\text{Adsorption} = -0.92 + 0.26\text{pH} + 0.03\text{SOC}; \text{ adjusted } r^2 = 0.90 \tag{9}$$

$$\text{Adsorption} = -0.57 + 0.17\text{pH} + 0.04\text{SOC} + 0.006\text{eCEC}; \text{ adjusted } r^2 = 0.92 \tag{10}$$

Multiple regression model for desorption

$$\text{Desorption} = 0.89 - 0.11\text{pH} - 0.03\text{SOC}; \text{ adjusted } r^2 = 0.70 \tag{11}$$

$$\text{Desorption} = 0.89 - 0.11\text{pH} - 0.03\text{SOC} - 0.00005\text{eCEC}; \text{ adjusted } r^2 = 0.69 \tag{12}$$

In general, the models were better in predicting adsorption as compared with desorption, and the use of more model parameters improved the prediction for desorption (but had the risk of overfitting). In these models, including eCEC improved the adjusted r^2 for adsorption very little and gave no improvement for desorption. This implies that in the studied soil, pH and SOC drive the adsorption-desorption process and help to determine the potentially available soil Zn for plant uptakes. These models help to understand and potentially estimate the amount of Zn adsorbed in the soil and desorbed from the soil and, therefore, the plant availability of Zn in these soils. They can also help to guide fertilizer recommendation schemes to improve crop Zn use efficiency as the soil pH and soil organic carbon were found to be influential soil parameters.

4. Discussion

4.1. Amount of Adsorbed and Desorbed Zn with Landscape Positions

The amount of adsorbed Zn (mg kg^{-1}) increased with increasing initial Zn concentrations of the stock solutions, and the amount of Zn desorption increased with increasing amounts of adsorbed Zn (Figure 2). This was as expected and due to the increasing mass transfer driving force from high concentrations (high stock solution Zn concentration or soil with high adsorbed Zn) to low concentrations (low Zn adsorption on the soil or the zero Zn blank solution), resulting in adsorption and desorption processes moving towards an equilibrium. Similar research findings have been reported on many different soils of Tigray [8]. However, the strength of the reactions differed considerably for the different soils.

In all landscape positions, an increase in the adsorption (Figure 2) of Zn was highly associated with an increase of clay content, soil pH, and eCEC of these soils (Tables 1 and 2). Along the landscape positions from upslope to footslope at Debre Mewi (pH, clay, eCEC) and Aba Gerima fields planted with maize (eCEC), and from footslope to upslope for Aba Gerima fields planted with teff (pH, eCEC), these soil parameters increased, leading to greater adsorption. However, no clear differences were observed among the different landscape positions at Markuma and this was probably because these soil parameters did not differ substantially along the landscape positions. The effects of these soil parameters in the adsorption of Zn has been well studied by several authors. Studies found out that increasing soil pH, soil organic carbon and eCEC significantly increase the amount of adsorbed Zn in soils [8,34–36].

Low soil pH in the upslope of Debre Mewi and Aba Gerima fields planted with maize and the footslope of Aba Gerima fields planted with teff could be a stronger driving factor for low adsorption rather than landscape positions. These low pH soils were associated with corresponding low eCEC, hence adsorption is low and Zn is more freely available and can be found in the soil solution [15,16]. In contrast, higher soil pH, usually accompanied by higher eCEC increases adsorption [17]. Except at Debre Mewi, no consistent trends were observed with these soil characteristics and clay content (Tables 1 and 2). However, increasing clay content increased Zn adsorption, and consequently the activity of Zn in the soil solution decreased with increasing clay content [15]. Furthermore, it seemed that a decrease or increase in soil organic carbon did not influence the adsorption patterns on these locations, possibly because the differences in soil organic carbon were too small and crop management practices are relatively similar within the location. This study aligns with several others that show the activity of Zn in the soil solution increases with decreasing soil pH and decreases with increasing the content of organic carbon and clay particles through adsorption [15–17].

Desorption is the opposite of adsorption; as adsorption increases, desorption decreases and vice versa. Desorption continually decreased from upslope to footslope in Aba Gerima fields planted with maize and Debre Mewi while in increased at Aba Gerima fields planted with teff (Figure 2). However at Markuma, desorption was relatively uniform across the landscape positions, most likely because most of the soil parameters such as soil pH, eCEC and even the total Zn content of these soils were similar (Table 1). It has been found that SOC, CEC, and soil pH are the most important factors controlling Zn desorption while calcium carbonate equivalent and clay content were not [37]. In addition, these authors found that soil pH had a negative relationship with Zn desorption. The multiple regression models developed for desorption align with these findings (Equations (3) and (4)).

4.2. Langmuir and Freundlich Isotherms

Adsorption and desorption isotherms can be used to describe the equilibrium relationship between the amounts of adsorbed and dissolved species at a given temperature considering the intensity, quantity and capacity factors, which are important for predicting the amount of soil nutrient required for maximum plant growth. In the Langmuir model it is assumed that even at maximum adsorption capacity, there is only a monomolecular layer on the surface. This means that there is no stacking of adsorbed molecules. The Freundlich model does not have this restriction and stacked cation layers are possible. Both models were applied in a number of studies investigating Zn availability in soils and found that soils with divergent characteristics showed good fit to either Langmuir or the Freundlich isotherms [38–40]. On calcareous Vertisols soils from Jordan [41] found that both fitted to the soil studied but Freundlich resulted in better fits as compared with Langmuir. Failure of Zn adsorption data to conform to the linear Langmuir equation has been attributed to the existence of more than one type of Zn adsorbing sites, such as occur on different types of clay. In Ethiopia, some of the soil data could not be described with the Langmuir isotherm [8]. The findings by [17] also showed poor fits of soil characteristics to Langmuir adsorption isotherms. The same author reported that the reason for the poor fit was unclear although even though low realistic Zn additions were used for their study. Similarly, the Zn additions in the current study were low representing the low soil Zn status of agricultural soils in Ethiopia. And as described by [17], the Langmuir isotherm did also not fit well to the soils we analyzed, possibly because this isotherm assumes a linear relationship between adsorbed vs. adsorbed/equilibrium Zn variables (Table 4). However, desorption conformed to the Langmuir isotherm (Table 4, Figure 4). In contrast to these results, [42] observed Langmuir isotherm best fits to addition of high Zn concentrations.

The Freundlich isotherm fitted very well to the observed Zn adsorption (Table 5, Figure 4) and desorption (Table 5, Figure 6) for the soils investigated. Similar findings have been reported by [38,39]. Because the Freundlich isotherm is applicable to adsorption and desorption processes that occur on heterogeneous surfaces [43], the good fit of this isotherm indicates that the soil characteristics such as soil pH, eCEC and clay content do vary with landscape positions and locations, significantly affecting the amount of adsorbed and desorbed Zn in the studied soils. The higher r^2 values for both adsorption and desorption suggested that the Freundlich isotherm is the better model for soils in the studied regions of Ethiopia.

4.3. Soil Factors Driving These Processes

Soil pH has been identified in many studies as one of the main factors affecting Zn mobility and sorption in soils [8,15,16]. Zn becomes more soluble as soil pH decreases, it is more mobile and increasingly available in low pH environments, especially below pH 5.0 [44]. As the soil pH at Markuma is classified as strongly acidic (below 4.9 and almost the same for all landscape positions, see Table 1), the rate of adsorption (Figure 3) is very low compared with the moderate Aba Gerima fields planted with Maize, and the less acidic soils at Aba Gerima fields planted with teff and Debre Mewi with soil pH values of 5.5, 6.0 and 6.2, respectively (Table 1, Figure 2).

Absorption and adsorption are two properties related to the surface area of clay minerals. Therefore, the bioavailability of trace elements, including Zn, decreases generally with the clay mineral content in soils [44,45]. Zn may even be irreversibly fixed by clay through isomorphous substitutions or solid-state diffusion into the crystal structure of layered silicates. However, although the soil texture at all the studied sites is classified as clay, the actual clay percentage varied considerably. The amount of clay in Markuma (39–44) was low compared with the other sites which is probably another reason for the low adsorption of Zn at this location. In contrast, high adsorption at Debre Mewi (50–70% clay) is related to the high amount of clay in the soils there, while Aba Gerima fields planted with Maize (47–37% clay) and teff (38–50% clay) had moderate clay content, contributing to the modest Zn adsorption and desorption (Table 2).

In addition, eCEC seems to affect the adsorption of Zn in the studied soils. Ref. [8] found that in his studies of many soils from Tigray region, this was one of the main soil factors affecting adsorption and fitting to the different isotherms. At Debre Mewi

(17.0–37.3 cMole kg^{-1}) and Aba Gerima fields planted with Maize (13.6–25.9 cMole kg^{-1}), the eCEC increased with landscape position which was associated with increasing adsorption and decreasing desorption at these locations. In contrast, at Aba Gerima fields planted with teff (29.8–12.4 cMole kg^{-1}) decreasing eCEC values reduced the adsorption and promoted the desorption process. The eCEC values for the Markuma site were similar across landscape positions (12.7–12.5 cMole kg^{-1}) and hence the adsorption and desorption processes were similar. eCEC is of course affected by pH, clay content, clay mineralogy and soil organic carbon content, and all these factors interact to produce the observed effects.

4.4. Implications for Zn in the Soil and Potential Availability for Crop Uptake

Generally, as soil pH changes across positions in the landscape, the solubility of native soil Zn differs, and adsorption increases with increasing soil pH and vice versa. For example, in the tef planted field at Aba Gerima, soil pH decreased from upslope to footslope while at Debre Mewi it increased with landscape position from upslope to footslope (Table 1). Therefore, in landscape positions with low soil pH, the native soil Zn solubility increased and this, coupled with low adsorption (Equation (1)) and relatively high desorption (Equation (2)), suggests that the application of Zn fertilizers can potentially improve net available soil Zn levels and hence enhance plant Zn uptake.

Soil factors such as soil pH and organic carbon in the studied soils do vary along landscape positions and play a vital role in determining the availability of Zn in the soil by affecting the solubility of native soil Zn reserves and/or added Zn from applied fertilizers through adsorption-desorption process. This will, in turn, determine the net soil Zn potentially available for the plant uptake and improve efficiency. Hence, using important soil factors helps to estimate the amount of Zn in the soil that could be available for crop uptake and would be useful for refining fertilizer recommendation schemes or for suggesting the introduction of Zn uptake efficient crops.

5. Conclusions

We conducted this study with the aim of better understanding the soil Zn characteristics along the different landscape positions in order to improve crop uptake through adsorption-desorption studies in Ethiopia. For this objective, we analyzed how well adsorption-desorption data fitted to the most common isotherms and identified the influential soil factors affecting the potentially available soil Zn for uptake by plants.

In general, adsorption fitted to the Freundlich isotherm only while desorption fitted both isotherms. Soil parameters such as pH and SOC were identified as the most factors governing the adsorption and desorption processes and we determined the potential available net soil Zn at different locations. From this study it can be concluded that the most probable reasons for the widespread Zn deficiency in the study is the high rate of Zn adsorption with little desorption. Hence, in areas where the soil has high adsorption capacity, high application rates of Zn fertilizer are needed while soils with low adsorption will need lower rates of Zn fertilizers, which would minimize expense and accumulation of Zn. The models will help to quantify the amount of potentially available soil Zn for crop uptake and can be used to devise stratified Zn fertilizer recommendations for these sites and different landscape positions.

Further studies linking the net potentially available Zn in the soil with plant uptake are needed to better understand uptake efficiencies of different crops and factors affecting plant uptake of Zn from the soil. This will help to improve our understandings on Zn uptake efficiencies on highly adsorptive soils and help in making a decision to select crops which are efficient in the different landscape positions and locations.

Author Contributions: Conceptualization, M.K.D., M.R.B., S.P.M., S.G., T.A. and S.M.H.; methodology, M.K.D., J.H.-A., K.L.H. and S.M.H.; validation, M.K.D., J.H.-A., K.L.H., and S.M.H.; formal analysis, M.K.D., M.R.B., S.P.M., K.L.H. and S.M.H.; investigation M.K.D., M.R.B., S.P.M., J.H.-A., K.L.H., S.G., T.A. and S.M.H.; resources, M.R.B., S.P.M., and S.M.H.; data curation, M.K.D., J.H.-A., K.L.H., and S.M.H.; writing—original draft preparation, M.K.D.; writing—review and editing, M.R.B., S.P.M., J.H.-A., K.L.H., S.G., T.A. and S.M.H.; visualization, M.K.D. and K.L.H.; supervision, M.K.D., M.R.B., S.P.M., K.L.H., S.G., T.A. and S.M.H.; project administration, M.R.B., S.P.M. and S.M.H. All authors have read and agreed to the published version of the manuscript.

Funding: This work was supported by the Nottingham-Rothamsted Future Food Beacon Studentships in International Agricultural Development and the GeoNutrition project funded by the Bill & Melinda Gates Foundation (BMGF) [INV-009129].

Institutional Review Board Statement: Not applicable.

Informed Consent Statement: The authors declare no conflict of interest.

Acknowledgments: Rothamsted Research is highly acknowledged for providing access to the laboratory facilities needed for this study.

Conflicts of Interest: The funders had no role in the design of the study; in the collection, analyses, or interpretation of data; in the writing of the manuscript, or in the decision to publish the results.

References

1. Hambridge, K.M. Human Zn deficiency. *J. Nutr.* **2000**, *130*, 1344–1349. [CrossRef] [PubMed]
2. Broadley, M.R.; White, P.J.; Hammond, J.P.; Zelko, I.; Lux, A. Zn in plants. *New Phytol.* **2007**, *173*, 677–702. [CrossRef] [PubMed]
3. Chasapis, C.T.; Loutsidou, A.C.; Spiliopoulou, C.A.; Stefanidou, M.E. Zn and human health: An update. *Arch. Toxicol.* **2012**, *86*, 521–534. [CrossRef] [PubMed]
4. Kumssa, D.; Edward, J.M.; Joy, E.; Ander, L.; Watts, M.J.; Young, S.D.; Walker, S.; Broadley, M.R. Dietary calcium and Zn deficiency risks are decreasing but remain prevalent. *Sci. Rep.* **2015**, *5*, 10974. [CrossRef] [PubMed]
5. EPHI—Ethiopian Public Health Institute. *Annual Report*; EPHI: Addis Ababa, Ethiopia, 2016.
6. Masresha, T.; De Groote, H.; Brouwer, I.D.; Feskens, E.J.M.; Belachew, T.; Zerfu, D.; Belay, A.; Demelash, Y.; Gunaratna, N.S. Soil Zn is associated with Serum Zn but not with linear growth of children in Ethiopia. *Nutrients* **2019**, *11*, 221.
7. Alloway, B.J. Soil factors associated with Zn deficiency in crops and humans. *Environ. Geochem. Health* **2009**, *31*, 537–548. [CrossRef]
8. Bereket, H. Assessment of Zn and Iron, Zn and Adsorption, and Effects of Zn and Iron on Yields and Grain Nutrient Concentration of Teff and Bread Wheat in Some Soils of Tigray, Northern Ethiopia. Ph.D. Thesis, Haramaya University, Dire Dawa, Ethiopia, 2018.
9. Teklu, B.; Suwanarit, A.; Osotsapar, Y.; Sarobol, E. Status of B, Cu, Fe, Mo and Zn of Soils of Ethiopia for Maize Production: Greenhouse Assessment. *Kasetsart J. Nat. Sci.* **2005**, *39*, 357–367.
10. Asgelil, D.; Taye, B.; Yesuf, A. The status of micro-nutrients in Nitisols, Vertisols, Cambisols and Fluvisols in major maize, wheat, teff and citrus growing areas of Ethiopia. In Proceedings of the Agricultural Research Fund Research Projects Completion Workshop, Addis Ababa, Ethiopia, 1–2 February 2007; pp. 77–96.
11. Yifru, A.; Kebede, M. Assessment on the Status of Some Micronutrients in Vertisols of the Central Highlands of Ethiopia. *Inter. Res. J. Agr. Sci. Soil Sci.* **2013**, *3*, 169–173.
12. Yifru, A.; Kassa, S. Status of Soil Micronutrients in Ethiopian Soils: A Review. *J. Envir. Earth Sci.* **2017**, *7*, 85–90.
13. Ashenafi, W.; Bedadi, B.; Mohammed, M. Assessment on the Status of Some Micronutrients of Salt Affected Soils in Amibara Area, Central Rift Valley of Ethiopia. *Acad. J. Agri. Res.* **2016**, *4*, 534–542.
14. Dandanmozd, F.; Hosseinpur, A.R. Thermodynamic parameters of Zn sorption in some calcareous soils. *J. Am. Sci* **2010**, *6*, 298–304.
15. Rutkowska, B.; Szulc, W.; Bomze, K. Soil factors affecting solubility and mobility of Zn in contaminated soils. *Int. J. Environ. Sci. Technol.* **2015**, *12*, 1687–1694. [CrossRef]
16. Dhanwinder, S.; McLaren, R.G.; Cameron, K.C. Effect of pH on Zn Sorption–Desorption by Soils. *J. Comm. Soil Sci. Plant. Anal.* **2008**, *39*, 2971–2984.
17. Muhammad, I.; Alloway, B.J.; Aslam, M.; Memon, M.Y.; Khan, P.; Siddiqui, S.-u.-H.; Shah, S.K.H. Zn Sorption in Selected Soils. *Comm. Soil Sci. Plant. Anal.* **2006**, *37*, 1675–1688.
18. Gaudalix, M.E.; Pardo, M.T. Zn sorption by acid tropical soil as affected by cultivation. *Eur. J. Soil Sci.* **1995**, *46*, 317–322. [CrossRef]
19. Tisdale, S.L.; Nelson, W.L.; Beaton, J.D.; Havlin, J.L. *Soil Fertility and Fertilizers*, 5th ed.; Macmillan Publishing Company: New York, NY, USA, 1993.
20. Available online: https://www.meteoblue.com (accessed on 24 September 2020).

21. Tiessen, H.; Bettany, J.R.; Stewart, J.W.B. An improved method for the determination of carbon in soils and soil extracts by dry combustion. *Commun. Soil Sci. Plant. Anal.* **1981**, *12*, 211–218. [CrossRef]
22. Olsen, S.R.; Cole, C.V.; Watanabe, F.S.; Dean, L.A. Estimation of available phosphorus in soils by extraction with sodium carbonate. *USDA Circ.* **1954**, *939*, 1–19.
23. McGrath, S.P.; Cunliffe, C.H. A Simplified Method for the Extraction of the metals Fe, Zn, Cu, Ni, Cd, Pb, Cr, Co and Mn from Soils and Sewage Sludges. *J. Sci. Food Agric.* **1985**, *36*, 794–798. [CrossRef]
24. Schwertmann, U. Differenzierung der Eisenoxide des Bodens durch Extraktion mit Ammoniumoxalat-Lösung. *Z. Pflanzenernahr. Bodenkd.* **1964**, *105*, 194–202. (In German) [CrossRef]
25. Ciesielski, H.; Sterckeman, T. Determination of cation exchange capacity and exchangeable cations in soils by means of cobalt hexamine trichloride. Effect of experimental conditions. *Agronomie* **1977**, *17*. [CrossRef]
26. Reyhanitabar, A.; Ardalan, M.; Gilkes, R.J.; Savaghebi, G. Zn sorption characteristics of some selected calcareous soils of Iran. *J. Agr. Sci. Tech.* **2010**, *12*, 99–110.
27. Tan, I.A.W.; Ahmed, A.L.; Hammed, B.H. Adsorption isotherms, kinetics, thermodynamics and desorption studies of 2,4,6-trichlorophenol on oil palm empty fruit bunch-based activated carbon. *J. Hazard. Mat.* **2009**, *164*, 473–482. [CrossRef] [PubMed]
28. Langmuir, I. The sorption of gases on plane surface of glass Mica and Platin. *J. Am. Chem. Sos.* **1918**, *40*, 1361–1403. [CrossRef]
29. Freundlich, H. *Kapillarchemie: Eine Darstellung der Chemie der Kolloide und verwandter Gebiete*; Akademische Verlagsgesellschaf: Leipzig, Germany, 1909.
30. Tekalign, T. *Soil, Plant, Water, Fertilizer, Animal Manure and Compost Analysis*; Working Document No. 13; International Livestock Research Center for Africa: Addis Ababa, Ethiopia, 1991.
31. Landon, J.R. *Booker Tropical Soil Manual: A Handbook for Soil Survey and Agricultural Land Evaluation in the Tropics and Subtropics*; Longman Scientific and Technical: Essex, NY, USA; John Wiley & Sons Inc.: New York, NY, USA, 1991; 474p.
32. Jones, J.B. *Agronomic Handbook: Management of Crops, Soils, and Their Fertility*; CRC Press LLC: Boca Raton, FL, USA, 2003; 482p.
33. FAO. *Plant Nutrition for Food Security: A Guide for Integrated Nutrient Management*; Food and Agriculture Organization, Fertilizer and Plant Nutrition Bulletin; FAO: Rome, Italy, 2006; p. 16.
34. Mohammed, H.S. Sorption Mechanisms of Zn in Different Clay Minerals and Soil Systems as Influenced by Various Natural Ligands. Ph.D. Thesis, Louisiana State University and Agricultural and Mechanical College, Baton Rouge, LA, USA, 2010.
35. Aysen, A.; Doulati, B. The Effect of Soil Properties on Zn Adsorption. *J. Int. Environ/ App. Sci.* **2012**, *7*, 151–160.
36. Fan, T.T.; Wang, Y.J.; Li, C.B.; He, J.Z.; Gao, J.; Zhou, D.M.; Friedman, S.P.; Sparks, D.L. Effect of Organic Matter on Sorption of Zn on Soil: Elucidation by Wien Effect Measurements and EXAFS Spectroscopy. *Environ. Sci. Technol.* **2016**, *50*, 2931–2937. [CrossRef]
37. Hamid Reza, B.; Najafi-Ghiri, M.; Amin, H.; Mirsoleimani, A. Zn desorption kinetics from some calcareous soils of orange (*Citrus sinensis* L.) orchards, southern Iran. *Soil Chem. Soil Min.* **2018**, 20–27. [CrossRef]
38. Ashraf, M.S.; Ranjha, A.M.; Yaseen, M.; Ahmad, N.; Hannan, A. Zn adsorption behaviour of different textured calcareous soils using Freundlich and Langmuir models. *Pakistan J. Agric. Sci.* **2008**, *45*, 6–10.
39. Hashemi, S.S.; Baghernejad, M. Zn sorption by acid, calcareous and gypsiferous soils as related to soil mineralogy. *Iran. Agric. Res.* **2009**, *28*, 1–16.
40. Gurpreet-Kaur; Sharma, B.D.; Sharma, P. Zn adsorption as affected by concentration, temperature, and time of contact in the presence of electrolytic and aqueous medium in benchmark Soils of Punjab in Northwest India. *Comm. Soil Sci. Plant Anal.* **2012**, *43*, 701–715.
41. Hararah, M.A.; Al-Nasir, F.; El-Hasan, T.; Al-Muhtaseb, A.H. Zn adsorption–desorption isotherms: Possible effects on the calcareous Vertisols from Jordan. *Environ. Earth Sci.* **2012**, *65*, 2079–2085. [CrossRef]
42. Maskina, M.S.; Randhawa, N.S.; Sinha, M.K. Relation of growth and Zn uptake of rice to quantity, intensity and buffering capacity factors of Zn in soils. *Plant. Soil* **1980**, *54*, 195–205. [CrossRef]
43. Nimibofa, A.S.S.; Wankasi, A.D.; Dikio, E.D. Synthesis, characterization and application of Mg/Al layered double hydroxide for the degradation of congo red in aqueous solution. *Open J. Phys. Chem.* **2015**, *5*, 56–70.
44. Shuman, L.M. The effect of soil properties on Zn adsorption by soils. *Soil Sci. Soc. Am. Proc.* **1975**, *39*, 454–458. [CrossRef]
45. Sipos, P.; Nemeth, T. Effect of clay mineralogy on trace metal geochemistry as reflected by the soil profiles from the Cserhat Mts. In Proceedings of the NE Hungary, MEC Conference, Stara Lesna, Slovakia, 1 October 2001; Volume 99.

Article

Beneficial Effect of Root or Foliar Silicon Applied to Cucumber Plants under Different Zinc Nutritional Statuses

José María Lozano-González, Clara Valverde, Carlos David Hernández, Alexandra Martin-Esquinas and Lourdes Hernández-Apaolaza *

Department of Agricultural Chemistry and Food Science, Universidad Autónoma de Madrid, Av. Francisco Tomás y Valiente 7, 28049 Madrid, Spain; josem.lozano@uam.es (J.M.L.-G.); clara.valverdes@estudiante.uam.es (C.V.); carlosd.hernandez@estudiante.uam.es (C.D.H.); alexandra.martin@uam.es (A.M.-E.)
* Correspondence: lourdes.hernandez@uam.es

Abstract: Zinc (Zn) is an essential micronutrient involved in a large variety of physiological processes, and its deficiency causes mainly growth and development disturbances, as well as oxidative stress, which results in the overproduction and accumulation of reactive oxygen species (ROS). A possible environmentally friendly solution is the application of silicon (Si), an element that has shown beneficial effects under abiotic and biotic stresses on many crops. Si could be applied through the roots or leaves. The aim of this work is to study the effect of Si applied to the root or shoot in cucumber plants under different Zn statuses (sufficiency, deficiency, and re-fertilization). Cucumber plants were grown in hydroponics, with 1.5 mM Si applied at the nutrient solution or sprayed on the leaves. During the different Zn statuses, SPAD index, fresh weight, ROS, and Si, Zn, P, Cu and B mineral concentration were determined. The results suggested that Si application had no effect during sufficiency and deficiency periods, however, during re-fertilization foliar application of Si, it showed faster improvement in SPAD index, better increment of fresh weight, and a decrease in ROS quantity, probably due to a memory effect promoted by Si previous application during the growing period. In summary, Si application to cucumber plants could be used to prepare plants to cope with a future stress situation, such as Zn deficiency, due to its prompt recovery after overcoming the stress period.

Keywords: silicon; Zn-deficiency; Zn-sufficiency; Zn re-fertilization

Citation: Lozano-González, J.M.; Valverde, C.; Hernández, C.D.; Martin-Esquinas, A.; Hernández-Apaolaza, L. Beneficial Effect of Root or Foliar Silicon Applied to Cucumber Plants under Different Zinc Nutritional Statuses. *Plants* **2021**, *10*, 2602. https://doi.org/10.3390/plants10122602

Academic Editor: Gokhan Hacisalihoglu

Received: 5 November 2021
Accepted: 25 November 2021
Published: 27 November 2021

Publisher's Note: MDPI stays neutral with regard to jurisdictional claims in published maps and institutional affiliations.

1. Introduction

Monosilicic acid (H_4SiO_4) is the plant-assimilable form of silicon (Si), which can be found in soils with a concentration ranging between 0.1 and 1.4 mM [1]. Although all soil-grown plants contain some Si in their tissues [2], plants have different capacities to accumulate it, with values varying between 0.1% and 10% Si (dry weight) [3]. This element has been classified as a beneficial but non-essential nutrient for higher plants [1,2,4,5], as some species are almost unaffected by silicon fertilisation compared to others [6], and with different expression and functionality of Si transporters [7]. However, the beneficial effect of Si on the plant growth promotion under biotic and abiotic conditions has been extensively studied [2,8–12] in both root [13–15] and shoot [16–18] application. As discussed above, many studies have been conducted on the beneficial effects of Si in plants, but the mechanism of action of this element is still under discussion. Coskun et al. [6] proposed that the effect of Si is due to Si deposits formed in the apoplast, triggering the activation of the plant's stress responses. Hernández-Apaolaza et al. [17] also noted that the application of Si appears to be related to the induction of the corresponding stress responses, at least under micronutrient deficiencies, and included the hypothesis of the activation of some kind of memory effect that was activated by the Si addition, which was evident under

109

resupply experiments. In any case, further studies are needed to clarify and prove these emerging theories.

Zinc (Zn) is one of the essential microelements for plants, which has a functional and structural role in enzymatic reaction, being part of numerous enzymes, such as: carbonic anhydrase, responsible for CO_2 fixation in photosynthesis; alcohol dehydrogenase, that converts acetaldehyde to ethanol in anaerobic respiration in roots; or superoxide dismutase (SOD), which protects the plant against oxidation by superoxide radicals [19]. It also is involved in the synthesis of tryptophan, a precursor of the hormone auxin (IAA) necessary for plant growth [20], has an essential role in maintaining the structure and permeability of the plasma membrane [21], and is involved in the transport of phosphorus through the plant [19]. The high pH in calcareous soils decreases the solubility of the metal, thus precipitating it in the form of carbonates or hydroxides [22], decreasing its bioavailability, and causing a deficiency of this element in plants. Zn deficiency generates a delay in the growth of the plant, decrease and malformations in the young leaves, shortening of the internodal distance and, in the case of a severe deficiency, chlorosis and necrotic spots [1,23]. Furthermore, Zn deficiency can give rise to oxidative stress, which triggers an over-production and accumulation of reactive oxygen species (ROS), species that, in excess, can behave as toxic compounds [24]. Additionally, Zn deficiency is related to a higher accumulation of phosphorus in the old leaves; the more severe the Zn deficiency, the higher the phosphorus concentration [25]. There are different strategies to alleviate Zn deficiency, such as the application of inorganic salts to the soil or synthetic chelates among others [26], however, these strategies are expensive and have a high environmental impact. Another environmentally friendly alternative to alleviate Zn deficiency symptoms could be Si fertilization.

Bityutskii et al. [13] tested the effect of root Si addition on Zn-deficient cucumber (*Cucumis sativus* L.); for this, plants were grown for 2 weeks in hydroponics with or without Zn, Si was added as 1.5 mM H_4SiO_4. They observed that Zn deficiency symptoms were only partially prevented because no necrotic spots were observed, which may be relayed to an enhanced antioxidant capacity. Mehrabanjoubani et al. [27] observed that the application of Si in the form of sodium metasilicate ($Na_2SiO_3 \cdot 5H_2O$) increased: nutrient uptake, shoot biomass, and grain yield in rice plants submitted to Zn deficiency, toxicity and optimal Zn levels. It is commonly known that Si is essential for rice growth. Moreover, silicon application increased Zn, as well as Ca^{2+}, K^+, P and B contents in plants supplied up to $50\ \mu g\ L^{-1}$ Zn. In the experiment conducted by Pascual et al. [28] 0.5 or 1.0 of $Na_2SiO_3 \cdot 5H_2O$ was added to the nutrient solution of Zn-deficiency soybean (*Glycine max* L.) plants; after each sampling, Zn concentration in root, stem and leaves were measured. It was observed that treatment with 0.5 mM of Si promoted Zn accumulation in the root apoplast, and its subsequent remobilization by shoot, ameliorating the Zn deficiency symptoms. These authors developed a preculture with Si addition before the Zn depletion from the nutrient solution.

The aim of this work is to evaluate, for the first time, as far as we know, the effect of foliar and root application of Si in cucumber plants under different Zn statuses: Zn sufficiency, Zn deficiency, and Zn re-fertilization. Moreover, the effect of Si in other nutrient content (Cu, P, B) was also studied under the three Zn statuses. P concentration was studied for the well-known antagonism between this element and Zn, and Cu was evaluated due to its presence in the Cu/Zn SOD enzyme, which among others is uncharged, to decrease ROS concentration generated by Zn deficiency (and the rest of biotic and abiotic stressors), and finally, B has been studied due to its chemical similarities to Si.

2. Results

Silicon addition to the root (+SiR) or shoot (+SiF) of cucumber plants has been evaluated under Zn sufficiency (+Zn), deficiency (−Zn), and re-fertilization (−Zn(+Zn)). Two samplings were carried out: sampling 1 at the end of the Zn deficiency period, and sampling 2 at the end of the re-fertilization period. In each sampling, plants were compared with their respective controls (−Si+Zn and −Si+Zn(+Zn)).

2.1. *Effect of Silicon on SPAD Index and Biomass under Different Zn Nutrient Statuses*

The SPAD index was assessed in new leaves after the Zn deficiency period (sampling 1, Figure 1a), and at different days (4, 6, 8 and 11) during the Zn resupply period (Figure 1b), and compared to their respective controls (+Zn for sampling 1 and +Zn(+Zn) for sampling 2). In sampling 1, +Zn treatments showed significantly higher SPAD index values than the −Zn treatments. Among the treatments with +Zn, no differences were observed with or without Si application, and the same occurred in the deficient treatments (−Zn). In sampling 2 (Zn resupply), treatments with a continuous Zn addition (+Zn(+Zn)) showed significantly higher SPAD index than plants that have suffered a previous period of Zn deficiency (−Zn(+Zn)), being the treatment without Si application (−Si+Zn(+Zn)), the one with the highest SPAD index. The application of Si to the leaves after a period of Zn deficiency followed by a Zn resupply (+SiF−Zn(+Zn)) showed significantly better results than the other Si treatments (−Si−Zn(+Zn) or +SiR−Zn(+Zn)), presenting similar levels to the treatments with a continuous Zn application (+SiR+Zn(+Zn) and +SiF+Zn(+Zn)), at day 8 and 11 after Zn resupply (Figure 1b).

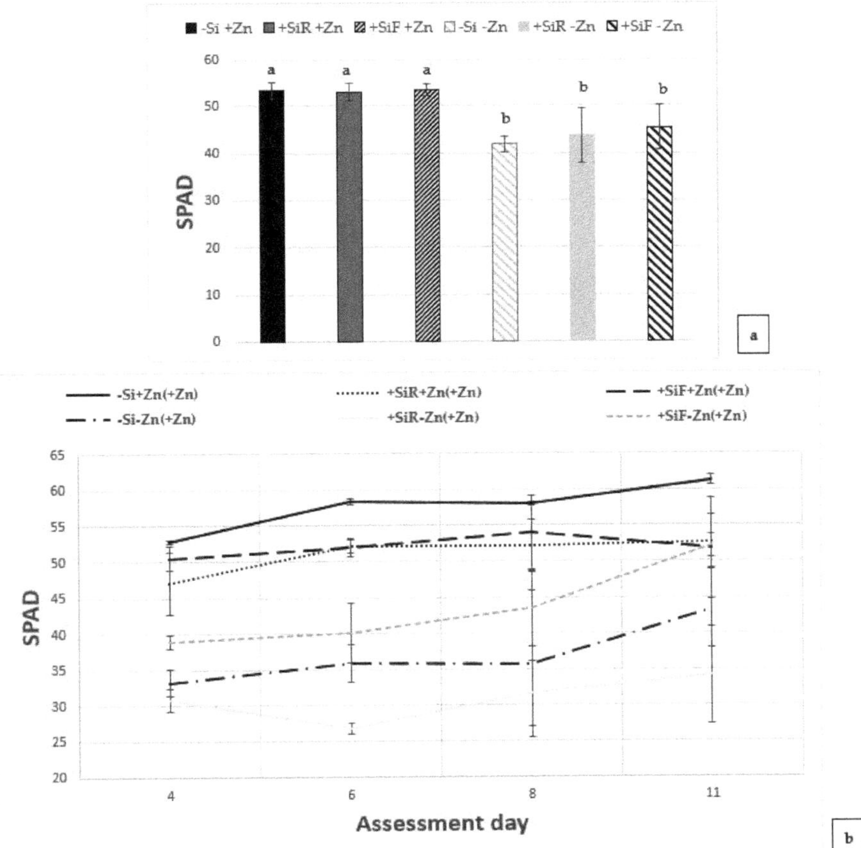

Figure 1. Effect of Si addition on the root (+SiR) or on the shoot (+SiF) of cucumber plants grown under different Zn nutritional statuses on SPAD values: (**a**) After a Zn deficiency period (−Zn) compared to the control plants with continuous Zn addition (+Zn) and (**b**) During a Zn re-fertilization period (−Zn(+Zn)) compared with their corresponding control plants (+Zn(+Zn)). The data are the mean ± SE (n = 5). Different letters indicate significant differences according to Duncan's test ($p < 0.05$).

With respect to plant biomass (Table 1), plants were divided into root, stem, old leaves (grown before the Zn depletion from the nutrient solution in sampling 1 or grown before the refertilization in sampling 2), and new leaves (grown during Zn deficiency and resupply periods). In sampling 1 (after Zn deficiency), control plants (+Zn) showed significantly higher fresh weight (FW) values than the −Zn ones. In both sufficiency (+Zn) and deficiency (−Zn) statuses, the application of Si (+SiR and +SiF) had no effect; also, in both cases, it was observed that the highest percentage of fresh weight was located in the root. After Zn resupply (sampling 2), control treatments with +Zn(+Zn) have a significantly higher fresh weight than treatments that have undergone a period of deficiency (−Zn(+Zn)). In the treatment (+Zn(+Zn)), the application of Si (+SiR and +SiF) significantly increases fresh weight compared to the treatment without Si (−Si), the foliar application (+SiF) being the treatment that presented the highest fresh weight values. With respect to the −Zn(+Zn) treatments, the application of Si (+SiR and +SiF) had no effect. As in sampling 1, most of the fresh weight was located in the root (Table 1). In addition, in the treatments with Zn-refertilization (−Zn(+Zn)), an important percentage of the fresh weight was also located in the new leaves (Table 1).

Table 1. Fresh weight distribution in plant tissues (%), and total fresh weight (FW, g) of cucumber plants with a root (+SiR) or shoot (+SiF) Si application and grown under different Zn nutritional statuses. Sampling 1: plants collected after a Zn deficiency period (−Zn) compared to the control plants grown with continuous Zn addition (+Zn), and sampling 2: plants collected after a Zn re-fertilization period (−Zn(+Zn)) and compared to their corresponding control plants (+Zn (+Zn)).

	Fresh Weight Distribution in Plant Tissues (%)				Total FW (g)
	Root	Stem	Old Leaves	New Leaves	
			Sampling 1		
−Si+Zn	47.40 [b]	9.52 [ab]	31.23 [a]	11.85 [ab]	11.56 [a]
+SiR+Zn	50.94 [b]	8.32 [b]	33.58 [a]	7.17 [b]	9.62 [a]
+SiF+Zn	55.26 [a]	7.33 [ab]	25.44 [a]	11.97 [a]	15.96 [a]
−Si−Zn	45.62 [c]	15.67 [c]	11.06 [b]	27.65 [b]	2.17 [b]
+SiR−Zn	43.70 [c]	14.96 [c]	4.72 [b]	36.61 [b]	2.54 [b]
+SiF−Zn	38.24 [c]	15.69 [c]	12.25 [b]	33.82 [b]	2.04 [b]
			Sampling 2		
−Si+Zn(+Zn)	35.62 [c]	25.12 [b]	14.81 [b]	24.46 [c]	31.81 [c]
+SiR+Zn(+Zn)	51.45 [b]	19.51 [a]	12.01 [a]	17.04 [d]	72.95 [b]
+SiF+Zn(+Zn)	57.48 [a]	15.25 [a]	11.28 [a]	15.98 [d]	96.1 [a]
−Si−Zn(+Zn)	37.15 [c]	22.55 [c]	7.95 [c]	32.35 [b]	5.41 [d]
+SiR−Zn(+Zn)	41.90 [c]	16.50 [c]	7.83 [c]	33.76 [b]	13.15 [d]
+SiF−Zn(+Zn)	34.74 [c]	18.36 [c]	4.49 [c]	42.40 [a]	7.57 [d]

The data are the mean $\pm$ SE (n = 3). Different letters in the same column for each sampling indicate significant differences according to Duncan's test ($p < 0.05$).

2.2. Effect of Silicon on ROS under Different Zn Nutrient Statuses

Reactive oxygen species (ROS) are an indication of oxidative stress in a plant; the more ROS there are, the greater the oxidative stress. A decrease in ROS indicates the activation of the plant's antioxidant defense capacity. After Zn deficiency, control plants treated with foliar Si (+SiF+Zn) seemed to be more stressed than control plants without Si application (−Si+Zn) (Figure 2), but when Si was applied to the root, no differences were observed. However, in Zn deficient plants (−Zn), Si addition, either to roots or leaves, increased ROS concentration in these tissues.

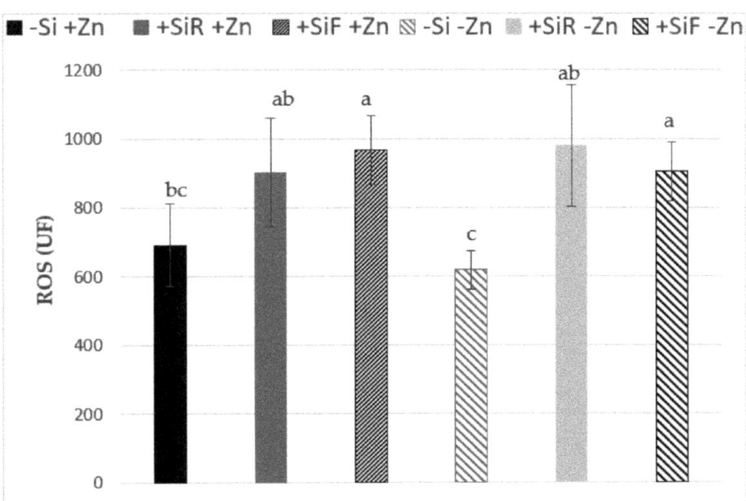

Figure 2. Effect of Si addition on the root (+SiR) or on the shoot (+SiF) of cucumber plants grown under different Zn nutritional statuses on reactive oxygen species (ROS) (FU). Plants were collected after a Zn deficiency period (−Zn) and compared to the control plants grown with continuous Zn addition (+Zn) after 16 days of growth. The data are the mean ± SE ($n = 9$). Different letters indicate significant differences according to Duncan's test ($p < 0.05$).

To study the effect of Si addition on Zn resupply response, the variation of ROS was calculated between both statuses (Zn deficiency and resupply) by subtracting the ROS after the Zn resupply period (Sampling 2) from the ROS, after the Zn deficiency period (Sampling 1). Moreover, this ROS increment was compared with the ROS variation in control plants which did not suffer a Zn deficiency period (+Zn(+Zn)) (Table 2). Both types of Si application reflected a significant decrease in ROS increment, being the Si foliar addition, the one which reduces ROS after Zn resupply in a greater amount. These results indicated that, after overcoming a period of Zn deficiency, the foliar application of Si (+SiF) helped the plant to decrease its ROS concentration.

Table 2. Effect of Si addition to the root (+SiR) or to the shoot (+SiF) of cucumber plants on reactive oxygen species increment (ΔROS (FU)) in the new leaves of plants harvested after a Zn deficiency period (sampling 1), followed by a Zn re-fertilization period (sampling 2), compared to their corresponding control plants, with a continuous Zn supply, and without Si (−Si).

	ΔROS (FU)
	Zn Sufficiency (Sampling 2-Sampling 1)
−**Si**	-182.47 ± 44.06 [e]
+**SiR**	-223.23 ± 39.52 [e]
+**SiF**	-220.93 ± 26.77 [e]
	Zn Resupply (Sampling 2)—Zn Deficiency (Sampling 1)
−**Si**	202.68 ± 13.42 [a]
+**SiR**	114.54 ± 21.77 [b]
+**SiF**	-43.33 ± 8.45 [c]

The data are the mean ± SE ($n = 3$). Different letters in the same column for sufficiency and resupply treatments, respectively, indicate significant differences, according to Duncan's test ($p < 0.05$).

2.3. Effect of Silicon on Mineral Concentration in Plant Tissues under Different Zn Nutrient Statuses

The concentrations of silicon (Si), zinc (Zn), copper (Cu), boron (B) and phosphorus (P) in plant tissues (root, old leaves and new leaves) were determined at the end of the Zn deficiency period (sampling 1) and Zn re-fertilization period (sampling 2).

Regarding the Si concentration at the end of the Zn deficiency period (Figure 3a), the −Si+Zn treatment showed the lowest concentrations, and the treatment with root application of Si (+SiR+Zn) showed higher concentrations in all plant organs than the foliar application (+SiF+Zn). The same trend could be observed in the treatments where Zn deficiency was generated (−Zn). The results obtained indicated that more Si was absorbed in Zn-deficient plants, although it was more poorly remobilized in the plant, since the concentration of Si in the root after root or foliar application of Si (+SiR and +SiF) in the Zn-deficient treatments (−Zn) was higher than in the counterpart treatments with sufficient Zn (+Zn); however, the opposite occurred with respect to the concentration of Si in the aerial part. On the other hand, these results indicated that Si was better absorbed and remobilized when applied by root (+SiR) than when applied foliar (+SiF). After Zn re-fertilization (sampling 2) (Figure 3b), the trend in Si concentration in the different plant tissues was similar to that observed after the deficiency period (sampling 1), Si concentration in the treatment without Si application (−Si) was the lowest, Si concentration in the treatments where it was applied to the root (+SiR) was higher in all tissues than in the treatments where it was applied to the leaves (+SiF); this trend was observed both in plants with a continuous supply of Zn (+Zn(+Zn)), and in those with a period of deficiency (−Zn(+Zn)).

In relation to the total Zn concentration in the whole plant at the end of the Zn deficiency period (sampling 1) (Table 3), in the control treatments with continuous Zn supply, no differences due to Si application have been found. However, Zn distribution was affected by the type of Si application. Compared to the control treatment (−Si+Zn), root Si addition (+SiR+Zn) decreased Zn percentage in root, but no differences were obtained in old or new leaves, although with foliar Si supply (+SiF+Zn), Zn significantly increased in new leaves. In sampling 2, plants with foliar Si and continuous Zn supply (+SiF+Zn(+Zn)) significantly increased the total Zn concentration with respect to −Si plants, although no differences with Si applied to the roots were obtained. According to Zn distribution in plant organs, Si enhanced Zn accumulation in root with both application types. Under Zn deficiency (sampling 1), only foliar Si application maintained similar Zn levels than in sufficiency (Table 3), and this Zn was mainly accumulated in new leaves. In the Zn resupply period (sampling 2), both Si treatments (+SiR−Zn(+Zn) and +SiF−Zn(+Zn)) enhanced Zn restoration to sufficiency levels compared with the control plants (−Si+Zn(+Zn)), and no statistical differences were observed between them. Interestingly, Zn was mainly in new leaves in +SiF treatment and in roots in +SiR treatment after re-fertilization. Related to P concentration, in sampling, 1 Zn depletion from the nutrient solution did not affect P concentration in the whole −Zn plants (Table 3). Phosphorous percentage in new leaves decreased in −Zn conditions without Si addition to the media but increased when Si was applied to the leaves. In sampling 2, after plant recovery, no differences in total P concentration in the entire plant were obtained in −Zn(+Zn). In root, +SiR−Zn(+Zn) treatment had the highest percentage of P; meanwhile, in the +SiF−Zn(+Zn) one, P was accumulated in new leaves (Table 3). In +Zn conditions at both samplings, data for roots showed that +SiR addition enhanced P percentage compared with the control plants (−Si+Zn and −Si+Zn(+Zn)); in old leaves, plants without Si drastically increased P percentage; although when Si was applied, this increment was not observed.

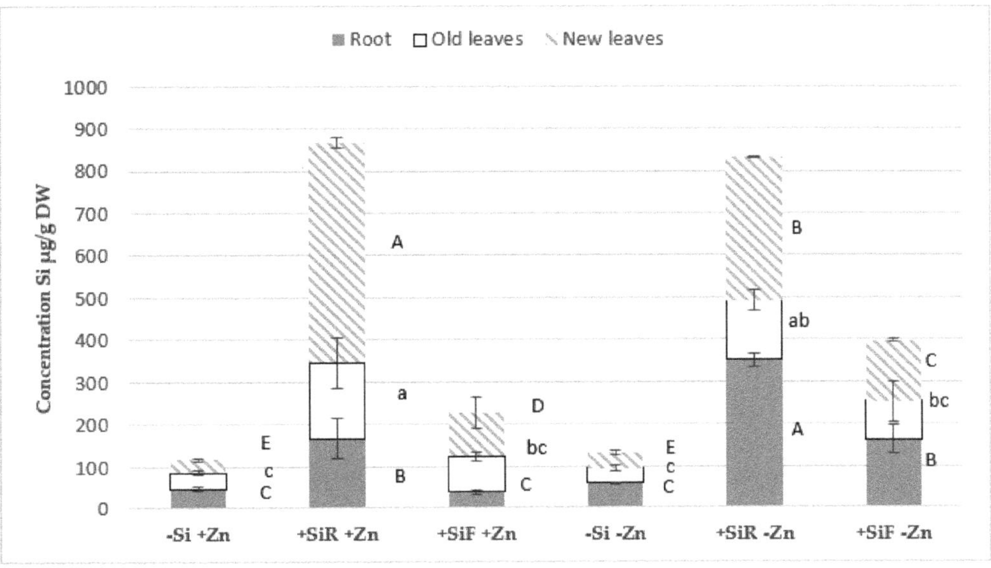

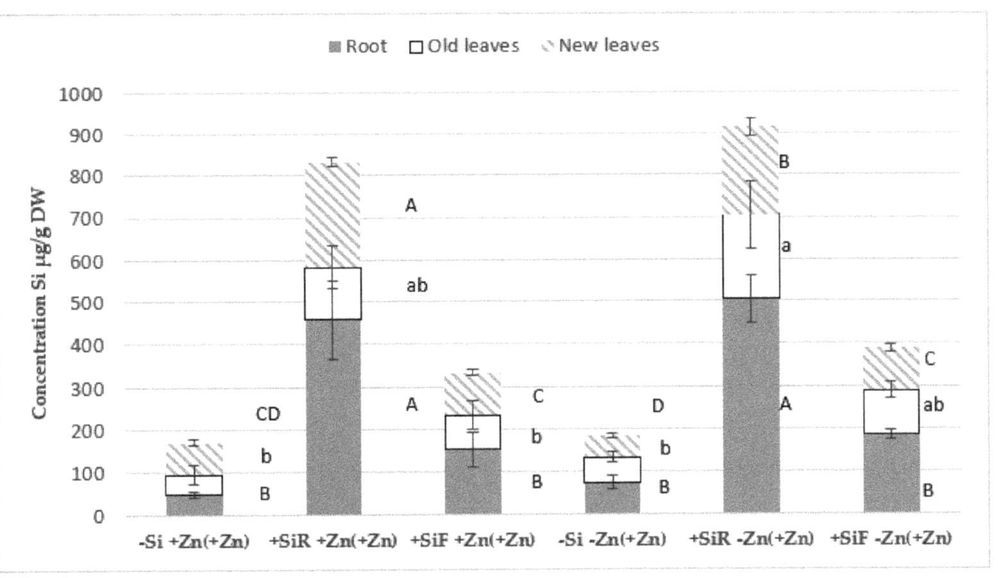

Figure 3. Effect of Si addition on the root (+SiR) or on the shoot (+SiF) on the Si concentration ($\mu g\ g^{-1}$ DW) in root, old and new leaves of cucumber plants (**a**) After a Zn deficiency period ($-$Zn) compared to the control plants with a continuous Zn addition (+Zn) and (**b**) After a Zn re-fertilization period ($-$Zn(+Zn)) compared with their corresponding control plants (+Zn(+Zn)). Data are mean $\pm$ SE ($n = 9$). Different letters indicate significant differences for each plant organ according to Duncan's test ($p < 0.05$).

Table 3. Zinc (Zn) and phosphorus (P) distribution in plant organs (%) and Zn (μg g^{-1} DW) and P concentration (mg g^{-1} DW) in the whole plant under different Zn and Si treatments (no Si ($-$Si), root (+SiR), and foliar (+SiF) application): (a) After a Zn deficiency period ($-$Zn) compared to the control plants with continuous Zn addition (+Zn) (sampling 1) and (b) After a Zn re-fertilization period ($-$Zn(+Zn)), compared to their corresponding control plants (+Zn(+Zn)) (sampling 2).

	Zn Distribution in Plant Tissues (%)				P Distribution in Plant Tissues (%)			
	Root	**Old Leaves**	**New Leaves**	**Total (µg/g)**	**Root**	**Old Leaves**	**New Leaves**	**Total (mg/g)**
Sampling 1								
$-$Si+Zn	51.59 [a]	21.28 [n.s]	27.14 [bc]	117.78 [a]	33.49 [cd]	20.36 [c]	46.14 [b]	4.18 [b]
+SiR+Zn	45.97 [bc]	18.13	35.90 [abc]	94.48 [ab]	32.381 [b]	48.56 [ab]	19.05 [c]	6.75 [ab]
+SiF+Zn	44.40 [ab]	15.02	40.58 [a]	118.57 [a]	25.53 [bcd]	31.99 [bc]	42.46 [c]	5.61 [ab]
$-$Si$-$Zn	40.49 [c]	29.50	30.00 [c]	78.57 [b]	34.60 [d]	51.40 [a]	13.99 [c]	5.89 [ab]
+SiR$-$Zn	42.65 [c]	24.79	32.56 [c]	80.37 [b]	36.43 [a]	47.75 [abc]	15.82 [c]	6.18 [ab]
+SiF$-$Zn	40.06 [bc]	20.67	39.28 [ab]	110.75 [a]	26.09 [bc]	21.10 [bc]	52.80 [a]	9.45 [a]
Sampling 2								
$-$Si+Zn(+Zn)	26.02 [b]	22.30 [ab]	51.68 [a]	145.64 [b]	14.36 [c]	64.07 [c]	21.55 [b]	1.96 [b]
+SiR+Zn(+Zn)	47.30 [a]	15.43 [b]	37.27 [a]	170.64 [ab]	59.82 [ab]	15.87 [c]	24.30 [b]	7.91 [a]
+SiF+Zn(+Zn)	37.33 [a]	23.51 [a]	39.16 [a]	183.89 [a]	23.27 [c]	50.79 [bc]	25.93 [b]	3.57 [b]
$-$Si$-$Zn(+Zn)	33.45 [c]	25.77 [b]	40.77 [b]	93.20 [c]	36.03 [bc]	48.68 [a]	15.27 [b]	7.71 [a]
+SiR$-$Zn(+Zn)	46.78 [a]	15.73 [b]	37.49 [a]	164.31 [ab]	41.42 [a]	43.99 [ab]	14.57 [b]	6.70 [a]
+SiF$-$Zn(+Zn)	34.80 [b]	18.01 [b]	47.19 [a]	128.68 [bc]	22.46 [c]	29.96 [bc]	47.57 [a]	8.00 [a]

The data are the mean $\pm$ SE (n = 3). Different letters in the same column for each sampling indicate significant differences according to Duncan's test ($p < 0.05$).

Copper (Cu) and boron (B) concentrations in the entire plant at the end of Zn sufficiency, Zn deficiency and Zn re-fertilization periods in all treatments were measured (Table 4). With respect to the total Cu concentration in the whole plant at sampling 1, among the +Si treatments of Zn sufficient plants, the +SiR+Zn one showed the highest concentration of Cu, and it was mainly located in new leaves. In sampling 2, the treatment with foliar application of Si (+SiF+Zn(+Zn)) presented the lowest concentration of Cu in the whole plant. In this sampling, the distribution of Cu in the Si treatments was mainly in the root, while in the control plants ($-$Si+Zn(+Zn)), Cu was found in the new leaves. The concentration of Cu in the treatments under Zn deficiency (sampling 1) was higher than in the Zn sufficiency plants. Under Zn deficiency, the treatment with root application of Si (+SiR$-$Zn) showed the lowest concentration. Plants without Si application ($-$Si$-$Zn) and with root application of Si (+SiR$-$Zn) presented a higher percentage of Cu in new leaves; however, after foliar application of Si (+SiF$-$Zn), Cu was found mainly in the root. In the Zn resupply period (sampling 2), Cu concentration attained much higher values than the control treatment ($-$Si+Zn(+Zn)), except for the root application of Si (+SiR$-$Zn(+Zn)), which showed similar values to the control. In all treatments, Cu was mainly located in the root. In relation to B concentration, in sampling 1 treatments, with or without Si application and with an optimal or deficient Zn nutritional level (+SiR+Zn; +SiF+Zn), they did not present significant differences with respect to control plants in the concentration of B in the entire plant (Table 4). Regarding the distribution of B in the plant, in all treatments B, was found mainly in the old leaves. In sampling 2, Si addition significantly increased B concentration in the plant compared to untreated plants in well fed plants. Under Zn resupply, +SiF$-$Zn(+Zn) presented the highest B concentration in the plant, and this element was found mainly in old leaves.

Table 4. Copper (Cu) and boron (B) distribution in plant organs (%), and Cu and B concentration (μg g^{-1} DW) in the whole plant under different Zn and Si treatments (no Si ($-$Si), root (+SiR), and foliar (+SiF) application: (a) After a Zn deficiency period ($-$Zn) compared to the control plants with continuous Zn addition (+Zn) (sampling 1) and (b) After a Zn re-fertilization period ($-$Zn(+Zn)), compared to their corresponding control plants (+Zn(+Zn)) (sampling 2).

	Cu Distribution in Plant Tissues (%)				B Distribution in Plant Tissues (%)			
	Root	Old Leaves	New Leaves	Total (μg/g)	Root	Old Leaves	New Leaves	Total (μg/g)
Sampling 1								
$-$Si+Zn	29.12 [bc]	27.42 [a]	43.46 [b]	84.31 [c]	17.40 [b]	68.22 [a]	14.39 [c]	49.51 [ab]
+SiR+Zn	28.90 [bc]	11.83 [c]	59.26 [a]	152.56 [b]	32.64 [a]	64.50 [a]	2.86 [d]	55.47 [a]
+SiF+Zn	26.79 [c]	14.91 [c]	58.30 [a]	77.85 [c]	15.71 [b]	62.41 [a]	21.88 [bc]	58.23 [a]
$-$Si$-$Zn	35.16 [b]	25.05 [a]	39.79 [c]	259.78 [a]	10.82 [b]	52.17 [b]	37.01 [a]	44.55 [b]
+SiR$-$Zn	39.73 [b]	12.93 [c]	47.34 [b]	168.67 [b]	12.73 [b]	51.61 [b]	35.67 [a]	43.72 [b]
+SiF$-$Zn	59.50 [a]	20.12 [b]	20.38 [d]	274.10 [a]	12.77 [b]	48.56 [b]	38.66 [a]	56.11 [a]
Sampling 2								
$-$Si+Zn(+Zn)	33.01 [d]	22.13 [ab]	44.86 [a]	81.78 [c]	32.64 [b]	46.31 [b]	21.05 [b]	70.29 [b]
+SiR+Zn(+Zn)	48.91 [c]	24.75 [a]	26.34 [c]	76.02 [c]	12.77 [c]	68.20 [a]	19.03 [b]	94.34 [a]
+SiF+Zn(+Zn)	49.79 [c]	22.79 [ab]	27.43 [c]	62.06 [d]	28.79 [b]	37.43 [c]	33.78 [a]	103.19 [a]
$-$Si$-$Zn(+Zn)	60.67 [b]	7.28 [c]	32.05 [b]	170.52 [a]	26.84 [c]	70.03 [a]	3.13 [c]	44.73 [c]
+SiR$-$Zn(+Zn)	56.80 [b]	18.51 [b]	24.70 [c]	86.43 [c]	26.85 [c]	36.95 [c]	36.20 [a]	57.93 [c]
+SiF$-$Zn(+Zn)	71.57 [a]	27.23 [a]	1.20 [d]	116.25 [b]	38.78 [a]	40.25 [bc]	20.97 [b]	108.87 [a]

The data are the mean $\pm$ SE ($n = 3$). Different letters in the same column for each sampling indicate significant differences according to Duncan's test ($p < 0.05$).

3. Discussion

Different effects could be observed in the parameters measured during the study, depending on the nutritional status of the cucumber plant and the treatment applied, so discussion has been divided into Si impact on Zn sufficient, deficient, and re-fertilized plant for each determination.

3.1. Effect of Silicon on SPAD Index and Biomass under Different Zn Nutrient Statuses

For the first time, the effects of Si applied to the root (+SiR) and leaves (+SiF) were tested on cucumber plants that underwent periods of Zn sufficiency, deficiency, and Zn deficiency recovery. In the plants that had a continuous supply of Zn in sampling 1 (16 days of growth), no differences have been observed in the SPAD index, although in sampling 2 (27 days of growth), the application of Si to both root (+SiR+Zn(+Zn)) and leaves (+SiF+Zn(+Zn)) provided a significantly lower SPAD index compared to the control ($-$Si+Zn(+Zn)) (Figure 1). Hernández-Apaolaza et al. [17], in a similar experiment in which cucumber plants were grown with Si applied both foliar and to the roots to well nourished seedlings, only found a decrease in the SPAD index with respect to the control plants when Si was applied to the roots (+SiR+Fe). These authors grew the plants under optimal nutrient conditions for 10 days, but in the present work, only 7 days of growth with a completed nutrient solution were applied. This fact may explain why the treatment with foliar Si did not suffer an SPAD decrease at that point. Several authors have argued that Si application enhances the Casparian Band of the root exodermis development [6,29–32]; this increase of apoplastic barriers may hinder nutrient uptake such as iron [33–35], and maybe Zn, which could explain the low SPAD levels compared to the Si-untreated plants, as it is well known that Zn deficiency promotes chlorosis in leaves [1]. This parameter may indicate that Si, when applied to the root, promoted Zn deficiency even when plants were submitted to a complete nutrient solution, as it was described before by Hernandez-Apaolaza et al. [17]. However, results obtained at this stage (Sampling 2) showed that plants with foliar Si (+SiF+Zn(+Zn)) significantly increased the total Zn concentration in the plant (Table 3) with respect to $-$Si plants, although no differences when Si was applied to the roots were obtained. However, the Zn distribution in plant organs showed that Si enhanced Zn accumulation in root with both application types, so this may explain the lower SPAD index found when Si was added. SPAD results of Zinc deficient plants (Figure 1a) indicated that Si addition did not have any influence on this parameter under $-$Zn. However, when Zn concentration is taken into account (Table 3), plants with foliar Si presented a significant higher amount of Zn than the other Si treatments, and were similar to control plants without Si and +Zn. In this case, Zn was equally distributed between roots and new leaves (Table 3). Likewise, Farias Guedes et al. [36] observed a significant increase in chlorophyll content, with foliar applications of Si to Zn-deficient Sorghum plants. Furthermore, the studies of Pascual et al. [28] and Bityutskii et al. [13] with soybean and cucumber plants, respectively, did not show significant increases in SPAD or chlorophyll content when Si was added to the nutrient solution (+SiR).

In plants that underwent a Zn deficiency period followed by a recovery one ($-$Zn(+Zn)) the treatment with foliar application of Si showed significantly higher SPAD levels than the other Si treatments (Figure 1), similar to that observed by Hernández-Apaolaza et al. [17] with Fe resupply of cucumber plants. These authors hypothesized that this quicker recovery from the chlorosis, was possibly related to plant memory effect. Srivastava et al. [37] observed that plants have to cope with different stresses all along their crop cycle, and retained 'memories' of previously encountered stresses as an adaptive mechanism that help them to confront forthcoming stresses more rapidly and efficiently. Such memories can be induced artificially, through preexposure to low-dose stressor, or by the addition of beneficial compounds like silicon. The induced stress memory is called 'acquired tolerance', and it can be retained either in the short term (somatic memory), or may be transferred to succeeding generations (intergenerational memory), or in some cases, inherited across generations (trans-generational memory) [37]. Ding et al. [38] applying successive dehydration stress/recovery treatments in a relatively short time in Arabidopsis plants, pointed out that leaf cells during recurring dehydration stresses displayed a transcriptional stress memory. During recovery (watered) states, trainable genes produce transcripts at basal (preinduced stress) levels, as expected, but remain associated with atypically high stress marks, that reinforced the response to the rewatering process. Our results may conduct a similar plant response related to Zn resupply. With respect to Zn concentration (Table 3), both +SiR$-$Zn(+Zn) and +SiF$-$Zn(+Zn) presented a tendency of Zn accumulation in the plant compared to $-$Si$-$Zn(+Zn), although only the +SiR treatment had a significant higher amount of Zn; even though Zn distribution in plant organs pointed to new leaves of +SiF having the greatest Zn percentage, which would explain this treatment having the highest SPAD index.

With respect to biomass (Table 1), in sampling 1, no significant differences were found in plants that had a correct Zn nutritional status (16 days +Zn); nevertheless, after 27 days of growth (sampling 2) the application of Si to the roots or shoots significantly increased the fresh weight compared to the treatment without Si application, +SiF being the treatment that produced the greatest increase. These results pointed to the beneficial effect of Si on biomass increase, especially if Si was applied to the leaves. A period of more than 16 days of growth was required to observe the beneficial effect of Si on biomass. Similar results were obtained by Mehrabanjoubani et al. [27] in rice plants, when different Zn concentrations (1, 10, 50 and 100 µg L^{-1}) were applied, and root Si was added as sodium silicate (1.5 mM) in hydroponics. They observed a significant increase in biomass in plants, with an optimal Zn nutritional status due to the application of Si after 105 days of treatment. Zn is an essential micronutrient necessary for optimal growth [1], and as expected, the observed increase in biomass was well correlated with significant increase in Zn concentration in the whole plant in +SiF+Zn(+Zn) plants, with respect to the control plants ($-$Si+Zn(+Zn)) (Table 3). As mentioned before, the apoplastic obstruction caused by Si and described by Coskun [6] was avoided when Si was applied to the leaves. This fact may explain why these plants presented the highest fresh weight values. On the contrary, Si application had no effect on fresh weight under Zn deficiency conditions (Table 1), which were consistent with the results obtained by Pascual et al. [28] in soybean plants. However, Mehrabanioubani et al. [39] observed that root Si addition increased total weight in −Zn rice plants. These differences among experiments were probably due to the different plant species used, and their ability to accumulate Si. Plants with active uptake of Si are classified as high accumulators, with passive silicon uptake are classified as intermediate accumulators, and with rejection mechanisms, are classified as non-accumulators [40]. Rice is known to be a high Si accumulator [41], and both soybean and cucumber are intermediate accumulators [42,43]. The total Zn concentration in the +SiF−Zn plants (Table 3) had a significantly higher concentration of Zn than the other treatments (−Si−Zn and +SiR−Zn), though a higher fresh weight was not recorded. In plants to which Zn was added back into the nutrient solution (−Zn(+Zn)), no significant differences in fresh weight were observed (Table 1), but fresh weight and Zn concentration (Table 3) values tend to be higher in the treatments with Si application. Maybe, if the period of re-fertilization were extended, significant differences in fresh weight could be observed.

3.2. Effect of Silicon on ROS under Different Zn Nutritional Statuses

Reactive oxygen species (ROS) are regulated naturally by the action of enzymes such as catalase and superoxide dismutase [44], or by non-enzymatic mechanisms, such as the accumulation of phenols [45,46]. Several authors have reported the improvement of the antioxidant defense capacity of the plant due to the application of Si to the root, as enzymatic activities such as SOD and CAT have been improved by the application of Si, which would cause a ROS reduction [47,48]. In addition, as with root application (+SiR), foliar application (+SiF) also stimulates the activation of antioxidant defenses [49–51]. Here, plants with continuous Zn supply and treated with foliar Si were more stressed than the −Si plants after 16 days of growth (sampling 1; Figure 2), but the Zn concentration in the whole plant was similar in all the treatments (Table 3). These data have been supported by Hacisalihoglu et al. [52], who observed no differences in shoot Zn concentrations between the Zn-inefficient and Zn-efficient bean genotypes grown under low−Zn conditions, where differences in Zn efficiency were exhibited. Zn efficiency is defined as the ability of plants to maintain high yield under Zn-deficiency stress. However, at the end of the experiment (27 days) no significant differences between treatments were found in ROS variation values, which decreased for all of them when Zn was resupplied to the nutrient solution (Table 2). Even though there were no stress situations, plants with Si application presented higher ROS (specially with Si foliar application) at the first sampling time. After that period, plants will be recovered (Table 2) and probably show the beneficial effect of this element at longer testing periods. Different results were obtained by de Farias Guedes et al. [36] in sorghum, where they observed that plants with a correct nutritional status sprayed with Si decreased their cell damage. In their experiment, the effect of Si on oxidative stress was studied 111 days after emergence, so the difference in the experiment duration, as well as the plant species used, may explain the disparity of the results. ROS levels in Zn-deficient plants with both Si applications (+SiR−Zn; +SiF−Zn) were significantly higher than in Si untreated plants (−Si−Zn) (Figure 2). It is well known that Zn deficiency produces an increment of ROS [20], and due to the results obtained, both types of Si application increased plant stress, since they presented significantly higher levels of ROS compared to the treatment without Si application (−Si). The results obtained for Zn sufficiency at first sampling time, and at Zn deficiency, were in agreement with the previously explained hypothesis of apoplastic obstruction. On the other hand, if after a period of Zn deficiency, Zn was added back to the nutrient solution, the treatments with Si application showed an improvement in the mitigation of ROS concentration in new leaves; specifically, the effect was more relevant in +SiF plants (Table 2). The data obtained suggest that there was a significant improvement in plant antioxidant defense capacity with foliar Si application, which was attributed to the "memory effect" of the plants [38], that "remember" the previous stress caused by the Si addition and were better prepared to cope with it than the non−Si-treated plants.

Silicon application improves the accumulation of Cu-ligands, such as proteins like Zn/CuSOD [53], which will contribute to ROS concentration reduction. There are three types of superoxide dismutase enzymes that are described in plants: FeSOD, MnSOD, and Cu/ZnSOD (where Cu is the redox active catalytic metal). However, the overproduction of superoxide dismutase only gives limited protection to abiotic stress, and does not remarkably improve plant performance [54]. Surprisingly, plant lines lack the most abundant Cu/ZnSOD or FeSOD activities performed, as well as the wild-type under most conditions tested, indicating that these superoxide dismutases were not limiting to the prevention of oxidative damage. As Cu was the catalytic metal involved in SOD activity, Cu was measured either for sufficient, deficient, or refertilized plants. In general, under Zn sufficiency, foliar Si decreased Cu concentration in plant (Table 4), and no Cu accumulation in roots was observed, as described by Bosnić et al. [53]. The expected increase in ROS concentration in the +SiF+Zn due to Cu decrease was obtained after 16 days of plant growth (Figure 2), although this has not reflected at the end of sampling 2 (Table 2), which supported the Pilon et al. [54] statement that SOD was not really a limiting factor to prevent oxidative damage. Comparing these data according to Zn concentration results (Table 3), in sampling 1 no differences in Zn concentration was observed, so the copper effect was the main fact that controlled Cu/Zn SOD activity. In sampling 2, +SiF significantly increased Zn concentration in the plant compared to −Si+Zn plants, so this may compensate for the decreased in Cu concentration found for this treatment, and finally show similar behavior according to ROS accumulation. Zinc-deficient plants showed that +SiR−Zn had the lowest Cu concentration in plant; subsequently, a high ROS concentration was found, although no statistical differences with +SiF−Zn were obtained. Moreover, the +SiF treatment was the one that accumulated the most Cu in the root. This may be due to the fact that foliar-applied Si enhances Cu binding capacity to the cell wall, and the accumulation of Cu-ligands, more than root-applied Si [55]. As mentioned above, Pilon et al. [54] indicated that the lack of the most abundant Cu/ZnSOD or FeSOD activities is not limiting, but a strong defect in chloroplast gene expression and development appeared in plants that lack the two minor FeSOD isoforms, which are expressed predominantly in seedlings, and that associate closely with the chloroplast genome. On the contrary, several authors have reported that Fe deficiency enhanced SOD activity [8,56,57], but the literature did not distinguish between the FeSOD isoforms altered. In refertilization, −Si plants presented the highest ROS increment, which means that ROS concentration produced due to −Zn was still present in the plants. Their Cu concentration was the highest among treatments, but its Zn concentration was the lowest, along with foliar Si treatment. This shows that the Zn concentration plays an important role in ROS scavenging, related to Cu/ZnSOD activity. As can be seen in the distribution of Cu in plant tissues (Table 4), the +SiF treatment had most of the Cu lodged in the root, which could provide evidence for an improvement in the Cu binding capacity of the root cell wall; this may have enhanced Zn/CuSOD accumulation, which would help to decrease oxidative stress. To verify this fact, future experiments could measure the amount of SOD isoforms present in the plant in such conditions.

3.3. Effect of Silicon on Mineral Concentration in Plant Tissues under Different Zn Nutritional Statuses

Concentrations of Si at the different Zn statuses were measured (Figure 3), and the first relevant thing observed was that, when Si was applied through the roots, a significantly higher amount of Si was measured than when applied to the shoot. It was necessary to note here that foliar Si addition tested was at the optimal levels reported in previous works (for example [12,17,58]) for different crops. Therefore, the Si amount added through root and leaf applications was not the same. When Si was added to the nutrient solution, the concentration was 1.5 mM, and it was renewed weekly, however, when Si was added to the leaves by spraying, three doses (125 μL each) per leaf per week were used. The difference was due to the instability of the Si solutions at the high concentrations required to spray shoots, if a similar amount of Si should be added through roots and shoots. When Si concentration in Zn-sufficient and -deficient plants were compared (sampling 1)

(Figure 3a), a significantly higher accumulation of Si in the root at Zn deficiency treatments was observed. In Figure 3b, resupply and sufficient plants only differ in the Si root treatment, in which Si in new leaves were lower in the resupply experiment than in the sufficient one.

Many attempts have been made to reveal the mechanisms of Zn-efficient plants in response to low$-$Zn, but they are still not well described [59]. As Hacisalihoglu [59] described in his review, several uptake Zn^{+2} studies in crop plants found no strong correlation between root influx and Zn efficiency, which indicates that Zn efficiency in higher plants is likely not a root-focused trait, but a shoot-focused trait, possibly related to shoot Zn compartmentation; because efficient plants maintain higher cytoplasmic Zn concentrations under Zn deficiency conditions, which provided enough elements to continue with the numerous cytoplasmatic localized physiological processes that require Zn [52]. As observed by Bityutskii et al. [13] in cucumber plants, root application of Si in Zn deficient plants had no effect on Zn concentration and mobility in plant tissues, which is confirmed in this study. In plants to which Zn was added back to the nutrient solution ($-$Zn(+Zn)), the concentration of Zn in the treatment with root application of Si (+SiR) showed significantly higher values than the treatment without Si application ($-$Si). No significant differences were observed between the Si treatments. These results suggested that root application of Si after a period of Zn deficiency, significantly increases the concentration of this element compared to the non-application of Si. Similar findings were observed by Hernández-Apaolaza et al. [17] in a Fe resupply assay.

It has been found that Si improves P uptake by the root, and a subsequent increase in soluble inorganic P concentration in leaves of wheat and corn, and the enhancement of the utilization of P within the plant tissues under P deficiency conditions was described [60,61]. In +Zn plants in sampling 2, root application of Si (+SiR) showed significantly higher concentration of P (Table 3) than the other treatments. However, +SiR treatment presented the highest fraction of P in the root, indicating that Si improved P uptake; although, no remobilization to the leaves was observed. Silicon stimulated root Pi acquisition by increasing the exudation of carboxylates [60], although the effect of exuded carboxylates on P-mobilization in the soil is still controversial [62]. Neither in deficiency nor re-fertilization periods were significant differences in P concentration found; although under Zn deficiency, an overaccumulation of P in the aerial part was expected according to the literature [63]. It is interesting to note that, at $-$Zn treatments, a higher accumulation in the shoot of the plant was detected, similar to what was described by [60]. Moreover, in Zn deficiency and resupply periods, the $-$Si and +SiR treatments had more P fraction in old leaves, and the +SiF treatment in new leaves. This may be due to the effect of foliar application of Si on the mobility of P to the new leaves.

Keller et al. [64] found that Si induces Cu accumulation in the root epidermal cells, thus limiting root-to-shoot Cu translocation in wheat (Triticum turgidum) seedlings. The author proposed an increase in the Cu adsorption onto the root surface and immobilization in the vicinity of root epidermis, and a limitation of translocation through the thickened Si-loaded endodermis areas. Moreover, Bosnić et al. [53] provided evidence that the binding of Cu to the Cu-chelating proteins, such as Zn/Cu SOD in roots and plastocyanin in leaves, are important components of the Si-alleviating mechanism in cucumber exposed to Cu excess. With respect to +Zn treatments, in sampling 1, Cu concentration (Table 4) was significantly higher when Si was applied to the root (+SiR), but this effect disappeared in sampling 2. However, a significantly higher fraction of Cu in the root was observed in the Si-applied treatments (sampling 2), which may be due to the improved binding capacity of Cu to the root cell wall previously described [53,55,64]. In Zn-deficient plants, no accumulation of Cu was observed, due to Si addition (Table 4). Cu concentration, in plants with Zn re-fertilization, was significantly higher in treatment without Si. Furthermore, the Cu concentration in +SiF treatment, it was significantly higher than in +SiR treatment. In the deficiency period, in the re-fertilization, the treatment with foliar application of Si (+SiF) also presented the highest Cu fraction in the root.

Several authors [65,66] proposed the existence of a certain degree of competition within the B transport system, favoring Si uptake, due to both elements showing considerable chemical similarities. The effect of Si in decreasing B accumulation has been reported in numerous species [67–69]. A possible mechanism could be the formation of Si complexes with B in soil/nutrient solution, thus reducing the availability of B [70]. On the other hand, Rogalla and Römheld [71] reported that the application of Si to cucumber exposed to high B had no effect on total B concentration. However, Loomis and Durst [72] supported that there were positive correlations between Si and B uptake within different barley genotypes. Thus, genotypes with lower Si uptake also showed lower B uptake. At the three different Zn statuses, Si foliar application significantly increased B concentration in the whole plant (Table 4). Although the effect of +SiR was not clear, insufficient B availability affects several physiological and metabolic processes in plants, such as cell wall and plasma membrane structure and function, phenolics and nitrogen metabolisms, secondary metabolism and oxidative stress, gene expression, shoot and root growth (see [73]). Therefore, Si application leads to higher B concentration in the whole plant, and these will cause the cell wall and plasma membrane to become more structured, and then more prepared to cope with different stresses (in this case, Zn deficiency). Moreover, as B controls the phenolic compounds biosynthesis, and these compounds are related to antioxidant activity (ROS decrease), a relationship between Si, ROS, and B could explain the Si effect on ROS scavenging in refertilization. Further research about the interaction between Si, ROS and B for different plant species and nutrition statuses is needed.

To sum up, Zn sufficiency plants, as expected, presented the highest SPAD index values at both sampling times, but Si addition significantly reduced this index. However, this fact was not related to the Zn concentration in the whole plant, as Zn concentration in the +SiF plants was significantly higher with respect to the −Si control plants. This Zn concentration was mainly located at the roots; when Si was added either to the root or to the shoot, which could explain the lowest SPAD index found when Si was applied. Si supply increased fresh weight; especially the foliar application. Silicon addition to plants grown under optimal nutritional conditions, did not present clear advantages or disadvantages according to ROS values as stress indicators; however, Si influenced P, Cu, and B concentrations. Then, analyzing the Zn-deficient plants results, it can be seen that the Si addition did not improve the SPAD index, the fresh weight, or the P concentration. Moreover, the plants treated with Si were more stressed than the control without this element, although the Zn and B concentration in +SiF−Zn showed levels like Zn sufficient plants. This fact did not support the apoplastic obstruction theory given for Fe, that explained the lower uptake of this element when Si was added to the nutrient solution. In conclusion, Si addition to Zn-deficient plants did not show clear benefits on ameliorating its symptoms. Finally, a distinctive benefit of Si addition has been observed on refertilized plants, as foliar Si supply improves SPAD index, drastically reduced ROS concentration, and increased Zn and B concentration in plants, to levels similar to those of plants which did not suffer from Zn deficiency. This effect was slightly lower when Si was applied through the root, and could probably be due to the "memory effect" of the plants. These results suggest that foliar Si application to cucumber plants could be used to prepare plants to cope with a future stress situation, such as Zn deficiency, and to give them a prompt recovery after overcoming the stress period.

4. Materials and Methods

4.1. Plant Material and Growing Conditions

Cucumber (*Cucumis sativus* L. cv. Ashley) seeds were sterilized and placed on top of a filter paper which was moistened with a solution of $CaSO_4$ 1 mM; then, the paper was rolled out and allowed to germinate in a growth chamber (Dycometal-type CCK) under controlled conditions (photoperiod 16/8 h (day/night), 25/20 °C, 40/60% relative humidity, and a photosynthetic photon flux density at the leaf of 1000 $\mu mol \cdot m^{-2} \cdot s^{-1}$), with a $CaSO_4$ 1 mM solution at the base of the paper. After one week of germination, uniform seeds were selected and transferred to 2 L plastic buckets, with nutrient solutions uninterruptedly aerated. The composition of the nutrient solution was: macronutrients (mM) $Ca(NO_3)_2$ (1.0), KNO_3 (0.9), $MgSO_4$ (0.1) and KH_2PO_4 (0.1); micronutrients (µM) NaCl (35), H_3BO_3 (10), Na_2MoO_4 (0.05), Na_2-EDTA (Ethylenediamine tetraacetic acid) (115.5), $MnSO_4$ (2.5), $CuSO_4$ (1.0), $ZnSO_4$ (1.0), $CoSO_4$ (1.0), $NiCl_2$ (1.0). Furthermore, 0.1 g/L of solid $CaCO_3$ and 0.1 mM of HEPES (4-(2-hydroxyethyl)-1-piperazineethanesulfonic acid) was added to simulate conditions of calcareous soil; the pH was buffered to 7.5 and checked daily.

Iron chelate solution N,N'-bis(2-hydroxybenzyl)ethylenediamine-N,N'-diacetic acid (HBED) was prepared as described by [74], HBED was purchased from Strem Chemicals, ligand was dissolved in NaOH 1:4 molar ratio; subsequently, $Fe(NO_3)_3$ solution was added, pH was adjusted to 7.0, and the mixture was left to stand overnight, filtered, and made up to volume. Silicic acid (H_4SiO_4) was freshly prepared as described by [75] passing on $Na_2SiO_3 \cdot 5\ H_2O$ (Sigma-Aldrich, Darmstadt, Germany), through a column containing a cation-exchange resin in its H+ form (Amberlite IR 120+, Sigma-Aldrich, Darmstadt, Germany).

Plants were grown in a full-strength nutrient solution continuously aerated for 1 week, and Si was applied as follows: A concentration of 1.5 mM of silicic acid was applied via foliar (+SiF) by spraying three doses (125 µL each) per leaf per week, dripping was not observed; the same concentration was applied through the root (+SiR) by adding silicic acid to the nutrient solution,; finally control plants without silicon (−Si) were also tested. Then, zinc deficiency (−Zn) conditions were induced in half of the plant material of each treatment, allowing them to grow during 9 days in zinc-free nutrient solution. Finally, Zn was re-added (−Zn(+Zn)) to the nutrient solution for 11 days more. Three replications of each treatment were performed; each replica had 3 plants. Two samplings were carried out: S1 at the end of the Zn deficiency period, and S2 at the end of the Zn re-fertilization period. In sampling 1: Zn sufficient plants [(+Zn): 16 days with Zn]; and Zn deficient plants [(−Zn): 7 days with Zn + 9 days without Zn] were collected; and in sampling 2: Zn sufficient plants [(+Zn(+Zn): 27 days with Zn] and Zn resupply plants [(−Zn(+Zn)): 7 days with Zn + 9 days without Zn + 11 days Zn re-fertilization] were obtained

4.2. Determinations

The degree of chlorosis of the leaves was quantified by a nondestructive method using the SPAD (Soil and Plant Analyzer Development) model 502 (Minolta Co., Osaka, Japan) digital chlorophyll meter.

Plant material was divided into root, stem, old leaves, and new leaves, then was washed with 0.1% Tween 80 (*v/v*) and 1% HCl (*v/v*), rinsed twice with distilled water, and fresh weight (FW) was determined. Subsequently, all the material was stored at −80 °C for stress analysis. Macronutrients and micronutrients concentration was quantified after microwave (CEM Corporation MARS 240/50; Matthews, NC, USA) digestion with HNO_3 65% and H_2O_2 30% by Thermo Scientific Inductively Coupled Plasma Optical Emission Spectroscopy (ICP-OES, Thermo Fisher Scientific, Waltham, MA, USA). Phosphorus concentration was determined by Bray I method [76].

Analysis of reactive oxygen species (ROS) was performed using fresh plant material (0.2 g); this material was chopped into 2 mL 50 mM HEPES at pH 7. Then, the extract (50 µL) with 150 µL 50 mM HEPES and 4 µL 5 µM H_2DCFDA (diacetate of 2',7'-diclorodihydrofluorescein) (Molecular Probes, Invitrogen, Carlsbad, CA, USA) was mixed and incubated for 30 min at 37 °C in agitation (100 rpm). Thereafter, the extract was centrifuged at 1000 rpm for 10 min; the pellet was resuspended in 0.2 mL HEPES and incubated for 10 min more at 37 °C fluorescence intensity of DCF was measured using a fluorescence spectrophotometer (Cary Eclipse Fluorescence, Varian, Australia) at room temperature, with an excitation wavelength of 488 nm and emission filter between 500 and 600 nm (excitation and emission slits width 5 nm). Relative ROS production was determined using the fluorescence intensity.

4.3. Statistical Analysis

Data were treated by one-way analysis of variance (ANOVA). Treatments were compared using Duncan's test for $p < 0.05$. Statistical analysis was performed by the SPSS software for Windows (V.24.0; SPSS, Chicago, IL, USA).

Author Contributions: Conceptualization, L.H.-A.; methodology L.H.-A. and J.M.L.-G.; formal analysis C.D.H., C.V., A.M.-E. and J.M.L.-G.; investigation L.H.-A., C.D.H., C.V., A.M.-E. and J.M.L.-G.; writing—original draft preparation, J.M.L.-G. and L.H.-A.; writing—review and editing J.M.L.-G. and L.H.-A.; project administration L.H.-A.; funding acquisition L.H.-A. All authors have read and agreed to the published version of the manuscript.

Funding: This research was funded by FEDER/Spanish Ministry of Science, Innovation, and Universities Project: RTI2018-096268-B-I00 and partially supported by Comunidad de Madrid (Spain) and Structural Funds 2014–2020 (ERDF and ESF) (project AGRISOST-CM S2018/BAA-4330).

Data Availability Statement: The data presented in this study are available in the article.

Acknowledgments: Group of chemical sensors and biosensors (SENSOUAM). Departamento de Química Analítica y Análisis Instrumental, Universidad Autónoma de Madrid, for letting us measure the ROS concentration in their equipment.

Conflicts of Interest: The authors declare no conflict of interest.

References

1. Marschner, H. *Marschner's Mineral Nutrition of Higher Plants*; Academic Press: Cambridge, MA, USA, 2011.
2. Hernandez-Apaolaza, L. Can Silicon Partially Alleviate Micronutrient Deficiency in Plants? A Review. *Planta* **2014**, *240*, 447–458. [CrossRef] [PubMed]
3. Epstein, E. The Anomaly of Silicon in Plant Biology. *Proc. Natl. Acad. Sci. USA* **1994**, *91*, 11–17. [CrossRef] [PubMed]
4. Castellanos González, L.; de Mello Prado, R.; Silva Campos, C.N. El Silicio En La Resistencia de Los Cultivos. *Cultiv. Trop.* **2015**, *36*, 16–24.
5. Liang, Y.; Nikolic, M.; Bélanger, R.; Gong, H.; Song, A. *Silicon in Agriculture*; Springer: Dordrecht, The Netherlands, 2015. [CrossRef]
6. Coskun, D.; Deshmukh, R.; Sonah, H.; Menzies, J.G.; Reynolds, O.; Ma, J.F.; Kronzucker, H.J.; Bélanger, R.R. The Controversies of Silicon's Role in Plant Biology. *New Phytol.* **2019**, *221*, 67–85. [CrossRef]
7. Pavlovic, J.; Kostic, L.; Bosnic, P.; Kirkby, E.A.; Nikolic, M. Interactions of Silicon with Essential and Beneficial Elements in Plants. *Front. Plant Sci.* **2021**, *12*, 1224. [CrossRef] [PubMed]
8. Carrasco-Gil, S.; Rodríguez-Menéndez, S.; Fernández, B.; Pereiro, R.; de la Fuente, V.; Hernandez-Apaolaza, L. Silicon Induced Fe Deficiency Affects Fe, Mn, Cu and Zn Distribution in Rice (*Oryza sativa* L.) Growth in Calcareous Conditions. *Plant Physiol. Biochem.* **2018**, *125*, 153–163. [CrossRef]
9. Etesami, H.; Jeong, B.R. Silicon (Si): Review and Future Prospects on the Action Mechanisms in Alleviating Biotic and Abiotic Stresses in Plants. *Ecotoxicol. Environ. Saf.* **2018**, *147*, 881–896. [CrossRef]
10. Ma, J.F. Role of Silicon in Enhancing the Resistance of Plants to Biotic and Abiotic Stresses. *Soil Sci. Plant Nutr.* **2004**, *50*, 11–18. [CrossRef]
11. Pavlovic, J.; Samardzic, J.; Kostic, L.; Laursen, K.H.; Natic, M.; Timotijevic, G.; Schjoerring, J.K.; Nikolic, M. Silicon Enhances Leaf Remobilization of Iron in Cucumber under Limited Iron Conditions. *Ann. Bot.* **2016**, *118*, 271–280. [CrossRef]
12. Peris-Felipo, F.J.; Benavent-Gil, Y.; Hernández-Apaolaza, L. Silicon Beneficial Effects on Yield, Fruit Quality and Shelf-Life of Strawberries Grown in Different Culture Substrates under Different Iron Status. *Plant Physiol. Biochem.* **2020**, *152*, 23–31. [CrossRef]
13. Bityutskii, N.; Pavlovic, J.; Yakkonen, K.; Maksimović, V.; Nikolic, M. Contrasting Effect of Silicon on Iron, Zinc and Manganese Status and Accumulation of Metal-Mobilizing Compounds in Micronutrient-Deficient Cucumber. *Plant Physiol. Biochem.* **2014**, *74*, 205–211. [CrossRef]
14. Martín-Esquinas, A.; Hernández-Apaolaza, L. Rice Responses to Silicon Addition at Different Fe Status and Growth PH. Evaluation of Ploidy Changes. *Plant Physiol. Biochem.* **2021**, *163*, 296–307. [CrossRef]
15. Pavlovic, J.; Samardzic, J.; Maksimović, V.; Timotijevic, G.; Stevic, N.; Laursen, K.H.; Hansen, T.H.; Husted, S.; Schjoerring, J.K.; Liang, Y.; et al. Silicon Alleviates Iron Deficiency in Cucumber by Promoting Mobilization of Iron in the Root Apoplast. *New Phytol.* **2013**, *198*, 1096–1107. [CrossRef]
16. Ahmad, A.; Afzal, M.; Ahmad, A.; Tahir, M. Effect of Foliar Application of Silicon on Yield and Quality of Rice (*Oryza sativa* L.). *Cercet. Agron. Mold.* **2013**, *46*, 21–28. [CrossRef]
17. Hernández-Apaolaza, L.; Escribano, L.; Zamarreño, Á.M.; García-Mina, J.M.; Cano, C.; Carrasco-Gil, S. Root Silicon Addition Induces Fe Deficiency in Cucumber Plants, but Facilitates Their Recovery After Fe Resupply. A Comparison with Si Foliar Sprays. *Front. Plant Sci.* **2020**, *11*, 1851. [CrossRef]

18. Hussain, S.; Mumtaz, M.; Manzoor, S.; Shuxian, L.; Ahmed, I.; Skalicky, M.; Brestic, M.; Rastogi, A.; Ulhassan, Z.; Shafiq, I.; et al. Foliar Application of Silicon Improves Growth of Soybean by Enhancing Carbon Metabolism under Shading Conditions. *Plant Physiol. Biochem.* **2021**, *159*, 43–52. [CrossRef] [PubMed]

19. Rattan, R.K.; Goswami, N.N. *Mineral Nutrition of Plants. Soil Science: An Introduction*; Indian Society of Soil Science: New Delhi, India, 2015; pp. 499–539.

20. Alloway, B.J. *Zinc in Soils and Crop Nutrition*; International Zinc Association Communications: Brussels, Belgium, 2004; p. 116.

21. Bettger, W.J.; O'Dell, B.L. A Critical Physiological Role of Zinc in the Structure and Function of Biomembranes. *Life Sci.* **1981**, *28*, 1425–1438. [CrossRef]

22. Malakouti, M.J. The Effect of Micronutrients in Ensuring Efficient Use of Macronutrients. *Turk. J. Agric. For.* **2008**, *32*, 215–220.

23. Karthika, K.S.; Rashmi, I.; Parvathi, M.S. Biological Functions, Uptake and Transport of Essential Nutrients in Relation to Plant Growth. In *Plant Nutrients and Abiotic Stress Tolerance*; Hasanuzzaman, M., Fujita, M., Oku, H., Nahar, K., Hawrylak-Nowak, B., Eds.; Springer: Singapore, 2018; pp. 1–49. [CrossRef]

24. Demidchik, V. Mechanisms of Oxidative Stress in Plants: From Classical Chemistry to Cell Biology. *Environ. Exp. Bot.* **2015**, *109*, 212–228. [CrossRef]

25. Webb, M.J.; Loneragan, J.F. Effect of Zinc Deficiency on Growth, Phosphorus Concentration, and Phosphorus Toxicity of Wheat Plants. *Soil Sci. Soc. Am. J.* **1988**, *52*, 1676–1680. [CrossRef]

26. Alloway, B.J. Micronutrients and Crop Production: An Introduction. In *Micronutrients Deficiencies in Global Crop Production*; Springer: Berlin/Heidelberg, Germany, 2008; pp. 1–39. [CrossRef]

27. Mehrabanjoubani, P.; Abdolzadeh, A.; Sadeghipour, H.R.; Aghdasi, M. Impacts of Silicon Nutrition on Growth and Nutrient Status of Rice Plants Grown under Varying Zinc Regimes. *Theor. Exp. Plant Physiol.* **2015**, *27*, 19–29. [CrossRef]

28. Pascual, M.B.; Echevarria, V.; Gonzalo, M.J.; Hernández-Apaolaza, L. Silicon Addition to Soybean (*Glycine max* L.) Plants Alleviate Zinc Deficiency. *Plant Physiol. Biochem.* **2016**, *108*, 132–138. [CrossRef] [PubMed]

29. Becker, M.; Ngo, N.S.; Schenk, M.K.A. Silicon Reduces the Iron Uptake in Rice and Induces Iron Homeostasis Related Genes. *Sci. Rep.* **2020**, *10*, 5079. [CrossRef]

30. Fleck, A.T.; Nye, T.; Repenning, C.; Stahl, F.; Zahn, M.; Schenk, M.K. Silicon Enhances Suberization and Lignification in Roots of Rice (*Oryza sativa*). *J. Exp. Bot.* **2011**, *62*, 2001–2011. [CrossRef] [PubMed]

31. Hinrichs, M.; Fleck, A.T.; Biedermann, E.; Ngo, N.S.; Schreiber, L.; Schenk, M.K. An ABC Transporter Is Involved in the Silicon-Induced Formation of Casparian Bands in the Exodermis of Rice. *Front. Plant Sci.* **2017**, *8*, 671. [CrossRef]

32. Kreszies, T.; Kreszies, V.; Ly, F.; Thangamani, P.D.; Shellakkutti, N.; Schreiber, L. Suberized Transport Barriers in Plant Roots: The Effect of Silicon. *J. Exp. Bot.* **2020**, *71*, 6799–6806. [CrossRef] [PubMed]

33. Ma, F.; Peterson, C.A. Current Insights into the Development, Structure, and Chemistry of the Endodermis and Exodermis of Roots. *Can. J. Bot.* **2003**, *81*, 405–421. [CrossRef]

34. Li, B.; Kamiya, T.; Kalmbach, L.; Yamagami, M.; Yamaguchi, K.; Shigenobu, S.; Sawa, S.; Danku, J.M.C.; Salt, D.E.; Geldner, N.; et al. Role of LOTR1 in Nutrient Transport through Organization of Spatial Distribution of Root Endodermal Barriers. *Curr. Biol.* **2017**, *27*, 758–765. [CrossRef] [PubMed]

35. Wang, P.; Calvo-Polanco, M.; Reyt, G.; Barberon, M.; Champeyroux, C.; Santoni, V.; Maurel, C.; Franke, R.B.; Ljung, K.; Novak, O.; et al. Surveillance of Cell Wall Diffusion Barrier Integrity Modulates Water and Solute Transport in Plants. *Sci. Rep.* **2019**, *9*, 4227. [CrossRef]

36. de Farias Guedes, V.H.; de Mello Prado, R.; Frazão, J.J.; Oliveira, K.S.; Cazetta, J.O. Foliar-Applied Silicon in Sorghum (*Sorghum bicolor* L.) Alleviate Zinc Deficiency. *Silicon* **2020**. [CrossRef]

37. Srivastava, A.K.; Suresh Kumar, J.; Suprasanna, P. Seed 'Primeomics': Plants Memorize Their Germination under Stress. *Biol. Rev.* **2021**, *96*, 1723–1743. [CrossRef] [PubMed]

38. Ding, Y.; Fromm, M.; Avramova, Z. Multiple Exposures to Drought "train" Transcriptional Responses in Arabidopsis. *Nat. Commun.* **2012**, *3*, 740. [CrossRef]

39. Mehrabanjoubani, P.; Abdolzadeh, A.; Sadeghipour, H.R.; Aghdasi, M. Silicon Affects Transcellular and Apoplastic Uptake of Some Nutrients in Plants. *Pedosphere* **2015**, *25*, 192–201. [CrossRef]

40. World Vegetable Center. The Possibility of Silicon as an Essential Element for Higher Plants. Available online: https://worldveg.tind.io/record/19607/ (accessed on 18 October 2021).

41. Ma, J.F.; Miyake, Y.; Takahashi, E. Chapter 2 Silicon as a Beneficial Element for Crop Plants. In *Studies in Plant Science*; Datnoff, L.E., Snyder, G.H., Korndörfer, G.H., Eds.; Silicon in Agriculture Series; Elsevier: Amsterdam, The Netherlands, 2001; Volume 8, pp. 17–39. [CrossRef]

42. Miyake, Y.; Takahashi, E. Effect of Silicon on the Growth of Solution-Cultured Cucumber Plant. *Soil Sci. Plant Nutr.* **1983**, *29*, 71–83. [CrossRef]

43. Deshmukh, R.K.; Vivancos, J.; Guérin, V.; Sonah, H.; Labbé, C.; Belzile, F.; Bélanger, R.R. Identification and Functional Characterization of Silicon Transporters in Soybean Using Comparative Genomics of Major Intrinsic Proteins in Arabidopsis and Rice. *Plant Mol. Biol.* **2013**, *83*, 303–315. [CrossRef]

44. Foyer, C.H.; Noctor, G. Oxidant and Antioxidant Signalling in Plants: A Re-Evaluation of the Concept of Oxidative Stress in a Physiological Context. *Plant Cell Environ.* **2005**, *28*, 1056–1071. [CrossRef]

45. Jacobo-Velázquez, D.A.; Martínez-Hernández, G.B.; del Rodríguez, S.C.; Cao, C.-M.; Cisneros-Zevallos, L. Plants as Biofactories: Physiological Role of Reactive Oxygen Species on the Accumulation of Phenolic Antioxidants in Carrot Tissue under Wounding and Hyperoxia Stress. *J. Agric. Food Chem.* **2011**, *59*, 6583–6593. [CrossRef]

46. Surjadinata, B.B.; Jacobo-Velázquez, D.A.; Cisneros-Zevallos, L. Physiological Role of Reactive Oxygen Species, Ethylene, and Jasmonic Acid on UV Light Induced Phenolic Biosynthesis in Wounded Carrot Tissue. *Postharvest Biol. Technol.* **2021**, *172*, 111388. [CrossRef]

47. Tripathi, D.K.; Singh, S.; Singh, V.P.; Prasad, S.M.; Dubey, N.K.; Chauhan, D.K. Silicon Nanoparticles More Effectively Alleviated UV-B Stress than Silicon in Wheat (*Triticum aestivum*) Seedlings. *Plant Physiol. Biochem.* **2017**, *110*, 70–81. [CrossRef]

48. Luyckx, M.; Hausman, J.-F.; Lutts, S.; Guerriero, G. Silicon and Plants: Current Knowledge and Technological Perspectives. *Front. Plant Sci.* **2017**, *8*, 411. [CrossRef]

49. Shahzad, S.; Ali, S.; Ahmad, R.; Ercisli, S.; Anjum, M.A. Foliar Application of Silicon Enhances Growth, Flower Yield, Quality and Postharvest Life of Tuberose (*Polianthes tuberosa* L.) under Saline Conditions by Improving Antioxidant Defense Mechanism. *Silicon* **2021**. [CrossRef]

50. El-Hady, N.A.A.A.; ElSayed, A.I.; El-saadany, S.S.; Deligios, P.A.; Ledda, L. Exogenous Application of Foliar Salicylic Acid and Propolis Enhances Antioxidant Defenses and Growth Parameters in Tomato Plants. *Plants* **2021**, *10*, 74. [CrossRef] [PubMed]

51. Fatemi, H.; Esmaiel Pour, B.; Rizwan, M. Foliar Application of Silicon Nanoparticles Affected the Growth, Vitamin C, Flavonoid, and Antioxidant Enzyme Activities of Coriander (*Coriandrum sativum* L.) Plants Grown in Lead (Pb)-Spiked Soil. *Environ. Sci. Pollut. Res.* **2021**, *28*, 1417–1425. [CrossRef]

52. Hacisalihoglu, G.; Hart, J.J.; Vallejos, C.E.; Kochian, L.V. The role of shoot-localized processes in the mechanism of Zn efficiency in common bean. *Planta* **2004**, *218*, 704–711. [CrossRef] [PubMed]

53. Bosnić, D.; Bosnić, P.; Nikolić, D.; Nikolić, M.; Samardžić, J. Silicon and Iron Differently Alleviate Copper Toxicity in Cucumber Leaves. *Plants* **2019**, *8*, 554. [CrossRef]

54. Pilon, M.; Ravet, K.; Tapken, W. The Biogenesis and Physiological Function of Chloroplast Superoxide Dismutases. *Biochim. Biophys. Acta BBA Bioenerg.* **2011**, *1807*, 989–998. [CrossRef]

55. Bosnić, D.; Nikolić, D.; Timotijević, G.; Pavlović, J.; Vaculík, M.; Samardžić, J.; Nikolić, M. Silicon Alleviates Copper (Cu) Toxicity in Cucumber by Increased Cu-Binding Capacity. *Plant Soil* **2019**, *441*, 629–641. [CrossRef]

56. Sun, B.; Jing, Y.; Chen, K.; Song, L.; Chen, F.; Zhang, L. Protective Effect of Nitric Oxide on Iron Deficiency-Induced Oxidative Stress in Maize (*Zea mays*). *J. Plant Physiol.* **2007**, *164*, 536–543. [CrossRef]

57. M'sehli, W.; Houmani, H.; Donnini, S.; Zocchi, G.; Abdelly, C.; Gharsalli, M. Iron Deficiency Tolerance at Leaf Level in *Medicago ciliaris* Plants. *Am. J. Plant Sci.* **2014**, *05*, 2541–2553. [CrossRef]

58. Pilon, C.; Soratto, R.P.; Moreno, L.A. Effects of Soil and Foliar Application of Soluble Silicon on Mineral Nutrition, Gas Exchange, and Growth of Potato Plants. *Crop Sci.* **2013**, *53*, 1605–1614. [CrossRef]

59. Hacisalihoglu, G. Zinc (Zn): The Last Nutrient in the Alphabet and Shedding Light on Zn Eficiency for the Future of Crop Production under Suboptimal Zn. *Plants* **2020**, *9*, 1471. [CrossRef] [PubMed]

60. Kostic, L.; Nikolic, N.; Bosnic, D.; Samardzic, J.; Nikolic, M. Silicon Increases Phosphorus (P) Uptake by Wheat under Low P Acid Soil Conditions. *Plant Soil* **2017**, *419*, 447–455. [CrossRef]

61. Soratto, R.P.; Fernandes, A.M.; Pilon, C.; Souza, M.R. Phosphorus and Silicon Effects on Growth, Yield, and Phosphorus Forms in Potato Plants. *J. Plant Nutr.* **2019**, *42*, 218–233. [CrossRef]

62. Wang, Y.; Krogstad, T.; Clarke, J.L.; Hallama, M.; Øgaard, A.F.; Eich-Greatorex, S.; Kandeler, E.; Clarke, N. Rhizosphere Organic Anions Play a Minor Role in Improving Crop Species' Ability to Take Up Residual Phosphorus (P) in Agricultural Soils Low in P Availability. *Front. Plant Sci.* **2016**, *7*, 1664. [CrossRef]

63. Cakmak, I.; Marschner, H. Mechanism of Phosphorus-Induced Zinc Deficiency in Cotton. I. Zinc Deficiency-Enhanced Uptake Rate of Phosphorus. *Physiol. Plant.* **1986**, *68*, 483–490. [CrossRef]

64. Keller, C.; Rizwan, M.; Davidian, J.-C.; Pokrovsky, O.S.; Bovet, N.; Chaurand, P.; Meunier, J.-D. Effect of Silicon on Wheat Seedlings (*Triticum turgidum* L.) Grown in Hydroponics and Exposed to 0 to 30 MM Cu. *Planta* **2015**, *241*, 847–860. [CrossRef]

65. Liang, Y.; Shen, Z. Interaction of Silicon and Boron in Oilseed Rape Plants. *J. Plant Nutr.* **1994**, *17*, 415–425. [CrossRef]

66. Celikkol Akcay, U.; Erkan, I. Silicon Induced Antioxidative Responses and Expression of BOR2 and Two PIP Family Aquaporin Genes in Barley Grown Under Boron Toxicity. *Plant Mol. Biol. Rep.* **2016**, *34*, 318–326. [CrossRef]

67. Gunes, A.; Inal, A.; Bagci, E.G.; Pilbeam, D.J. Silicon-Mediated Changes of Some Physiological and Enzymatic Parameters Symptomatic for Oxidative Stress in Spinach and Tomato Grown in Sodic-B Toxic Soil. *Plant Soil* **2007**, *290*, 103–114. [CrossRef]

68. Soylemezoglu, G.; Demir, K.; Inal, A.; Gunes, A. Effect of Silicon on Antioxidant and Stomatal Response of Two Grapevine (*Vitis vinifera* L.) Rootstocks Grown in Boron Toxic, Saline and Boron Toxic-Saline Soil. *Sci. Hortic.* **2009**, *123*, 240–246. [CrossRef]

69. Kaya, C.; Tuna, A.L.; Guneri, M.; Ashraf, M. Mitigation Effects of Silicon on Tomato Plants Bearing Fruit Grown at High Boron Levels. *J. Plant Nutr.* **2011**, *34*, 1985–1994. [CrossRef]

70. Nozawa, S.; Sato, T.; Otake, T. Effect of Dissolved Silica on Immobilization of Boron by Magnesium Oxide. *Minerals* **2018**, *8*, 76. [CrossRef]

71. Rogalla, H.; Römheld, V. Effects of Silicon on the Availability of Boron. In *Boron in Plant and Animal Nutrition*; Goldbach, H.E., Brown, P.H., Rerkasem, B., Thellier, M., Wimmer, M.A., Bell, R.W., Eds.; Springer: Boston, MA, USA, 2002; pp. 205–211. [CrossRef]

72. Loomis, W.D.; Durst, R.W. Chemistry and Biology of Boron. *BioFactors* **1992**, *3*, 229–239.

73. Camacho-Cristóbal, J.J.; Martín-Rejano, E.M.; Herrera-Rodríguez, M.B.; Navarro-Gochicoa, M.T.; Rexach, J.; González-Fontes, A. Boron Deficiency Inhibits Root Cell Elongation via an Ethylene/Auxin/ROS-Dependent Pathway in *Arabidopsis* Seedlings. *J. Exp. Bot.* **2015**, *66*, 3831–3840. [CrossRef]
74. Gonzalo, M.J.; Lucena, J.J.; Hernández-Apaolaza, L. Effect of Silicon Addition on Soybean (*Glycine max*) and Cucumber (*Cucumis sativus*) Plants Grown under Iron Deficiency. *Plant Physiol. Biochem.* **2013**, *70*, 455–461. [CrossRef] [PubMed]
75. Nikolic, M.; Nikolic, N.; Liang, Y.; Kirkby, E.A.; Römheld, V. Germanium-68 as an Adequate Tracer for Silicon Transport in Plants. Characterization of Silicon Uptake in Different Crop Species. *Plant Physiol.* **2007**, *143*, 495–503. [CrossRef]
76. Sparks, D.L. *Methods of Soil Analysis Part 3: Chemical Methods*; Soil Science Society of America, American Society of Agronomy: Madison, WI, USA, 1996.

Article

Biofortification of Silage Maize with Zinc, Iron and Selenium as Affected by Nitrogen Fertilization

Djordje Grujcic [1], Atilla Mustafa Yazici [2], Yusuf Tutus [2], Ismail Cakmak [2] and Bal Ram Singh [1,*]

[1] Faculty of Environmental Sciences and Natural Resource Management, Norwegian University of Life Science, 1432 Ås, Norway; djgrujcic@yahoo.com

[2] Faculty of Engineering and Natural Sciences, Sabanci University, Istanbul 34956, Turkey; ayazici@sabanciuniv.edu (A.M.Y.); ytutus@sabanciuniv.edu (Y.T.); cakmak@sabanciuniv.edu (I.C.)

* Correspondence: balram.singh@nmbu.no

Abstract: Agronomic biofortification is one of the main strategies for alleviation of micronutrient deficiencies in human populations and promoting sustainable production of food and feed. The aim of this study was to investigate the effect of nitrogen (N)fertilization on biofortification of maize crop (*Zea mays* L.) with zinc (Zn), iron (Fe) and selenium (Se) grown on a micronutrient deficient soil under greenhouse conditions. Factorial design experiment was set under greenhouse conditions. The experiment consisted of two levels of each N, Zn, Fe and Se. The levels for N were 125 and 250 mg N kg^{-1} soil; Zn were 1 and 5 mg Zn kg^{-1} soil; levels of Fe were 0 and 10 mg Fe kg^{-1} soil; levels of Se were 0 and 0.02 mg Se kg^{-1} soil. An additional experiment was also conducted to study the effect of the Zn form applied as a ZnO or ZnSO$_4$ on shoot growth, shoot Zn concentration and total shoot Zn uptake per plant. Shoot Zn concentrations increased by increasing soil Zn application both with ZnSO$_4$ and ZnO treatments, but the shoot Zn concentration and total Zn uptake were much greater with ZnSO$_4$ than the ZnO application. Under given experimental conditions, increasing soil N supply improved shoot N concentration; but had little effect on shoot dry matter production. The concentrations of Zn and Fe in shoots were significantly increased by increasing N application. In case of total uptake of Zn and Fe, the positive effect of N nutrition was more pronounced. Although Se soil treatment had significant effect, N application showed no effect on Se concentration and accumulation in maize shoots. The obtained results show that N fertilization is an effective tool in improving the Zn and Fe status of silage maize and contribute to the better-quality feed.

Keywords: biofortification; maize; micronutrients; nutrient uptake; plant nutrition

Citation: Grujcic, D.; Yazici, A.M.; Tutus, Y.; Cakmak, I.; Singh, B.R. Biofortification of Silage Maize with Zinc, Iron and Selenium as Affected by Nitrogen Fertilization. *Plants* **2021**, *10*, 391. https://doi.org/10.3390/plants10020391

Academic Editor: Georgia Ntatsi
Received: 23 January 2021
Accepted: 11 February 2021
Published: 18 February 2021

Publisher's Note: MDPI stays neutral with regard to jurisdictional claims in published maps and institutional affiliations.

1. Introduction

Agronomic biofortification of field crops with micronutrients is one of the main strategies in sustainable production of healthy and nutrient rich food and feed [1]. Human health problems are associated with micronutrient deficiency worldwide, especially in developing countries, but also the productivity of farm animals grown in these countries is negatively affected [2–4]. Low amount of phytoavailable micronutrients in cultivated soils and commonly consumed food and feed crops are main reason of the high prevalence of micronutrient deficiencies in humans [5,6]. According to Hill and Shannon (2019) [7], grazing animals are often exposed to high risk of reduced Zn intake because the pastures usually contain inadequate Zn concentration for a proper animal nutrition that is associated with low amount of phytoavailable Zn in soils. Consequently, there is an increasing trend for biofortification of feed crops with Zn to contribute to better Zn nutrition of livestock. Similarly, enrichment of feed crops with Se is of great importance for animal nutrition and health [8]. Selenium and selenoproteins, such as selenomethionine and selenocysteine, play a role in several critical biological functions in human and animal body and prevent development of various important diseases [9,10]. In essence, micronutrient deficiency affects all phases of food and feed production chain, from field to the final consumer.

Thanks to successful impacts of green revolution, farmers have managed to grow more high-yielding cereal crops, leading to the increased feed and food production and decline in cereal prices. However, this trend had a negative side effect, and resulted in dilution in the concentrations of micronutrients in the food and feed and unintentionally enhanced hidden hunger problem [11–13].

Zinc, Fe and Se deficiencies in soils are common in both developed and developing countries [5,8,14]. Soils of Western Balkan countries differ greatly in the concentration and availability of micronutrients, such as Zn, Fe and Se, as their availability is affected by soil factors, such as pH, soil organic matter, fertilization application, micronutrient concentration [15]. Manojlovic and Singh (2012) (15) also found that some fodder crop samples contained Zn and Se below the critical deficiency level and dietary requirement for ruminants.

Dairy cattle nutrition is highly affected by Zn, Fe and Se presence in the feed, because these micronutrients have crucial role in different metabolic processes [4,16]. Many different enzymes in animal body either contain Zn or are activated by Zn. Zinc is required for up to 10% of the proteins in biological systems for their functioning and structural stability [17]. Proteins bind Zn tightly with very high affinity, from picomolar to femtomolar range, to maintain their cellular functions and interactions [18]. Zinc is involved in cell replication, hormone production and immune system and electrolyte balance [19]. Iron makes 90% part of proteins, e.g. hemoglobin. A host of biochemical reactions, especially the enzymes of the electron transport chain (cytochromes) are activated by Fe [19].

Selenium, which acts as an antioxidant, makes an integral part of several enzymes. Selenium as selenocysteine (Se-Cys) is incorporated in the active center of at least 25 seleno-proteins [8,20]. Analysis of 105 sheep and 160 cow blood samples collected from different Western Balkan countries indicated low Se nutrition in animals, and therefore the need to improve animal feeds with Se to ensure a better Se nutrition of animals was highlighted [21].

Maize (*Zea mays* L.), the most grown field crops worldwide, provide dietary staple food for > 200 million people with about 20% of their calories needs [22,23]. Furthermore, 67% of maize produced globally is utilized, either as grain or as silage, for livestock feed [24].

In Europe Union, green maize production is also increasing, especially as silage crop, and it was grown on more than 6.4 million hectares in the EU-28 in 2019. The area increased by 0.5 million hectares (+10.9%) compared with 2011 [25]. Its production amounted to 248.6 million tons, nearly 48 million tons more than in 2010 (+23.9%) [25]. This significant increase in production of silage maize is mainly because: (i) it extends the area of tolerable climatic conditions for maize growth, (ii) has high biomass yield, (iii) it represents main component of domestic ruminants diet and (iv) recently it is used as source for biogas production in developed countries [26]. Micronutrients play an important role in producing high-quality maize silage with respect to its mineral status and therefore, to improve crop productivity and its nutritive value, adequate micronutrient concentrations are needed [27]. Maize is extremely sensitive to deficiency of Zn and Fe and therefore, farmers have made a regular agronomic practice of using micronutrient fertilizers [24,28].

New agronomic approaches have been developed to improve capacity of maize plants to absorb more Zn from soils, such as localized ammonium sulphate and superphosphate applications [29]. The form of Zn fertilizers has a significant effect on plant growth and Zn accumulation in plants, especially in high pH soils. Previously, Mortvedt (1992) [30] and Gangloff et al. (2002) [31] have highlighted that the Zn fertilizers applied in high pH soils should have at least 50% water solubility to improve growth and Zn concentrations of plants in high pH calcareous soils.

Recent research on foliar and soil application of Zn and Fe showed that agronomic biofortification is efficient in reduction of these micronutrient deficiencies, particularly in wheat and rice, as main staple crops in human nutrition [32–34]. In recent years, it has been shown that foliar spray of a mixture of micronutrient solution containing simultaneously Zn, Se, Fe and iodine to wheat and rice grown in different countries greatly increased grain concentrations of Zn, Se and iodine [35,36]. In these studies, the effect of foliar sprayed

Fe on grain Fe was not adequately high compared to Zn, Se and iodine. In previous studies, it has been suggested that N nutritional status of plants has an important role in increasing root uptake, shoot transport and grain deposition of Fe as well as Zn [1,37]. It was interesting to notice that increasing rate of Fe application had little effect on grain Fe; but at a given Fe application rate increasing N application increased grain Fe [32]. Kutman et al. (2010) and Erenoglu et al. (2011) [38,39] showed that increasing N application promotes root uptake of Zn and Fe and increases shoot and grain concentration of these nutrients. Literature reports also suggest that, in case of maize, N application positively affects maize shoot and grain micronutrient concentrations to certain extent under different field conditions [40–43]. Agronomic biofortification, a widely accepted approach in preventing micronutrient deficiency in several food and feed crops, is not a well- known practice in Western Balkan countries [15].

Considering that, maize silage is one of basic feed component for the dairy cattle in many Balkan countries [44], it is important to investigate the effect of N fertilization on uptake and concentration of Zn, Fe and Se in maize plant. Since N, fertilization is shown to stimulate the uptake and concentration of Zn and Fe and perhaps Se in plants, this may help to produce silage of higher nutritional value, with respect to daily needs of dairy cattle. Practically this fortified silage could lead to reduce the use of different micronutrients supplements.

The present study was planned to examine the effect of N and Zn application on: (1) maize shoot dry matter yield; (2) Zn, Fe and Se concentration and uptake in maize; and (3) the relationship between the concentration of N and that of Zn and Fe in maize plants. Furthermore, the effect of the Zn form applied as a ZnO or $ZnSO_4$ on shoot growth, shoot Zn concentration and total shoot Zn uptake was also investigated.

2. Materials and Methods

2.1. Greenhouse Experiment

A climate-controlled greenhouse experiment was conducted at the Sabanci University campus (40°53′24.5″ N; 029°22′46.7″ E) Turkey.

The soil used for this experiment originated from the Zn-deficient region of Central Anatolia and has clay loam texture with pH 7.6 (H_2O), 1.5% organic matter, and 18% $CaCO_3$. The diethylentriamine pentacetic acid (DTPA)-extractable Zn and Fe concentration was 0.1 mg kg^{-1} and 2.1 mg kg^{-1} soil, respectively, and 0.1 M KH_2PO_4 extractable Se was 0.002 mg kg^{-1} soil.

Two experiments were established. In the first one, the experiment consisted of two levels of each N and Zn. The levels for N were 125 and 250 mg N kg^{-1} soil and that of Zn were 1 and 5 mg Zn kg^{-1} soil. The sources of N and Zn were Ca $(NO_3)_2$ $4H_2O$ and $ZnSO_4$ $7H_2O$, respectively. Similarly, there were two levels of Fe, i.e., 0 and 10 mg Fe kg^{-1} soil and two levels of Se, i.e., 0 and 0.02 mg kg^{-1}. Iron and Se fertilizers were applied in form of Fe-sequestrene and Na_2SeO_4, respectively.

Basic fertilizers added were phosphorus (P) 100 mg P kg^{-1} soil in the form of KH_2PO_4, sulfur (S) 50 mg S kg^{-1} in the form of K_2SO_4, and potassium (K) in the form of KH_2PO_4 and K_2SO_4. All these nutrients and basic fertilizers were homogeneously mixed with 3 kg soil prior to putting the mixture into plastic pots. The experiment was laid out as factorial randomized design with 4 replicates. Twelve seeds of maize (*Zea mays* L.cv. Shemal) were sown in each plastic pot. Shortly after emergence, number of plants was reduced to 6 plants/pot. The pots were irrigated daily with deionized water. When plants were 25 days old, their shoot parts were harvested.

In the 2nd experiment, two different forms of Zn (i.e., $ZnSO_4$ and ZnO) were used to study how these Zn sources affect the shoot growth and shoot concentration of Zn of maize plants. Zinc was applied at the rates of 0, 0.5, 1, 2.5 and 7.5 mg kg^{-1} soil in the forms of ZnO and $ZnSO_4$ at the sowing time. When the plants were 35 days old, plants were harvested and analyzed for shoot production and shoot Zn concentration.

2.2. Plant Sampling and Chemical Analysis

Plant shoot materials, after washing with deionized water, were dried at 70 °C for the dry mater determination of shoot weight and analysis of micronutrients. Dried and ground plant samples (0.2 g) were digested with acid [a mixture containing 2 mL of 30% (v/v) H_2O_2 and 5 mL of 65% (v/v) HNO_3] in a closed-vessel microwave system (Mars Express; CEM Corp., Matthews, NC, USA). The digested solution was diluted with DI water. For the determination of Zn, Fe and Se in the digested solution, inductively coupled plasma optical emission spectrometry (ICP-OES) (Vista-Pro Axial, Varian Pty Ltd., Mulgrave, Australia) was used. A dry combustion method (950 °C) using a LECO Tru-Spec C/N Analyzer (Leco Corp., St Joseph, MI, USA) was used for the determination of N concentration in the samples. The certified standard reference materials, obtained from the National Institute of Standards and Technology (Gaithersburg, MD, USA), were used for checking the precision in mineral nutrient analysis.

As standard sample, the SRM 1547 peach leaves, were used and the deviation was below 2%. The total uptake per plant was calculated by multiplying the concentration with dry mater yield.

2.3. Statistical Analysis

The R commander program was used for statistical analysis of the results obtained. Analysis of variance (ANOVA) was used to assess the significance level of the effects of treatments and their interactions on the reported traits. Tukey test at 5% level ($p \leq 0.05$) was used for significant difference among means, whenever ANOVA (general linear model) indicated significant effect of treatments. The relationship between the treatments was assessed by linear regression model.

3. Results

Zinc deficiency symptoms (i.e., development of yellow or yellowish–white stripes along the midrib of younger leaves) appeared in plants grown under low supply of Zn (Figure 1). The symptoms started to develop first following 2 weeks of growth without Zn application under given conditions. Expression of these symptoms was more severe at low N than at adequate N supply. In case of Fe deficiency, uniform chlorosis on younger leaves appeared. Under low N supply, older leaves turn uniformly pale green and then whole shoot look slightly yellowish. The effect of soil N and Zn applications on maize shoot dry matter (DM) yield, shoot Zn and Fe concentrations, and their total uptake per plant was significant.

By contrast, soil N and Zn applications did not significantly affect shoot Se concentration and total Se uptake (Table 1).

Table 1. Analysis of variance (ANOVA) of effects of soil N and Zn applications on shoot dry matter, Zn, Fe and Se concentration and Zn, Fe and Se uptake in 25-days old plants [a,b].

Source of Variation (Treatments)	df	Shoot Dry Matter		Shoot Zn Concentration		Shoot Zn Uptake		Shoot Fe Concentration		Shoot Fe Uptake		Shoot Se Concentration		Shoot Se Uptake	
		SS	F Pr	SS	F Pr	SS	F Pr	SS	F Pr	SS	F Pr	SS	F Pr	SS	F Pr
Soil N (A)	1	0.047	0.461	332.3	<0.001	3802.4	<0.001	6088.3	<0.001	67,088	<0.001	0.0463	0.586	0.221	0.703
Soil Zn (B)	1	1.981	<0.001	1539.2	<0.001	19,448.5	<0.001	1838.8	<0.001	6056	0.039	0.0019	0.911	0.192	0.722
A × B	1	0.058	0.749	34.2	0.172	564.7	0.093	41.5	0.576	32	0.878	0.0299	0.661	0.342	0.635
Experimental error	44	3.756		783.6		8472.5		5752.5		59,400		6.7965		66.2	

[a] Data of 25 days old maize (*Zea mays*) plants grown under greenhouse conditions. [b] ANOVA test values: df, SS and F Pr.

Variation in N supply did not show a significant impact on the shoot dry matter yield of the plants at both low and adequate Zn treatment (Table 2), indicating that low N supply was still enough at this growth phase of maize under given experimental conditions.

On the other hand, variation in Zn soil application both under low and adequate N supply significantly affected shoot dry mater production (Table 2). Table 2 shows that the shoot dry matter of plants at low Zn and adequate N supply was 10% higher than the shoot dry matter of plants produced under low Zn and low N treatments. The shoot dry matter of plants under adequate Zn and N supply was 12% higher than those plants grown under adequate Zn but low N supply.

Both Zn concentration and total Zn uptake by shoots were increased by increasing soil Zn application (Table 3). At the low Zn application rate, the plants grown under adequate N supply showed significantly higher concentration and shoot uptake of Zn than the plants grown under low N supply. The increased soil N supply showed positive effect on the shoot Zn concentration and uptake in plants grown with adequate Zn soil application (Table 3).

Figure 1. Zinc deficiency symptoms on young leaves of maize plants.

Table 2. Effect of varied N supply on shoot dry matter production of 25-days-old maize grown at low and adequate Zn levels.

N Treatment [a]	Shoot Dry Matter (g plant^{-1}) [b,c]	
	Low Zn	Adequate Zn
Low	2.99 ± 0.28 Aa	3.27 ± 0.29 Ab
Adequate	3.02 ± 0.12 Aa	3.37 ± 0.41 Ab

[a] N treatments: low (125 mg of N kg^{-1} of soil) and adequate (250 mg of N kg^{-1} of soil). [b] Data for 25 days old maize plants grown at low (1 mg of Zn kg^{-1} of soil) and adequate (5 mg of Zn kg^{-1} of soil) Zn supply on a Zn-deficient soil under greenhouse conditions. [c] The average of 4 independent replicates makes the mean values presented and those in column followed by different uppercase letters and in a row followed by different lowercase letters are significantly different by Turkey test at the 5% level.

Selenium soil treatment had significant effect on Se concentration and shoot Se uptake in maize shoot (Table 4), while both N and Zn application did not affect plant Se status (Table 3). Although, not statistically significant, a decrease of the Se concentration could be observed with increase of both N and Zn soil application (Table 3).

Table 3. Effect of varied N and Zn supply on the shoot concentration and uptake of Zn, Fe and Se in 25-days-old maize plants.

Micronutrient	Zn Treatment [a]	Shoot Concentration [c,e]		Shoot Uptake [d,e]	
		Low N [b]	Adequate N [b]	Low N [b]	Adequate N [b]
Zn	Low	7.2 ± 0.5 Aa	10.8 ± 0.5 Ab	21.4 ± 2.3 Aa	32.5 ± 2.4 Ab
	Adequate	16.8 ± 1.1 Ba	23.8 ± 1.2 Bb	54.9 ± 4.3 Ba	80.1 ± 11.8 Bb
Fe	Low	51.4 ± 5.5 Aa	80.0 ± 7.3 Ab	153.5 ± 16.1 Aa	241.6 ± 25.3 Ab
	Adequate	36.3 ± 2.5 Ba	57.2 ± 3.0 Bb	118.3 ± 11.8 Ba	192.8 ± 27.1 Bb
Se	Low	0.83 ± 0.06 Aa	0.80 ± 0.12 Aa	2.4 ± 0.30 Aa	2.5 ± 0.32 Aa
	Adequate	0.88 ± 0.03 Aa	0.71 ± 0.07 Aa	2.8 ± 0.25 Aa	2.4 ± 0.37 Aa

[a] Data for 25 days old maize plants grown at low (1 mg of Zn kg^{-1} of soil) and adequate (5 mg of Zn kg^{-1} of soil) Zn supply on a Zn-deficient soil under greenhouse conditions. [b] N treatments: low (125 mg of N kg^{-1} of soil) and adequate (250 mg of N kg^{-1} of soil). [c] Concentration measured in mg kg^{-1} per micronutrient. [d] Uptake measured as µg of micronutrient/plant. [e] The average of 4 independent replicates makes the mean values presented and those in column followed by different uppercase letters and in a row followed by different lowercase letters are significantly different by Turkey test at the 5% level.

Table 4. Effect of varied N supply on the shoot concentration and uptake of Fe and Se in 25-days-old maize plants grown under different Fe and Se soil treatments.

Micronutrient	Micronutrient Treatment [a]	Shoot Concentration [c,e]		Shoot Uptake [d,e]	
		Low N [b]	Adequate N [b]	Low N [b]	Adequate N [b]
Fe	No treatment	29.2 ± 2.8 Aa	47.2 ± 3.7 Ab	92.3 ± 4.4 Aa	147.4 ± 17.9 Ab
	Adequate	42.6 ± 2.5 Ba	71.0 ± 2.9 Bb	137.7 ± 9.6 Ba	222.3 ± 25.6 Bb
Se	No treatment	0.04 ± 0.01 Aa	0.05 ± 0.01 Aa	0.1 ± 0.0 Aa	0.2 ± 0.0 Aa
	Adequate	0.85 ± 0.03 Ba	0.88 ± 0.08 Ba	2.7 ± 0.2 Ba	2.8 ± 0.4 Ba

[a] Data for 25 days old maize plants grown at different Fe (0 and 10 mg of Fe kg^{-1} of soil) and Se (0 and 0.02 mg of Se kg^{-1} of soil) supply on micronutrient deficient soil under greenhouse conditions. [b] N treatments: low (125 mg of N kg^{-1} of soil) and adequate (250 mg of N kg^{-1} of soil). [c] Concentration measured in mg kg^{-1} per micronutrient. [d] Uptake measured as µg of micronutrient/plant. [e] Values are means of four independent replicates. [e] The average of 4 independent replicates makes the mean values presented and those in column followed by different uppercase letters and in a row followed by different lowercase letters are significantly different by Turkey test at the 5% level.

Shoot Fe concentration and uptake were greatly affected by N and Zn supply (Table 3). Both Zn and Fe uptake were positively affected by increasing soil N supply (Tables 3 and 4). Shoot Fe concentration and uptake at adequate N application were significantly higher compared to low N application, while the shoot Fe concentration and uptake at adequate Zn supply significantly decreased compared to the low Zn treatment (Table 3). Although increase in soil N supply promoted the Zn and Fe concentration in maize shoot, the ratio between Zn and Fe was mainly affected by the level of soil Zn application (Figure 2).

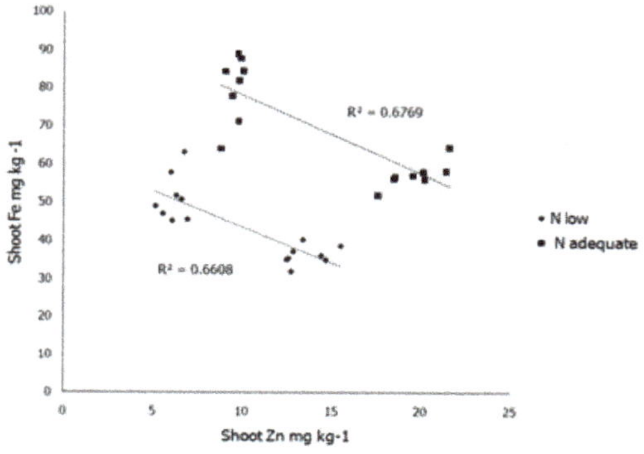

Figure 2. Correlation between shoot Zn and Fe concentrations under different N soil supply. Data for 25. days old maize plants. N rates: low (125 mg of N/kg of soil), (—, ■), and adequate (250 mg of N/kg of soil), (—, ◆).

The relationship between N-Zn and N-Fe concentrations in shoot is shown in Figure 3. The positive impact of N nutrition on Zn and Fe concentrations in shoot in relation to N shoot concentration is shown both under low and adequate Zn supply.

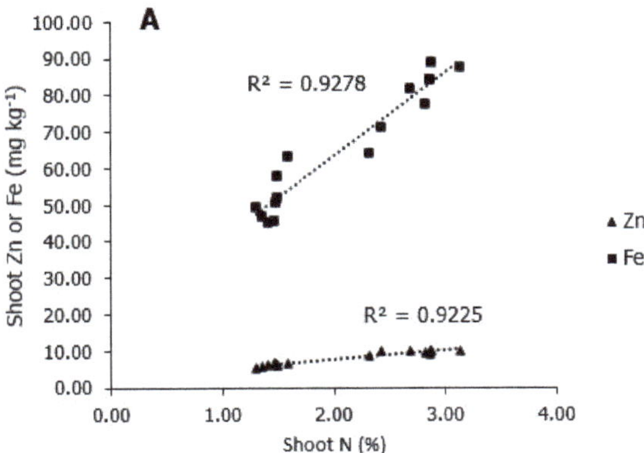

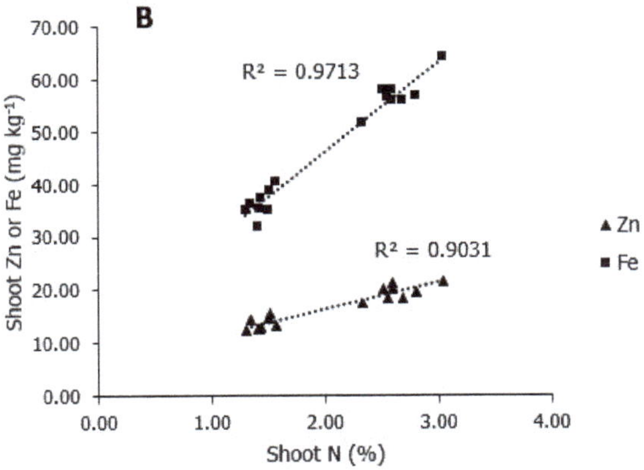

Figure 3. Correlations between Zn, Fe and N concentrations in maize shoot under different Zn soil supply. (**A**) Low Zn soil supply. (**B**) Adequate Zn soil supply. Data for 25 days old maize plants. Fe maize shoot concentration (—, ■), Zn maize shoot concentration (—, ▲).

The relation between N-Zn in shoot is highly dependent on availability of N in the soil irrespective of Zn soil availability. Although Zn application decreased shoot Fe concentration (Figure 2), still increase in N supply significantly affected N-Fe relationship, and as in case of N-Zn relationship, N-Fe relationship is strengthened by the increase of N soil availability (Figure 3).

The well-documented effects of Zn nutrition on growth of plants and shoot Zn concentrations and uptake were more distinct with the soil $ZnSO_4$ application when compared to the soil ZnO application (Table 5). A progressive increase in shoot dry matter production was observed with increasing soil Zn application, and this increase was stronger and more pronounced in case of soil ZnSO4 application. These much better effects of $ZnSO_4$ application on shoot growth than the ZnO application are also shown in Figure 4a,b.

Table 5. Effects of increasing soil Zn supply as ZnO and ZnSO$_4$ on shoot dry matter production, shoot Zn concentration and total Zn uptake of 34-days-old maize plants grown on a Zn-deficient calcareous soil. Zinc has been applied at the rates of 0, 0.5, 1, 2.5 and 7.5 mg Zn kg^{-1} soil in form of ZnO and ZnSO$_4$.

Zn Source	Zn Supply mg/kg	Dry Matter (g plant^{-1})	Shoot Zn (mg kg^{-1})	Shoot Zn Uptake (µg plant^{-1})
Control	No Zn	0.60 ± 0.02	5.65 ± 0.10	3.4 ± 0.1
	0.5	0.98 ± 0.11	5.01 ± 0.65	4.9 ± 1.1
ZnO	1	1.28 ± 0.32	6.32 ± 0.29	8.1 ± 1.9
	2.5	2.52 ± 1.35	8.09 ± 0.36	20.7 ± 11.9
	7.5	4.64 ± 1.97	9.24 ± 1.10	44.5 ± 23.3
	0.5	2.21 ± 0.48	6.88 ± 0.77	15.3 ± 4.3
ZnSO$_4$	1	3.97 ± 0.60	7.24 ± 0.54	28.7 ± 4.6
	2.5	6.16 ± 0.24	8.36 ± 0.27	51.5 ± 2.7
	7.5	7.43 ± 0.30	13.55 ± 1.32	99.6 ± 8.6

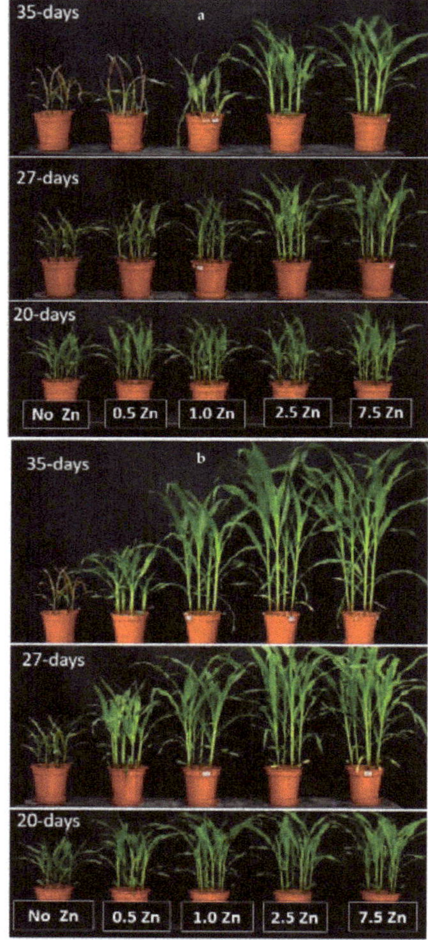

Figure 4. (**a**) Effects of increasing soil Zn supply **in form of ZnO** on growth of maize plants on a Zn-deficient calcareous for 20 days (bottom), 27 days (middle) and 35 days (top). (**b**) Effects of increasing soil Zn supply **in form of ZnSO$_4$** on growth of maize plants on a Zn-deficient calcareous for 20 days (bottom), 27 days (middle) and 35 days (top). Zinc has been applied at the rates of 0, 0.5, 1, 2.5 and 7.5 mg Zn kg^{-1} soil in form of ZnO and ZnSO4.

The plants receiving ZnO could not develop well when sol Zn application was <1 mg Zn kg^{-1} soil. However, shoot Zn concentrations showed a clear increase by increasing soil Zn application both with $ZnSO_4$ and ZnO treatments. Since the enhancements in shoot dry matter by increasing soil Zn supply was more obvious by $ZnSO_4$ application, the shoot total Zn uptake was accordingly much greater with $ZnSO_4$ than the ZnO applications (Table 5).

4. Discussion

Maize is known to be highly sensitive to low Zn in soils [24]. In good agreement with this, decreases in Zn soil application showed detrimental effects on the shoot growth of maize plants (Tables 2–5; Figure 4a,b). Very positive response of maize plants to increasing soil Zn application was pronounced in case of $ZnSO_4$ application. This indicates clearly that use of water-soluble Zn fertilizers in high pH soils is highly desirable. The results presented here are similar those presented for maize plants grown in calcareous soils in the previous studies [45,46]. ZnO is known to be nearly water insoluble (e.g., 0.0016 g per liter) while ZnSO4 shows very high-water solubility (i.e., 580 g per liter) [47]. Use of water-soluble Zn fertilizer is, therefore, highly desirable for high pH soils. At least 40 to 50% of Zn in each granular fertilizer should be water soluble to achieve a positive agronomic impact on plant growth by using Zn-containing compound fertilizers in high pH soils [30,46,48]. Recently, Degryse et al. (2020) [48] emphasized importance and relevance of water-soluble Zn in each granular fertilizer and highlighted that water-soluble Zn rather than the total Zn should be considered in the fertilizer labeling regulations.

Studies dealing with the use of nanoparticle ZnO in Zn fertilization of plants is growing with controversial results and debates [49,50]. One critical debate is related to the very poor solubility of Zn existing in nanoparticulated ZnO. Published evidence shows that solubility and diffusion of Zn from a granular fertilizer which is coated by a bulk or nanoparticulated ZnO are not affected from the size of ZnO used [51]. Therefore, use of nanoparticulated ZnO-containing granular fertilizers in high pH soils may have a very minimal agronomic impact on plant growth and plant Zn uptake when compared to use of fertilizers containing higher percentage of water-soluble Zn.

The present study shows that besides the form of the Zn fertilizer used, the N nutritional level of the plants also influences plant Zn concentration. Although varied soil N supply only slightly affected the shoot dry matter production of plants under given experimental conditions, shoot Zn concentration as well as Zn accumulation (i.e., total Zn uptake by shoot) of plants were significantly increased through the increase in N fertilization of plants (Table 2). These results are like those published by LeBlanc et al. (1997) [40], and Xue et al. (2014) [43] for maize and Cakmak et al. (2010b) [52] for wheat grown under field conditions. Positive effects of increasing N fertilization on plant Zn concentrations have been also shown under greenhouse conditions [53].

In a short-term experiment, Erenoglu et al. (2011) [39] showed that Zn transport through root uptake and root to shoot was significantly promoted by N fertilization. An increased N nutrition of plants showed probably positive effects on abundance of Zn-chelating ligands (such as amino acids and amines) and transporter proteins involved in the root uptake and root to shoot transport of Zn in the plants such as nicotianamine, ZIP family proteins and YSL transporters [37,38]. The presented positive effect of higher soil N supply on Zn concentration and uptake is more emphasized under adequate soil Zn application. This result compares well with those found by Kutman et al. (2010) [38] in wheat. According to Kutman et al. (2010) [38] Zn and N are synergistic in their effects on increasing plant Zn concentrations and their levels in growth medium should be at enough levels to achieve the synergistic effect of N on root Zn uptake.

Like Zn, increasing soil N supply also positively affected shoot Fe concentrations, even at much higher level than Zn (Tables 3 and 4). Losak et al. (2011) [41] in maize and Aciksoz et al. (2011a) [32] in wheat also found the positive effects of N fertilization on plant Fe concentrations. Interestingly, increasing soil Fe fertilization in the form of $FeSO_4$ and FeEDTA had no clear effect on shoot and grain Fe concentrations of wheat plants; but at a given Fe dose, increasing N fertilization significantly improved shoot and grain Fe [32]. The N-dependent increases in plant Fe concentrations were ascribed to N-dependent increases in level of Fe-chelating and transporting nitrogenous compounds such nicotianamine and phytosiderohores. Accordingly, it was shown that improving plants N nutritional status increased root release of phytosiderophores from roots [54]. It is known that phytosiderophores play an important role in mobilization of Fe and Zn from sparingly soluble Fe and Zn sources in soils as well as in root uptake and shoot transport of Zn and Fe within plants [55–57].

In contrast to Zn and Fe, increasing N fertilization did not affect plant Se concentration (Table 4), which shows that the increasing effect of N fertilization on plant Zn and Fe seems to be specific. Even, an increasing N fertilization has been found to have an inhibitory effect on Se concentrations of vegetables [58]. Similarly, the concentration and uptake of Se in maize plants was not significantly affected by varied Zn supply (Table 1), which could be explained by different root uptake mechanisms for Se and Zn.

Plants are very responsive to soil Se fertilization and show substantial increases in shoot and grain concentrations of Se as shown in field-grown maize plants by Chilimba et al. (2012) [59] in Malawi, Mao et al. (2014) [60] in China and Ngigi et al. (2019) [61] in Kenya. The results presented in Table 4 compare well with those results from the field trials.

Considering the average concentrations of observed micronutrients in experimental maize plants and dietary requirement of the dairy cattle, we can suggest that biofortified maize grown on the deficient soils can meet the dairy cattle requirements to certain extent. Average Se maize shoot DM levels ranged between 0.04–0.88 mg kg^{-1} (Table 4), while the defined selenium requirement is 0.3 mg/kg of dietary dry matter for all categories of dairy cattle [62].

This indicates that maize silage could be very successfully Se biofortified after optimization of the application rates. In case of Zn, required dietary concentration for milking cows is 63 mg kg^{-1} DM, for heifers 31 mg kg^{-1} DM and dry cows 22.8 mg kg-1 DM [63], while the average maize shoot DM values ranged between 7–24 mg kg^{-1}. According to this comparison, Zn biofortified silage could meet the requirements of dry cows and heifers but not those of milking cows. Required dietary concentration of Fe for milking cow is 24 mg kg^{-1} DM [63]. Thus, average range of maize Fe shoot concentration of 36–80 mg kg^{-1} DM could easily meet the dietary requirements of all cattle categories.

The present study showed that N nutritional status represents a key factor in biofortification of silage maize with Zn and Fe. It is suggested that an optimum N fertilization of feed and food crops are required to contribute to better human and animal dietary intake of Zn and Fe. The results also highlighted importance of use of water-soluble Zn fertilizers in Zn biofortification of plants grown on high pH soils.

Author Contributions: Methodology, D.G.; formal analysis, D.G., A.M.Y. and Y.T.; investigation D.G., A.M.Y. and Y.T.; resources, B.R.S. and I.C.; D.G. writing—original draft, D.G.; writing—review and editing, B.R.S. and I.C.; supervision, B.R.S. and I.C.; project administration, B.R.S. All authors have read and agreed to the published version of the manuscript.

Funding: The research was funded by the Norwegian State Education Loan Fund for supporting PhD fellowship to senior author, the Norwegian Royal Ministry of Foreign Affairs under the Program for Higher Education, Research and Development (HERD) in Western Balkan ((project number 09/1548) (grant number 332160UA) and the COST Action FA 0905 'Mineral Improved Crop Production for Healthy Food and Feed' project, supported by the European Commission. This project has also received a partial support from the HarvestZinc Project (www.harvestzinc.org) (accessed on 2 February 2021).

Institutional Review Board Statement: No approval was required for it.

Informed Consent Statement: No such statement was needed.

Data Availability Statement: The data presented in this study are available on request from the corresponding author.

Conflicts of Interest: No potential conflict of interest was reported by the authors.

References

1. Cakmak, I.; Kutman, U. Agronomic biofortification of cereals with zinc: A review. *Eur. J. Soil. Sci.* **2018**, *69*, 172–180. [CrossRef]
2. Bouis, H.E.; Hotz, C.; McClafferty, B.; Meenakshi, J.; Pfeiffer, W.H. Biofortification: A new tool to reduce micronutrient malnutrition. *Food Nutr. Bull.* **2011**, *32*, S31–S40. [CrossRef] [PubMed]
3. Oliver, M.A.; Gregory, P.J. Soil, food security and human health: A review. *Eur. J. Soil. Sci.* **2015**, *66*, 257–276. [CrossRef]
4. Paul, S.; Dey, A. Nutrition in health and immune function of ruminants. *Indian J. Anim. Sci.* **2015**, *85*, 103–112.
5. Cakmak, I. Enrichment of cereal grains with zinc: Agronomic or genetic biofortification? *Plant Soil.* **2008**, *302*, 1–17. [CrossRef]
6. Welch, R.M.; Graham, R.D.; Cakmak, I. Linking agricultural production practices to improving human nutrition and health. In Proceedings of the ICN2 Second International Conference on Nutrition Preparatory Technical Meeting, Rome, Italy, 13–15 November 2013.
7. Hill, G.M.; Shannon, M.C. Copper and Zinc Nutritional Issues for Agricultural Animal Production. *Biol. Trace Elemres* **2019**, *188*, 148–159. [CrossRef]
8. Schiavon, M.; Nardi, S.; dalla Vecchia, F.; Ertani, A. Selenium biofortification in the 21st century: Status and challenges for healthy human nutrition. *Plant Soil.* **2020**, *453*, 245–270. [CrossRef]
9. Dalgaard, T.S.; Briens, M.; Engberg, R.M.; Lauridsen, C. The influence of selenium and selenoproteins on immune responses of poultry and pigs. *Anim. Feed Sci. Technol.* **2018**, *238*, 73–83. [CrossRef] [PubMed]
10. Kang, D.; Lee, J.; Wu, C.; Guo, X.; Lee, B.J.; Chun, J.S.; Kim, J.H. The role of selenium metabolism and selenoproteins in cartilage homeostasis and arthropathies. *Exp. Mol. Med.* **2020**, *52*, 1198–1208. [CrossRef]
11. Welch, R.M.; Graham, R.D. A new paradigm for world agriculture: Meeting human needs and productive, sustainable, nutritious. *Field Crops Res.* **1999**, *60*, 1–10. [CrossRef]
12. Pingali, P.; Mittra, B.; Rahman, A. The bumpy road from food to nutrition security—Slow evolution of India's food policy. *Glob. Food Sec.* **2017**, *15*, 77–84. [CrossRef]
13. Rengel, Z.; Römheld, V.; Marschner, H. Uptake of zinc and iron by wheat genotypes differing in tolerance to zinc deficiency. *J. Plant Physiol.* **1998**, *152*, 433–438. [CrossRef]
14. White, P.J.; Broadley, M.R. Biofortification of crops with seven mineral elements often lacking in human diets—Iron, zinc, copper, calcium, magnesium, selenium and iodine. *New Phytol.* **2009**, *182*, 49–84. [CrossRef]
15. Manojlović, M.; Singh, B.R. Trace elements in soils and food chains of the Balkan region. *Acta Agric. Scand. B Soil. Plant Sci.* **2012**, *62*, 673–695. [CrossRef]
16. Kincaid, R.L. Assessment of trace mineral status of ruminants: A review. *J. Anim. Sci.* **2000**, *77*, 1–10. [CrossRef]
17. Andreini, C.; Banci, L.; Bertini, I.; Rosato, A. Zinc through the three domains of life. *J. Proteome Res.* **2006**, *5*, 31–73. [CrossRef]
18. Kluska, K.; Adamczyk, J.; Krężel, A. Metal binding properties, stability and reactivity of zinc fingers Coordination. *Chem. Rev.* **2018**, *367*, 18–64.
19. McDonald, P.; Edwards, R.; Greenhalgh, J.; Morgan, C.; Sinclair, L.A.; Wilkinson, R.G. *Animal Nutrition*, 7th ed.; Prentice-Hall: London, UK, 2007; Volume 11, pp. 264–271.
20. Rayman, M.P. Selenium and human health. *Lancet* **2012**, *379*, 1256–1268. [CrossRef]
21. Ademi, A.; Govasmark, E.; Bernhoft, A.; Bytyqi, H.; Djikic, M.; Manojlović, M.; Loncaric, Z.; Drinic, M.; Filipovic, A.; Singh, B.R. Status of selenium in sheep and dairy cow blood in Western Balkan countries. *Acta Agric. Scand. A Anim. Sci.* **2015**, *65*, 9–16. [CrossRef]
22. Shiferaw, B.; Prasanna, B.; Hellin, J.; Bänziger, M. Crops that feed the world 6. Past successes and future challenges to the role played by maize in global food security. *Food Sec.* **2011**, *3*, 307–327. [CrossRef]
23. Ranum, P.; Peña-Rosas, J.P.; Garcia-Casal, M.N. Global maize production, utilization, and consumption. *Ann. N. Y. Acad. Sci.* **2014**, *13121*, 105–112. [CrossRef] [PubMed]
24. Alloway, B.J. Micronutrients and Crop Production: An Introduction. In *Micronutrient Deficiencies in Global Crop Production*; Springer: Dordrecht, The Netherlands, 2008; pp. 1–39.
25. Eurostat. Available online: https://ec.europa.eu/eurostat (accessed on 20 January 2021).
26. La Frano, M.R.; de Moura, F.F.; Boy, E.; Lönnerdal, B.; Burri, B.J. Bioavailability of iron, zinc, and provitamin A carotenoids in biofortified staple crops. *Nutr. Rev.* **2014**, *72*, 289–307. [CrossRef] [PubMed]
27. Grujcic, D.; Drinic, M.; Zivanovic, I.; Cakmak, I.; Singh, B.R. Micronutrient availability in soils of Northwest Bosnia and Herzegovina in relation to silage maize production. *Acta Agric. Scand. B Soil. Plant Sci.* **2018**, *68*, 301–310. [CrossRef]
28. Subedi, K.D.; Ma, B.L. Assessment of some major yield-limiting factors on maize production in a humid temperate environment. *Field Crops Res.* **2009**, *110*, 21–26. [CrossRef]
29. Ma, Q.; Wang, X.; Li, H.; Li, H.; Cheng, L.; Zhang, F.; Rengel, Z.; Shen, J. Localized application of NH4+-N plus P enhances zinc and iron accumulation in maize via modifying root traits and rhizosphere processes. *Field Crops Res.* **2014**, *164*, 107–116. [CrossRef]
30. Mortvedt, J.J. Crop response to level of water-soluble zinc in granular zinc fertilizers. *Fertil. Res.* **1992**, *33*, 249–255. [CrossRef]
31. Gangloff, W.J.; Westfall, D.G.; Peterson, G.A.; Mortvedt, J.J. Relative availability coefficients of organic and inorganic fertilizers. *J. Plant Nutr.* **2002**, *25*, 259–273. [CrossRef]
32. Aciksoz, S.B.; Yazici, A.; Ozturk, L.; Cakmak, I. Biofortification of wheat with iron through soil and foliar application of nitrogen and iron fertilizers. *Plant Soil.* **2011**, *349*, 215–225. [CrossRef]

33. Zou, C.Q.; Zhang, Y.Q.; Rashid, A.; Ram, H.; Savasli, E.; Arisoy, R.Z.; Ortiz-Monasterio, I.; Simunji, S.; Wang, Z.H.; Sohu, V Biofortification of wheat with zinc through zinc fertilization in seven countries. *Plant Soil.* **2012**, *361*, 119–130. [CrossRef]

34. Phattarakul, N.; Rerkasem, B.; Li, L.J.; Wu, L.H.; Zou, C.Q.; Ram, H.; Sohu, V.S.; Kang, B.S.; Surek, H.; Kalayci, M. Biofortification of rice grain with zinc through zinc fertilization in different countries. *Plant Soil.* **2012**, *361*, 131–141. [CrossRef]

35. Zou, C.; Du, Y.; Rashid, A.; Ram, H.; Savasli, E.; Pieterse, P.J.; Ortiz Monasterio, I.; Yazici, A.; Kaur, C.; Mahmood, K.; et al. Simultaneous biofortification of wheat with zinc, iodine, selenium, and iron through foliar treatment of a micronutrient cocktail in six countries. *J. Agric. Food Chem.* **2019**, *67*, 8096–8106. [CrossRef]

36. Prom-u-thai, C.; Rashid, A.; Ram, H.; Zou, C.; Guilherme, L.; Roberto, G.; Corguinha, A.P.B.; Guo, S.; Kaur, C.; Naeem, A.; et al. Simultaneous Biofortification of Rice With Zinc, Iodine, Iron and Selenium Through Foliar Treatment of a Micronutrient Cocktail in Five Countries. *Front. Plant Sci.* **2020**, *11*, 15–16. [CrossRef]

37. Cakmak, I.; Pfeiffer, W.H.; McClafferty, B. Biofortification of durum wheat with zinc and iron. *Cereal Chem.* **2010**, *87*, 10–20. [CrossRef]

38. Kutman, U.B.; Yildiz, B.; Ozturk, L.; Cakmak, I. Biofortification of durum wheat with zinc through soil and foliar applications of nitrogen. *Cereal Chem.* **2010**, *87*, 1–9. [CrossRef]

39. Erenoglu, E.B.; Kutman, U.B.; Ceylan, Y.; Yildiz, B.; Cakmak, I. Improved nitrogen nutrition enhances root uptake, root-to-shoot translocation and remobilization of zinc (65Zn) in wheat. *New Phytol.* **2011**, *189*, 438–448. [CrossRef]

40. LeBlanc, P.V.; Gupta, U.C.; Christie, B.R. Zinc nutrition of silage corn grown on acid podzols. *J. Plant Nutr.* **1997**, *20*, 345–353. [CrossRef]

41. Losak, T.; Hlusek, J.; Martinec, J.; Jandak, J.; Szostkova, M.; Filipcik, R.; Manasek, J.; Prokes, K.; Peterka, J.; Varga, L. Nitrogen fertilization does not affect micronutrient uptake in grain maize (*Zea mays* L.). *Acta Agric. Scand. B Soil. Plant Sci.* **2011**, *61*, 543–550.

42. Wang, J.; Mao, H.; Zhao, H.; Huang, D.; Wang, Z. Different increases in maize and wheat grain zinc concentrations caused by soil and foliar applications of zinc in Loess Plateau, China. *Field Crops Res.* **2012**, *135*, 89–96. [CrossRef]

43. Xue, Y.; Yue, S.; Zhang, W.; Liu, D.; Cui, Z.; Chen, X.; Ye, Y.; Zou, C. Zinc, iron, manganese and copper uptake requirement in response to nitrogen supply and the increased grain yield of summer maize. *PLoS ONE* **2014**, *9*, e93895. [CrossRef]

44. Randjelovic, V.; Prodanovic, S.; Tomic, Z.; Simic, A. Genotype x year effect on grain yield and nutritive values of maize (*Zea mays* L.). *J. Anim. Vet. Adv.* **2011**, *10*, 835–840.

45. Amrani, M.; Westfall, D.G.; Peterson, G.A. Influence of water solubility of granular zinc fertilizers on plant uptake and growth. *J. Plant Nutr.* **1999**, *22*, 1815–1827. [CrossRef]

46. Shaver, T.M.; Westfall, D.G.; Ronaghi, M. Zinc fertilizer solubility and its effects on zinc bioavailability over time. *J. Plant Nutr.* **2007**, *30*, 123–133. [CrossRef]

47. ZINC. Available online: https://crops.zinc.org/ (accessed on 18 January 2021).

48. Degryse, F.; da Silva, R.C.; Baird, R.; Cakmak, I.; Yazici, M.A.; McLaughlin, M.J. Comparison and modelling of extraction methods to assess agronomic effectiveness of fertilizer zinc. *J. Plant Nutr. Soil Sci.* **2020**, *183*, 248–259. [CrossRef]

49. Monreal, C.M.; DeRosa, M.; Mallubhotla, S.C.; Bindraban, P.S.; Dimkpa, C. Nanotechnologies for increasing the crop use efficiency of fertilizer-micronutrients. *Biol. Fertil. Soils* **2016**, *52*, 423–437. [CrossRef]

50. Kopittke, P.; Lombi, E.; Wang, P.; Schjørring, J.K.; Husted, S. Nanomaterials as fertilizers for improving plant mineral nutrition and environmental outcomes. *Environ. Sci. Nano* **2019**, *6*, 3513–3524. [CrossRef]

51. Milani, N.; McLaughlin, M.J.; Stacey, S.P.; Kirby, J.K.; Hettiarachchi, G.M.; Beak, D.G.; Cornelis, G. Dissolution kinetics of macronutrient fertilizers coated with manufactured zinc oxide nanoparticles. *J. Agric. Food Chem.* **2012**, *60*, 3991–3998. [CrossRef]

52. Cakmak, I.; Kalayci, M.; Kaya, Y.; Torun, A.A.; Aydin, N.; Wang, Y.; Arisoy, Z.; Erdem, H.; Yazici, A.; Gokmen, O.; et al. Biofortification and localization of zinc in wheat grain. *J. Agric. Food Chem.* **2010**, *58*, 9092–9102. [CrossRef] [PubMed]

53. Kutman, U.B.; Yildiz, B.; Cakmak, I. Effect of nitrogen on uptake, remobilization and partitioning of zinc and iron throughout the development of durum wheat. *Plant Soil.* **2011**, *342*, 149–164. [CrossRef]

54. Aciksoz, S.B.; Ozturk, L.; Gokmen, O.O.; Romheld, V.; Cakmak, I. Effect of nitrogen on root release of phytosiderophores and root uptake of Fe(III)- phytosiderophore in Fe-deficient wheat plants. *Physiol. Plant* **2011**, *142*, 287–296. [CrossRef]

55. Römheld, V.; Marschner, H. Evidence for a specific uptake system for iron phytosiderophores in roots of grasses. *Plant Physiol.* **1986**, *80*, 175–180. [CrossRef]

56. Suzuki, M.; Tsukamoto, T.; Inoue, H.; Watanabe, S.; Matsuhashi, S.; Takahashi, M.; Nakanishi, H.; Mori, S.; Nishizawa, N.K. Deoxymugineic acid increases Zn translocation in Zn-deficient rice plants. *Plant Mol. Biol.* **2008**, *66*, 609–617. [CrossRef] [PubMed]

57. Masuda, H.; Suzuki, M.; Morikawa, K.C.; Kobayashi, T.; Nakanishi, H.; Takahashi, M.; Saigusa, M.; Mori, S.; Nishizawa, N.K. Increase in iron and zinc concentrations in rice grains via the introduction of barley genes involved in phytosiderophore synthesis. *Rice* **2008**, *1*, 100–108. [CrossRef]

58. Li, S.; Banuelos, G.S.; Min, J.; Shi, W. Effect of continuous application of inorganic nitrogen fertilizer on selenium concentration in vegetables grown in the Taihu Lake region of China. *Plant Soil.* **2015**, *393*, 351–360. [CrossRef]

59. Chilimba, A.D.C.; Young, S.D.; Black, C.R.; Meacham, M.C.; Lammel, J.; Broadley, M.R. Agronomic biofortification of maize with selenium (Se) in Malawi. *Field Crop Res.* **2012**, *125*, 118–128. [CrossRef]

60. Mao, H.; Wang, J.; Zan, Y.; Lyons, G.; Zou, C. Using agronomic biofortification to boost zinc, selenium, and iodine concentrations of food crops grown on the loess plateau in China. *J. Soil. Sci. Plant Nutr.* **2014**, *14*, 459–470. [CrossRef]

61. Ngigi, P.B.; Lachat, C.; Masinde, P.W.; Du Laing, G. Agronomic biofortification of maize and beans in Kenya through selenium fertilization. *Environ. Geochem. Health* **2019**, *41*, 2577–2591. [CrossRef] [PubMed]
62. Hendriks, S.J.; Laven, R.A. Selenium requirements in grazing dairy cows: A review. *N. Z. Vetj.* **2020**, *68*, 13–22. [CrossRef]
63. Council, N.R. *Nutrient Requirements of Dairy Cattle*, 7th ed.; The National Academies Press: Washington, DC, USA, 2001; p. 405.

Article

Combined Selenium and Zinc Biofortification of Bread-Making Wheat under Mediterranean Conditions

Dolores Reynolds-Marzal [1], Angelica Rivera-Martin [1], Oscar Santamaria [2] and Maria J. Poblaciones [1,*]

[1] Department of Agronomy and Forest Environment Engineering, University of Extremadura, Avenida Adolfo Suárez s/n, 06007 Badajoz, Spain; lolreymar@gmail.com (D.R.-M.); angelicarm@unex.es (A.R.-M.)

[2] Department of Construction and Agronomy, University of Salamanca, Avenida Cardenal Cisneros 34, 49029 Zamora, Spain; osantama@usal.es

* Correspondence: majops@unex.es; Tel.: +34-92-428-6201

Abstract: Millions of people worldwide have an inadequate intake of selenium (Se) and zinc (Zn), and agronomic biofortification may minimise these problems. To evaluate the efficacy of combined foliar Se and Zn fertilisation in bread making wheat (*Triticum aestivum* L.), a two-year field experiment was established in southern Spain under semi-arid Mediterranean conditions, by following a split-split-plot design. The study year (2017/2018, 2018/2019) was considered as the main-plot factor, soil Zn application (50 kg Zn ha^{-1}, nor Zn) as a subplot factor and foliar application (nor Se, 10 g Se ha^{-1}, 8 kg Zn ha^{-1}, 10 g Se ha^{-1} + 8 kg Zn ha^{-1}) as a sub-subplot factor. The best treatment to increase both Zn and Se concentration in both straw, 12.3- and 2.7-fold respectively, and grain, 1.3- and 4.3-fold respectively, was the combined foliar application of Zn and Se. This combined Zn and Se application also increased on average the yield of grain, main product of this crop, by almost 7%. Therefore, bread-making wheat seems to be a very suitable crop to be used in biofortification programs with Zn and Se to alleviate their deficiency in both, people when using its grain and livestock when using its straw.

Keywords: sodium selenate; zinc sulfate; cereal; rainfed conditions; forage yield

Citation: Reynolds-Marzal, D.; Rivera-Martin, A.; Santamaria, O.; Poblaciones, M.J. Combined Selenium and Zinc Biofortification of Bread-Making Wheat under Mediterranean Conditions. *Plants* **2021**, *10*, 1209. https://doi.org/10.3390/plants10061209

Academic Editor: Gokhan Hacisalihoglu

Received: 14 May 2021
Accepted: 8 June 2021
Published: 14 June 2021

Publisher's Note: MDPI stays neutral with regard to jurisdictional claims in published maps and institutional affiliations.

1. Introduction

Cereals are the most important crops for both animal feed and human nutrition, supplying between 25% and 90% of their daily energy needs. Among these, wheat is one of the most important, being grown in 120 countries, China, India and Russia as the main producers, with a harvested area of around 220 million ha and a production of more than 770 million Mg [1]. Its relevance lies in the fact that around 82% of wheat grain is made up of carbohydrates, with more than 60% being starch, with an adequate protein content [2]. However, the vitamin and mineral content is generally low [3]. This low mineral content is aggravated by the relatively high content of the anti-nutrient phytate that wheat has, which hinders the absorption of nutrients such as Ca, Fe, Mg, Se and Zn [4,5].

Nowadays, mineral malnutrition, or hidden hunger, is a global problem affecting around 60% of the world's population, with Fe, I, Se and Zn deficiencies being the most pronounced [6]. The main cause of these deficiencies is the low bioavailability of these nutrients into soil, which produce crops with an inadequate amount of these nutrients in their edible parts. Soils in the semi-arid Mediterranean area have generally low concentrations of both Se and Zn, especially in those of the Southwest of the Iberian Peninsula, which are classified according to [7,8] as deficient to marginal in available Se (<27 µg Se kg^{-1}) [9–11] and deficient in available Zn (<0.5 mg Zn-DTPA kg^{-1}) [12].

Selenium is not considered an essential nutrient for angiosperm plants, but it is for animals and humans, where it is a key component of more than 30 selenoproteins or selenoenzymes [13]. It is involved in cell protective processes and is related to the proper functioning of the immune and endocrine systems [14,15]. Its deficiency is linked

143

to oxidative stress, epilepsy, asthma, reduced male fertility, depression of the immune system and the increased risk of certain cancers, such as rectal, liver, prostate and colon cancer [13,16]. On the other hand, Zn is one of the most important trace elements for all living organisms, including plants, in which it plays a particularly important role during periods of rapid growth [17,18]. In animals and humans, it is present in high concentrations in all body tissues, and is involved in many vital functions [19,20]. Its deficiency is associated with diseases such as anemia, anorexia, cancer, gastrointestinal and kidney problems, immune system dysfunction, delayed bone and sexual maturation and DNA damage, as well as being linked to certain types of cancer [21,22]. Due to the antiviral and immune-boosting properties of Se and Zn, recent studies have linked Se levels to the severity of the infection of SARS-CoV-2 [23], and have proposed Se [24] and Zn supplementation [25,26] as treatments to alleviate their symptoms.

One of the most effective remediation strategies to alleviate this problem is agronomic biofortification, i.e., increasing the bioavailable concentration of nutrients in the edible parts of plants through agronomic intervention [27]. While for Se there is some consensus that foliar application of 10 g Se ha^{-1} applied as sodium selenate at anthesis is the most efficient in semi-arid conditions [9,10,28,29], for Zn, there is more controversy. In general, it is considered that zinc sulphate ($ZnSO_4$-$7H_2O$) is the most widely used fertiliser. However, while the Zn soil incorporation before sowing at a rate of 50 kg ha^{-1} increases mainly grain productivity, the foliar application of 4–8 kg ha^{-1} at the start of flowering seems to be more efficient in increasing Zn concentrations in grain. Therefore, the combined soil and foliar application is considered as a suitable option by [12,30].

However, the information regarding the combined biofortification of Se and Zn, which may allow to alleviate their intake deficiency simultaneously and might reduce application costs for farmers, is very limited, and mainly based on trials carried out under greenhouse conditions [31]. Under in-field conditions, such combined application of Zn and Se under semiarid Mediterranean climate has already demonstrated a high accumulation of those micronutrients in forage peas [32]. However, the effect of the combined Zn and Se biofortification on bread wheat, a crop of global importance, remains unknown under these semiarid conditions, where the irregularity of rainfall could substantively influence its efficiency. The general aim of this study is to contribute to achieve a basic crop in human nutrition with a high enough content of both nutrients able to reduce Zn and Se deficiencies, obtaining a functional crop with added value for farmers. Therefore, the present study aims to evaluate the effect of the biofortification with Zn and Se, both individually and in combination, on the accumulation of these minerals in the edible parts of wheat (grain for humans and straw for animals), and on the yield and nutritional quality of such parts. Likewise, in the plots with Zn treatment in the soil, the evolution and permanence of Zn into soil was also evaluated to analyse its residual effect.

2. Results

2.1. Evolution of Soil Zn-DTPA in the Soil

The split-plot ANOVA performed to evaluate the residual effect of the Zn applications by analysing the concentration of Zn into the topsoil showed the main effects 'sampling time' (degree of freedom, df = 4, *F* value = 20.03, $p < 0.001$), 'Zn application' (df = 2, *F* value = 75.04, $p < 0.001$) and their interaction (df = 8, *F* value = 5.49, $p < 0.001$) to be all significant variables. The Zn-DTPA concentration increased significantly since its application. Such an increment was lower in 50SZn + 0FZn, with an average of 1.00 ± 0.10 mg Zn-DTPA ha^{-1}, than in 50SZn + 8FZn, with 1.25 ± 0.12 mg Zn-DTPA ha^{-1}, but only in the first measurement after its application. Afterwards, no significant variation was observed between both treatments, including Zn (Figure 1).

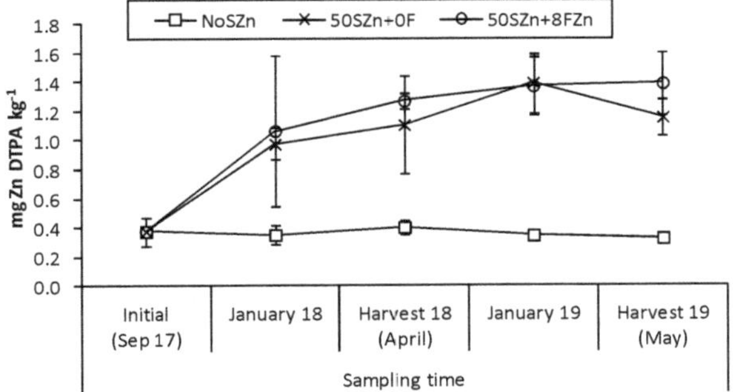

Figure 1. Zn DTPA concentration into topsoil of the study area as affected by the interaction 'sampling time (five times) * Zn application (3 treatments: NoSZn, 50SZn + 0F and 50SZn + 8FZn)'. Error bars indicate standard error (n = 3). Different letters mean significant differences between means according to LSD test ($p \leq 0.05$).

2.2. Zn and Se Concentrations and Contents in Straw and Its Bioavailability

Both the Zn concentration in the straw, and its bioavailability, measured through the molar ratio phytate/Zn, were significantly affected by the main effects 'Study year' and 'Foliar application', as well as by their interaction (Table 1). The foliar treatments containing Zn increase the Zn concentration in the straw almost 10-fold on average, although such differences took place mainly in the study year 2018/2019 (almost 16-fold) in comparison with 2017/2018 (more than 6-fold). The foliar Se application did not produce any effect on the concentration of Zn in the wheat straw (Figure 2). The molar ratio phytate/Zn in the straw was much lower in the treatments containing Zn, especially in the study year 2018/2019 (Table 2). When the total Zn content per ha was considered as response variable, only the foliar application had a significant influence (Table 1). In this case, the treatments containing Zn increased the total Zn content per ha about 8-fold (Figure 2).

Table 1. Summary of the split-split-plot ANOVAs showing the effect of the main-plot factor (year), subplot factor (Zn soil application), sub-subplot factor (foliar application) and their interactions on each parameter evaluated in straw and grain. DF, degree of freedom; F values, including the level of significance (* $p \leq 0.05$, ** $p \leq 0.01$, *** $p \leq 0.001$) are shown in the rest of the rows.

	Part	Year (Y)	Zn Soil Applic. (S)	Foliar Applic. (F)	Y*S	Y*F	S*F	Y*S*F
DF		1	1	3	1	3	3	3
Zn	Straw [1]	35.92 *	0.23	185.9 ***	1.25	13.70 ***	0.26	1.33
(mg kg⁻¹)	Grain	1.36	7.30 *	26.71 ***	0.00	3.75 *	0.10	0.49
Se	Straw [1]	0.07	0.06	27.78 ***	0.04	0.22	0.34	0.37
(µg kg⁻¹)	Grain [2]	75.01 **	0.33	55.93 ***	0.17	2.36	0.05	0.70
TZn	Straw [1]	5.39	6.31	100.6 ***	1.48	0.69	0.53	1.09
(g kg⁻¹)	Grain	101.2 **	24.01 **	25.92 ***	15.83 **	13.78 ***	0.50	0.26
TSe	Straw [1,2]	29.45 *	1.04	47.86 ***	0.03	0.91	0.81	0.85
(mg kg⁻¹)	Grain [2]	365.4 ***	0.50	53.62 ***	1.23	2.34	0.07	0.92
Phytic acid	Straw [1]	1.04	0.30	0.24	0.11	2.18	1.26	1.47
(g kg⁻¹)	Grain	0.04	3.09	0.33	0.93	0.45	0.20	0.14
Ph/Zn	Straw [1]	49.29 *	1.62	96.10 ***	0.46	6.61 **	0.90	1.09
	Grain	0.03	5.09	39.30 ***	0.02	4.29 *	0.68	0.61
Ph/Se	Straw [1]	0.05	0.04	57.46 ***	0.02	0.85	0.10	0.30
	Grain [2]	61.83 **	0.24	54.70 ***	0.13	1.68	0.06	0.79
Yield	Straw	195.4 ***	28.55 **	0.64	0.27	1.58	1.70	0.78
(kg ha⁻¹)	Grain	394.7 ***	2.23	1.70	8.01 **	2.19	1.42	0.55

Table 1. *Cont.*

	Part	Year (Y)	Zn Soil Applic. (S)	Foliar Applic. (F)	Y*S	Y*F	S*F	Y*S*F
1000 gw (g)	Grain	60.75 **	0.16	2.72	0.30	1.24	0.08	1.67
Hect. weight (kg hL^{-1})	Grain	31.38 *	0.29	2.34	3.16	0.91	0.82	0.78
NDF (%)	Straw	1766 ***	1.46	0.88	0.18	0.42	0.60	0.12
ADF (%)	Straw	937.1 ***	2.19	0.23	0.69	0.24	0.63	0.09
ADL (%)	Straw	2.14	8.13 *	1.22	7.65*	1.52	0.88	1.53
Ashes (%)	Straw [2]	102.1 **	0.07	0.32	3.35	0.38	1.74	1.09
Mg	Straw [1]	16.92	14.71*	0.56	3.77	0.48	0.87	0.79
(mg kg^{-1})	Grain	0.35	0.02	1.03	1.98	1.58	0.58	1.35
Ca	Straw [1]	8.25	85.14 ***	0.44	71.44 **	0.37	1.05	1.58
(mg kg^{-1})	Grain	2.73	0.49	1.17	1.94	2.10	1.07	2.33
Fe	Straw [1]	101.8 **	1.18	0.92	0.17	1.15	2.41	0.41
(mg kg^{-1})	Grain	14.56 *	0.40	0.69	5.87	3.85 *	0.89	0.49
Ph/Mg	Straw [1]	18.60 *	10.54 *	0.96	1.41	1.09	0.51	0.67
	Grain	0.00	0.18	1.31	3.00	1.82	1.45	1.57
Ph/Ca	Straw [1]	9.94	18.00 *	0.93	15.13 *	1.53	0.95	1.80
	Grain	0.14	1.10	1.57	2.33	2.08	1.52	2.55
Ph/Fe	Straw [1]	403.9 **	2.34	1.86	0.12	1.71	2.45	0.30
	Grain	25.57 *	0.31	0.54	9.97*	4.21*	0.90	0.51

TZn: total Zn content = Zn*yield; TSe: total Se content = Se*yield; Yield: grain yield; 1000 gw: thousand grain weight; NDF: neutral detergent fiber; ADF: acid detergent fiber; ADL: acid detergent lignin; Ph/mineral: molar ratio Phytate/each mineral.[1] In these parameters: n = 3; in the rest: n = 4. [2] These parameters were transformed by following: Ln(x + 1).

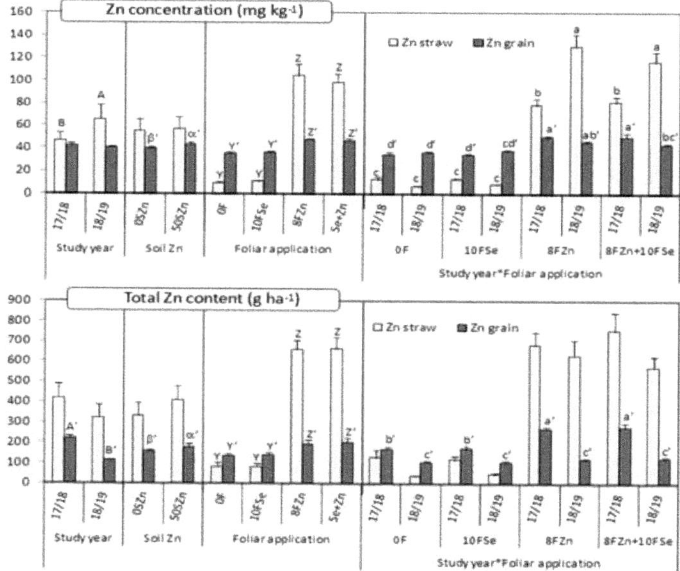

Figure 2. Concentration of Zn and total Zn content in the straw and grain as affected by the main effects 'Study year (Y)', 'Soil Zn application (S)', 'Foliar application (F)', and by the interaction 'Y*F'. Charts indicate means (for straw n = 3; for grain n = 4) and error bars indicate standard error. Within each factor and plant part, different letters mean significant differences between means according to LSD test ($p \leq 0.05$). To make the differences clearer, a different set of letters was assigned to each factor and plant part (lowercase letters for 'Y*F', Greek letters for 'S', uppercase letters [A, B] for 'Y' and uppercase letters [Z, Y] for 'F'). Letters follow by apostrophe (') for grain.

Table 2. Molar ratio phytate: mineral (Zn and Se) in the straw and grain, expressed as mean value ± standard error (*n* = 3 for straw and *n* = 4 for grain) as affected by the main effects 'Study year (Y)', 'Foliar application (F)' (in bold) and by their interaction 'Y*F'.

		Factor	Treatment	Study Year		
				2017/2018	2018/2019	Average
Phytate:Zn	Straw	Foliar application	0F	52.9 ± 10.1 bc	84.5 ± 8.4 a	68.7 ± 7.8 Z
			10FSe	45.0 ± 5.5 c	66.5 ± 4.7 b	55.8 ± 4.7 Y
			8FZn	7.2 ± 0.5 d	4.4 ± 0.3 d	5.8 ± 0.5 X
			8FZn + 10FSe	7.1 ± 0.5 d	4.9 ± 0.3 d	6.0 ± 0.4 X
			Average	28.0 ± 5.1 B	40.1 ± 8.0 A	
	Grain	Foliar application	0F	17.3 ± 1.0 a	16.4 ± 0.8 ab	16.9 ± 0.6 Z
			10FSe	17.1 ± 0.8 ab	15.5 ± 0.4 b	16.3 ± 0.5 Z
			8FZn	11.7 ± 0.3 d	12.9 ± 0.5 cd	12.3 ± 0.3 Y
			8FZn + 10FSe	12.0 ± 0.7 d	13.6 ± 0.4 c	12.8 ± 0.4 Y
			Average	14.5 ± 0.6	14.6 ± 0.4	
Phytate:Se	Straw	Foliar application	0F	19.9 ± 2.9	22.5 ± 1.8	21.2 ± 1.6 Z
			10FSe	8.1 ± 1.8	6.2 ± 0.5	7.2 ± 0.9 Y
			8FZn	20.9 ± 1.3	21.8 ± 3.1	21.4 ± 1.5 Z
			8FZn + 10FSe	8.2 ± 0.7	7.7 ± 0.5	8.0 ± 0.4 Y
			Average	14.3 ± 1.5	14.5 ± 1.8	
	Grain	Foliar application	0F	24.0 ± 3.1	35.0 ± 5.7	29.5 ± 3.4 Z
			10FSe	5.5 ± 0.8	14.0 ± 1.3	9.8 ± 1.4 Y
			8FZn	20.4 ± 4.0	37.3 ± 5.5	28.9 ± 3.9 Z
			8FZn + 10FSe	4.5 ± 0.8	11.0 ± 1.2	7.7 ± 1.1 Y
			Average	13.6 ± 2.0 B	24.3 ± 2.8 A	

Within each parameter and factor, different letters mean significant differences between means according to LSD test ($p \leq 0.05$). If letters do not appear, this factor did not have a significant effect according to split-split-plot ANOVA. To make the differences clearer, a different set of letters was assigned to each factor (lowercase letters [a, b, c, d] for 'Y*F', uppercase letters [Z, Y, X] for 'F' and uppercase letters [A, B] for 'Y'.

Regarding Se, its concentration in the straw and the molar ratio Phytate/Se were only affected by the foliar treatment (Table 1). The treatments containing Se, regardless of the Zn application, produced on average almost a 3-fold increase in the Se concentration in comparison with the rest of treatments (Figure 3). The same pattern was observed for the molar ratio phytate/Se (Table 2). The Se content per ha in the straw, besides the foliar application, was also affected by the study year (Table 1), being 44% higher in 2017/2018 than in 2018/2019. Furthermore, the treatments containing Se produced the highest values of total Se content in the straw (Figure 3).

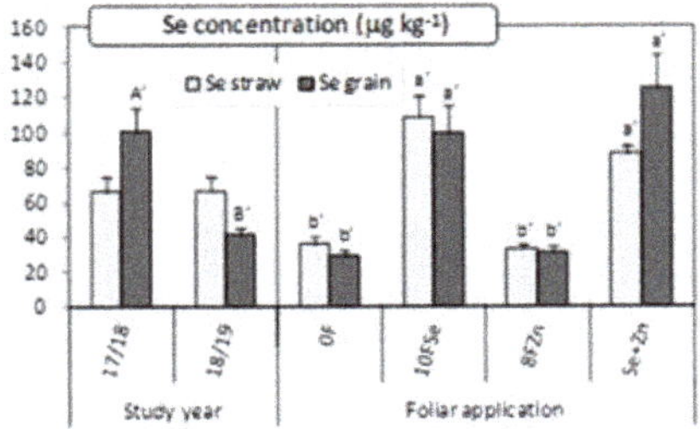

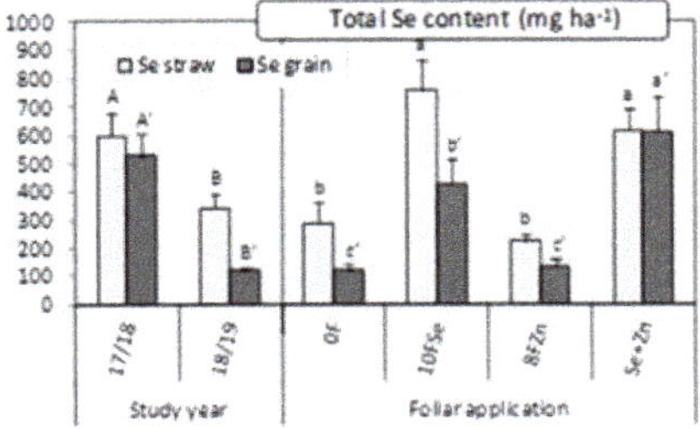

Figure 3. Concentration of Se and total Se content in the straw and grain as affected by the main effects 'Study year (Y)' and 'Foliar application (F)'. Charts indicate means (for straw *n* = 3; for grain *n* = 4) and error bars indicate standard error. Within each factor and plant part, different letters mean significant differences between means according to LSD test ($p \leq 0.05$). To make the differences clearer, a different set of letters was assigned to each factor and plant part (lowercase letters for 'F', uppercase letters for 'Y'). Letters follow by apostrophe (') for grain. Although the LSD test was performed on the transformed variable, back-transformed values are represented to ease interpretation.

2.3. Zn and Se Concentrations and Contents in Grain and Its Bioavailability

The 'soil Zn application' and 'foliar application' as main effects, and the interaction 'study year*foliar application' significantly affected the Zn concentration in grain. The same pattern was observed for the Zn bioavailability (measured through the molar ratio phytate/Zn), excepting for the 'soil Zn application', which did not have significant influence (Table 1). Such as in the case of the straw, the foliar treatments containing Zn produced the highest Zn accumulation in the grain, but the magnitude of the increase was much lower in this case, at on average around 24% (Figure 2). In 2018/2019, those differences were even lower. The soil application of Zn also increased the Zn concentration in grain to around 9%. When considering the molar ratio phytate/Zn, again, the treatments containing Zn showed the lowest values, especially in 2017/2018 (Table 2). The total Zn content in grain was affected by the three main effects of study year (Y), soil Zn application (S) and foliar application (F), and by the interactions 'Y*S' and 'Y*F' (Table 1). The treatments including Zn, either soil or foliar applied, showed the highest values, but only in 2017/2018, the year with the highest total Zn content in the grain (Figure 2). The Se application did not have any effect on the Zn accumulation in the grain.

The concentration of Se in the grain, the total Se content and the molar ratio phytate/Se, were all affected by the main effects 'study year' and 'foliar Zn application' (Table 1). For Se concentration and total Se content, the pattern was almost the same in all the significant factors. The highest values were obtained in 2017/2018 in the treatments containing Se (Figure 3). The treatments containing Se produced the lowest values of the ratio Phytate/Se (Table 2).

2.4. Effect of Zn and Se Application on Grain and Straw Yield and Straw Nutritive Parameters

While in the case of the grain, yield was affected by the main effect 'study year (Y)' and by the interaction 'Y*soil Zn application (S)', the straw yield was significantly influenced by the main effects 'Y' and 'S' (Table 1). Grain yield was much higher in 2017/2018, the most humid growing season, than in 2018/2019 (almost 2-fold higher). The Zn application in soil caused an increase in the grain yield of around 10%, but only in 2017/2018. In 2018/2019, the soil Zn application did not have any effect in the grain yield (Figure 4). Straw was yielded more also in 2017/2018 than in 2018/2019 (more than 1.7-fold), and when Zn was applied to soil (more than 21%) in comparison with the non-fertilised control. In both cases, the Se application did not have any effect on the yield (Figure 4). The thousand grain weight and the hectolitre weight were only affected by the study year (Table 1). In both cases, values were higher in 2017/2018 than in 2018/2019: for thousand grain weight $35.8 \pm 0.7\%$ vs. 25.8 ± 0.4 g, and for hectolitre weight $79.7 \pm 0.3\%$ vs. 76.1 ± 0.4 kg hL^{-1}, respectively.

Fibres (both neutral detergent, NDF, and acid detergent, ADF) and ashes were all only affected by the study year (Table 1). In all cases, the values were higher in 2017/2018 than in 208/2019 (for NDF: $72.8 \pm 0.4\%$ vs. $63.2 \pm 0.4\%$; for ADF: $43.3 \pm 0.3\%$ vs. $34.2 \pm 0.4\%$; for ashes: $1.9 \pm 0.1\%$ vs. $0.5 \pm 0.0\%$, respectively. Lignin (LAD) was affected by the main effect 'soil Zn application (S)' and by the interaction 'Study year*S' (Table 1). In this case, the highest values were obtained when Zn was applied into soil, but only in 2017/2018 (Figure 5). Regarding the mineral status, the influence of the main factors studied and their interactions on their concentration and their bioavailability (measured through the molar ratio phytate/mineral) in the straw and grain can be observed in Table 1. Within those, the most significant results are shown in Table 3. While the study year influenced the Fe concentration and its bioavailability in both the straw and grain and the molar ratio phytate/Mg in the straw, the soil Zn application affected the Mg and Ca concentration in the straw and their bioavailability. The foliar application, although significant in some interactions with the study year, did not present a clear pattern (Table 3). In general, the highest concentration values for Mg, Ca and Fe were obtained in 2017/2018. In the case of Mg and Ca, such highest values were reached when soil Zn was not applied, and in the case of Fe, when foliar Zn and Se were applied in combination, but only in 2017/2018 (Table 3).

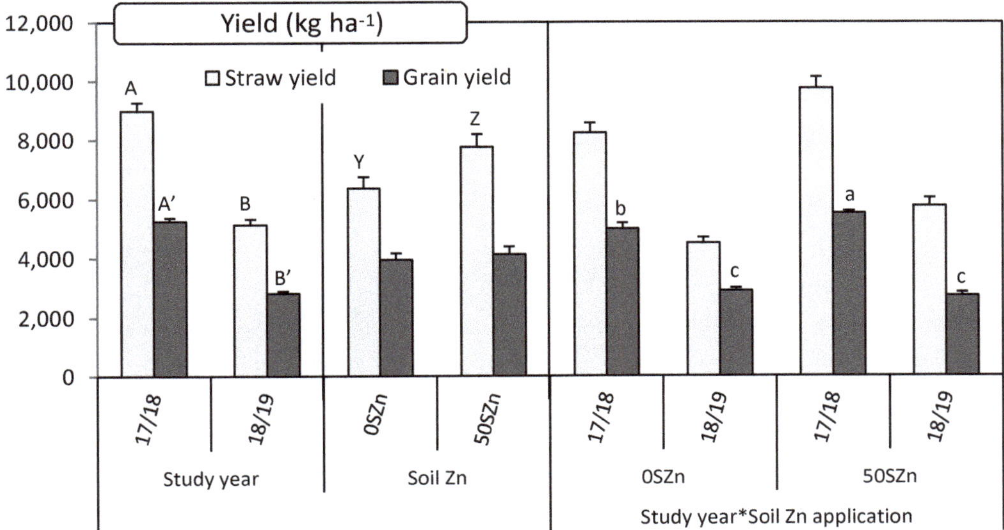

Figure 4. Influence of the main effects 'Study year (Y)', 'Soil Zn application (S)' and their interaction 'Y*S' on both straw and grain yield. Charts indicate means (*n* = 4) and error bars indicate standard error. Within each parameter and factor, different letters mean significant differences between means according to LSD test ($p \leq 0.05$). To make the differences clearer, a different set of letters was assigned to each factor (lowercase letters (a, b, c) for the interaction, uppercase letters [A, B] for 'Y' and uppercase letters (Z, Y) for 'S'. Letters follow by apostrophe (') for grain.

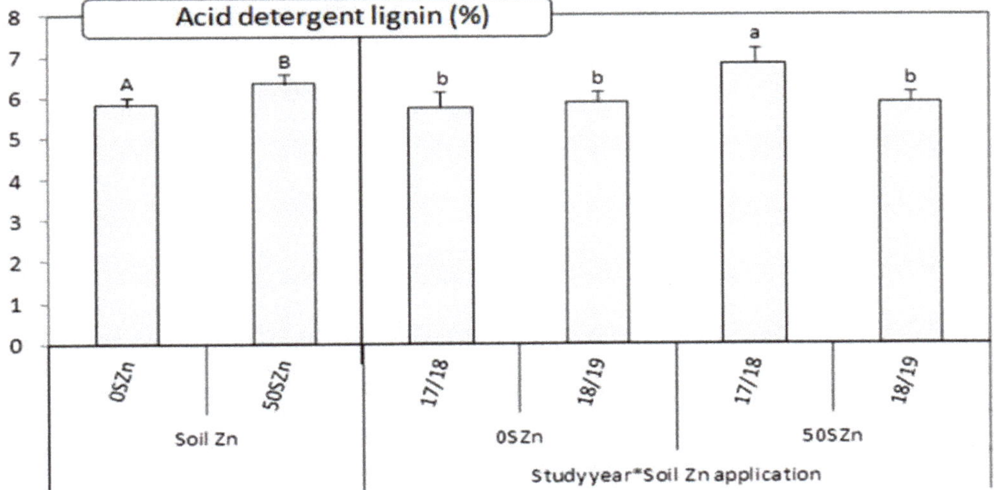

Figure 5. Acid detergent lignin for straw as affected by the main effect 'Soil Zn application (S)' and by the interaction 'Study year (Y)*S'. Charts indicate means (*n* = 4) and error bars indicate standard error. Within each factor, different letters mean significant differences between means according to LSD test ($p \leq 0.05$). To make the differences clearer, a different set of letters was assigned to each factor (lowercase letters for 'Y*S' and uppercase letters for 'S').

Table 3. Concentration of Mg, Ca and Fe, and their molar ratio phytate:mineral in either the straw or grain, expressed as mean value ± standard error (*n* = 3 for straw and = 4 for grain) as affected by the main effects 'Study year (Y)', 'Zn soil application (S)' and/or 'Foliar application (F)' (in bold) and by the interactions 'Y*S' and/or 'Y*F'.

	Factor	Treatment	Study Year		
			2017/2018	**2018/2019**	**Average**
Straw	Mg (g kg^{-1})	Soil Zn application	0SZn	1.05 ± 0.05	0.77 ± 0.03	0.91 ± 0.04 Z
			50SZn	0.85 ± 0.03	0.71 ± 0.02	0.78 ± 0.02 Y
			Average	0.95 ± 0.03	0.74 ± 0.02	
	Ph/Mg	Soil Zn application	0SZn	0.20 ± 0.01	0.27 ± 0.01	0.24 ± 0.01 Y
			50SZn	0.25 ± 0.01	0.30 ± 0.01	0.27 ± 0.01 Z
			Average	0.22 ± 0.01 B	0.28 ± 0.01 A	
	Ca (g kg^{-1})	Soil Zn application	0SZn	3.73 ± 0.15 z	2.67 ± 0.06 x	3.20 ± 0.14 Z
			50SZn	3.05 ± 0.13 y	2.64 ± 0.08 x	2.84 ± 0.09 Y
			Average	3.39 ± 0.12	2.65 ± 0.05	
	Ph/Ca	Soil Zn application	0SZn	0.09 ± 0.00 x	0.13 ± 0.00 z	0.11 ± 0.00 Y
			50SZn	0.12 ± 0.00 y	0.13 ± 0.00 z	0.12 ± 0.00 Z
			Average	0.10 ± 0.00	0.13 ± 0.00	
Grain	Fe (mg kg^{-1})	Foliar application	0F	36.3 ± 0.9 bc	35.0 ± 2.6 bcd	35.6 ± 1.3
			10FSe	39.3 ± 1.9 ab	34.7 ± 2.9 bcd	37.0 ± 1.7
			8FZn	37.8 ± 1.8 ab	32.0 ± 0.6 cd	34.9 ± 1.2
			8FZn + 10FSe	42.3 ± 2.9 a	30.6 ± 0.9 d	36.4 ± 2.1
			Average	38.9 ± 1.0 A	33.1 ± 1.0 B	
	Ph/Fe	Soil Zn application	0SZn	1.24 ± 0.02 x	1.65 ± 0.02 z	1.44 ± 0.04
			50SZn	1.38 ± 0.05 y	1.44 ± 0.05 y	1.41 ± 0.04
		Foliar application	0F	1.38 ± 0.04 cd	1.47 ± 0.09 bc	1.43 ± 0.05
			10FSe	1.29 ± 0.06 de	1.49 ± 0.09 abc	1.39 ± 0.06
			8FZn	1.34 ± 0.07 cde	1.56 ± 0.03 ab	1.45 ± 0.05
			8FZn + 10FSe	1.21 ± 0.08 e	1.64 ± 0.05 a	1.43 ± 0.07
			Average	1.31 ± 0.03 B	1.54 ± 0.03 A	

Within each parameter and factor, different letters mean significant differences between means, according to LSD test (*p* ≤ 0.05). If letters do not appear, this factor did not have a significant effect according to split-split-plot ANOVA. To make the differences clearer, a different set of letters was assigned to each factor (lowercase letters [a, b, c, d, e] for 'Y*F', lowercase letters (z, y, x) for 'Y*S', uppercase letters [Z, Y] for 'S' and uppercase letters [A, B] for 'Y'.

3. Discussion

The present study was designed to perform the soil Zn application only once at the beginning of the experiment, and with the minimum amount possible to reduce the total inputs that satisfy the crop requirements. In this research, after the soil application, the Zn-DTPA concentration in soil increased up to more than 1.00 mg kg^{-1}, remaining always above 0.5 mg kg^{-1}, which is a critical value to meet the crop needs [30]. This fact confirmed then the assumption that the used soil fertilisation rate was high enough to reach the values of available Zn into soil above the crop requirements in both the application year and at least in the following cropping year. This result agreed with the stated in previous studies [12,30], where an important Zn residual effect into soil after a Zn sulfate fertiliser application was reported. However, more years are needed to establish, in semi-arid conditions, what is the duration of the effect of this application.

Without biofortification, the values of Zn and Se concentration in the straw were on average 8.5 mg kg^{-1} and 33.0 μg kg^{-1}, respectively, while in the grain were 33.1 mg Zn kg^{-1} and 26.3 μg Se kg^{-1}. Considering that the required amount of Zn and Se by humans is about 15 mg Zn day^{-1} and 55 μg Se day^{-1}, respectively [33,34], and that livestock requires about 35 mg Zn kg^{-1} and 0.1–0.5 mg Se kg^{-1} feed DM, respectively [35], without biofortification, the levels reached in the different parts of the wheat plant might be under these values. This fact supports the idea that low levels of Zn and Se into soil, such as it was in this case, might produce plants with inadequate Zn and Se concentration in their edible parts to accomplish the required necessities in humans and livestock. Under these soil conditions, the implementation of strategies like agronomic biofortification, which allow alleviating such deficiencies, might make much more sense than in other situations. To get a general application, while the obtained plant-derived products do not reach higher prices because of the Zn and Se enrichment, public policies should fund to farmers the extra costs generated by the application.

Regarding the situation without fortification, two results are interesting to be remarked. The first is that while for Se, the concentration was quite similar in the straw and in the grain, for Zn, it was much higher in the grain than in the straw. Because Zn is an essential nutrient for plants involved in many physiological and metabolic processes [36], the Zn accumulated in senescent tissues might be transported to younger sinks still in development to be again used in the cellular activity of the novel part. This might be supported by Longnecker and Robson [37], who indicated that Zn concentrations are usually higher in growing tissues than in those that are mature. In the case of Se, because it is essential for mammals [38,39], but not for plants [40], although different positive effects have been reported, such as an increase on the chlorophyll content accumulation on the leaves [41] with an improvement in the photosynthetic system [42], alleviated adverse effects of drought stress in different species [43], maintaining under heat or drought stress and grain yield in cereals [44]. The second aspect to be remarked is that for Se, its concentration was higher in the most humid year (2017/2018) in the grain, but for Zn the effect was opposite, i.e., the highest values were obtained in the growing season with the lowest rainfall (2018/2019) in the straw (Figure 6). This apparently contradictory result can be explained by a dilution effect, as a result of the different yield obtained. When the total mineral content (multiplying the mineral concentration by the yield) is considered to take into account this effect, in both cases, for straw and grain, and for both micronutrients, the values obtained in the most humid year (2017/2018) were almost 2-fold of those in 2018/2019. Therefore, as stated previously [45,46], water availability might enormously favour the uptake of these micronutrients.

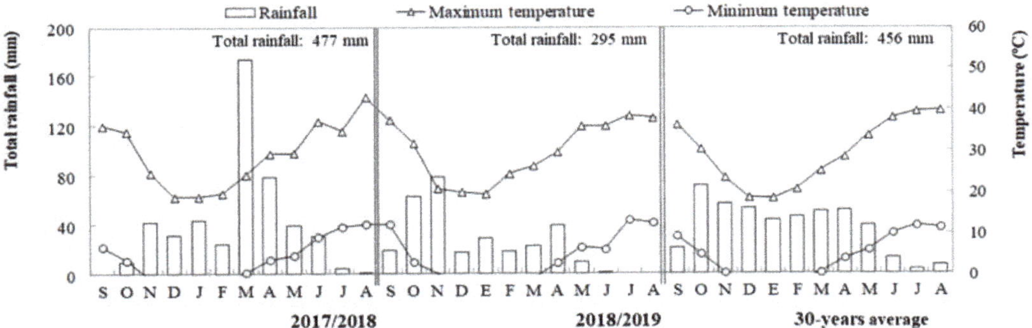

Figure 6. Monthly and annual rainfall and mean maximum and minimum temperatures in 2017/2018, 2018/2019 and in an average year from a 30-year period in Badajoz (Spain).

When biofortification was performed, the best treatment to increase both Zn and Se concentration in both straw and grain was the combined foliar application of 8 kg zinc sulfate ha^{-1} and 10 g sodium selenate ha^{-1}. With this application, the values of Zn and Se concentration in the straw were on average 98.8 mg Zn kg^{-1} (an increase of 12.3-fold in comparison with that of controls) and 87.6 µg Se kg^{-1} (2.7-fold that of controls), respectively, while in the grain were 46.4 mg Zn kg^{-1} (1.32-fold that of controls) and 124.9 µg Se kg^{-1} (4.3-fold that of controls), respectively. According to these results, the importance of the biofortification effect, although always effective, was quite inconsistent, depending on the plant tissue analysed and the mineral considered. To explain this inconsistency, data should be analysed considering each study year separately.

In 2018/2019, a year with an unusual very low precipitation (Figure 6), the foliar application affected at a lower extent the Zn concentration in grain. However, such an application had a very important influence in the Zn concentration in the straw, resulting in values 16.7-fold higher when both nutrients were applied in comparison with the non-fertilised controls. Therefore, under this situation, it seemed that the Zn absorbed by leaves after application was not so effectively transported to the grain, remaining mostly in the foliar system. This might be supported by the fact that phloem transport, main Zn fed source for young sink tissues such as developing grains [47], it is known to fail, or at least decrease, during drought [48]. In 2017/2018, the Zn concentration increased more importantly in the straw than in the grain (6.5-fold vs. 1.5-fold, respectively). Although the rainfall was higher in this year, it might not be still enough for an adequate transport to the grain. Therefore, under the semiarid Mediterranean climate, characterised by scarce and very irregular precipitations, the effectiveness of the Zn biofortification might be clear and positively linked with the amount of rainfall. Nevertheless, further studies, including a higher number of study years or designing specific experiments with different water regimes, should be performed in order to clarify the exact influence of the water availability in the efficiency of the Zn biofortification.

Regarding the Se accumulation after its foliar application, while in the straw the importance of the increase was quite similar regardless of the study year (2.1-fold in 2017/2018 and 2.9 in 2018/2019), in the grain the influence of the year was determinant. In the driest year, 2018/2019, the foliar application increased the Se concentration in grain around 2.8-fold, while in the most humid year, 2017/2018, it increased almost 5.4-fold. Again, as in the case of Zn, precipitation seems to substantively affect the efficiency of the Se biofortification, such as it was also found in previous studies on bread-making wheat [9,11,29]. The quite greater plant vegetative growth in 2017/2018, figured out after observing the greater straw yield of this year, and an increased opening of the leaf stomata, the main route for foliar nutrients entrance into the plant [49], as a consequence of the higher rainfall, might explain this highest efficiency. In any case, even in the most effective case, i.e., in the year 2017/2018 when Zn and Se were simultaneously applied, the Se concentration in grain was quite lower than that obtained previously for bread-making wheat in a very close area [9,11,29], which accounted for almost 800 μg kg^{-1} (vs. the 182 μg kg^{-1} of the present study). The higher amount of rainfall and the higher initial total Se in the topsoil (6 μg extractable Se kg^{-1} vs. 1.27 μg extractable Se kg^{-1}) of that study could explain the observed differences. Another factor which has been regarded to affect the Se uptake in bread-making wheat grain is the rain fallen during the days before the Se application [29]. There it was found that the lower the precipitation in this period, the higher the Se accumulation in grain. Considering all of these aspects, under the semiarid conditions of the Mediterranean climate, special attention should be paid to rainfall, especially in the days prior to the fertiliser application, in order to maximise the success of the combined biofortification with Zn and Se in the driest years.

In terms of biofortification, not only the concentration of the target micronutrients (Zn and Se in this case) is important, which increased by the way to values close to the recommended under the studied conditions, but also how they are bioavailable for the organism. In this regard, phytates are phosphorous-containing compounds that reduce the nutrient absorption, especially for Ca, Fe, Mg and Zn [50]. Considering that a phytate:Zn molar ratio lower than 15 is associated with high Zn bioavailability [51], foliar treatments containing Zn, especially for the straw, caused a decrease in such values below this threshold. Furthermore, biofortification is important to be analysed in terms of productivity and nutritional quality of the edible parts. As the application of Zn in deficient soil increased photosynthesis by increasing chlorophyll a and b concentrations, transpiration and stomatal conductance rate [52,53], a yield increase is expected. In this experiment, the combined soil fertilisation with 50 kg Zn ha^{-1} and the foliar application of 8 kg Zn ha^{-1} + 10 g Se ha^{-1} increased on average the yield of grain, the main product of this crop, by almost 7%. Although less important in the global farm incomes, the straw yield also increased by around 26% after that combined application. Both increases, especially that of the grain yield, might give solid arguments able to persuade farmers to implement these programs, besides the increase in the Zn and Se concentration in their products which, although beneficial for society and stockbreeders, are still not compensated in their selling prices. Future trials should aim at facilitating and reducing costs to make agronomic biofortification with Se and Zn even more attractive for farmers. A good example is that reported by Wang et al. [54], who combined Zn application plus pesticide, showing it as a cost-effective ready-to-use strategy to fight human Zn deficiency in wheat-dominated regions around the world.

The rest of parameters analysed, such as straw fibres, lignin, or mineral status (Mg, Ca, Fe), resulted in either unaffected or very few affected by the biofortification with Zn and Se. Once again, this is a good result to try to successfully implement these programs among farmers, as its application might not be detrimental for the quality of their productions.

4. Materials and Methods

4.1. Site, Experimental Design and Crop Management

The field experiment was conducted in Badajoz, southern Spain (38°54′ N, 6°44′ W, 186 m above sea level), in a Xerofluvent soil, according to Soil Taxonomy, under rainfed Mediterranean conditions in 2017/2018 and 2018/2019 growing seasons. Weather-related parameters for this area for the concerned years, as well as for the average year obtained from a 30-year period, are shown in Figure 6. All climate data were taken from a weather station located at the study site. In the first study year, rainfall was like the average year but much higher (40%) than in 2018/2019. The months of March and April were exceptionally rainy, with ~175 and ~80 mm, respectively, which is much higher than in the average year. The second year was extraordinarily dry, with a very different seasonal distribution, being autumn the wettest season, to the extent that it caused a two-month delay in sowing compared to the first year, and with important drought periods between February and March and May and June.

The experiment was designed as a split-split plot arrangement with four replications randomly distributed. The main plots were the study year (2017/2018 and 2018/2019), subplots were Zn soil application (without any application [0SZn] and with a soil application of 50 kg ZnSO$_4$-7H$_2$O ha^{-1} [50SZn]) (equivalent to 11.4 kg of Zn ha^{-1}) and sub-subplots were foliar application with four treatments (without any application [0F]; two foliar applications of 4 kg ZnSO$_4$-7H$_2$O ha^{-1} each (equivalent to 0.91 kg of Zn ha^{-1}) at the start of flowering and two weeks later [8FZn]: a foliar application of 10 g Se ha^{-1} as Na$_2$SeO$_4$ (equivalent to 24 g ha^{-1} of Na$_2$SeO$_4$) at the start of flowering [10FSe]: a combination of 8FZn and 10FSe [8FZn + 10FSe]). The crop area for each treatment was 15 m^2 (3 m × 5 m). Zinc soil treatment was only made at the beginning of the first season, in October 2017, before the sowing, sprayed as a solid in the soil surface and incorporated into the soil by tillage.

The foliar Zn application treatment consisted of two times of foliar Zn application at the start of flowering, and it was repeated two weeks later as described by Gomez-Coronado et al. [12]. At each time, 0.5% (w/v) of aqueous solution of ZnSO$_4$·7H$_2$O ha^{-1} with 800 L per hectare were sprayed until most of the leaves were covered at the very late afternoon to avoid burning in plants. For the Se treatment 10 g of Se applied as Na$_2$SeO$_4$ ha^{-1} was diluted in 800 L H$_2$O ha^{-1} to obtain a 0.003% (w/v) solution, and applied as in the case of foliar Zn, as described by Poblaciones et al. [9].

The bread wheat (*Triticum aestivum* L.) cultivar used was "Antequera". Conventional tillage treatment was used to prepare a proper seedbed before sowing. The sowing was in late October in the first year (2017) and late December in 2018 (due to the intense rainfall in autumn in the second year). The sowing rate was of 350 seeds m^2; each plot had six rows of

20 cm apart. A N-P-K fertiliser (15-15-15) was applied before sowing at a 200 kg ha^{-1} dose in all plots. Weed control was carried out by applying Trigonil (concentrated in suspension to 400 g L^{-1} of chlortoluron and 25 g L^{-1} of diflufenican) in the sowing.

4.2. Soil Analysis

To characterise the experimental soil, four representative soil samples of 30 cm depth were taken in September 2017 from the experimental site. Soil samples were air dried and sieved to <2 mm using a roller mil. Texture was clay loam, determined gravimetrically; soil pH was slightly acid, with 6.4 ± 0.2 (mean ± standard error) using a calibrated pH meter (ratio, 10-g soil:25-mL deionised H$_2$O), soil organic matter was very low with 1.31 ± 0.09% determined by oxidation with potassium dichromate [55], total N was medium with 0.12 ± 0.007% [56], P Olsen with 4.9 ± 0.05 g P kg^{-1} was low, measured following the Olsen procedure, and assimilable K was low with 321 ± 8 mg kg $^{-1}$, determined with ammonium acetate (1N) and quantified by atomic absorption spectrophotometry.

Soils contained low concentrations of Ca with 1248 ± 134 mg kg^{-1}, medium of Mg with 1455 ± 145 mg kg^{-1}. They were extracted according to the method of [57] by extraction with DTPA (diethylenetriamine pentaacetic acid) and measured by inductively coupled plasma optical emission spectroscopy (ICP-MS) (Agilent 7500ce, Agilent Technologies, Palo Alto, CA, USA). Extractable Se was very low (with 1.27 ± 0.01 μg Se kg^{-1}) determined by using KH$_2$PO$_4$ (0.016 mmol L^{-1}, pH 4.8) at a ratio of 10 g dry weight soil: 30 mL KH$_2$PO$_4$ (w/v) [58]. The Se concentration in the extracts was determined by ICP-MS, as described below. All the results were reported on a dry weight basis.

To evaluate the residual effect along the experiment of the Zn soil treatments, i.e., NoSZn, 50SZn + 0F and 50SZn + 8FZn, four sampling, as well as the initial sampling, was taken in January in both study years and before each harvest (therefore in September 2017, January 2018, May 2018, January 2019 and May 2019). Zinc-DTPA were determined in each moment.

4.3. Plant Analysis

Harvesting was done at maturity in early July. Straw and grain samples were thoroughly washed with tap water, and then with distilled water to avoid the eventual presence of residues from foliar applications. Afterwards, samples were dried at 70 °C until constant weight, and their dry matter yield was then recorded. Thousand grain weight and hectolitre weight were determined from the grain samples. Official procedures [59] were followed to determine neutral detergent fiber (NDF), acid detergent fiber (ADF) and acid detergent lignin (ADL) by means of a fiber analyser (ANKOM8–98, ANKOM Technology, Macedon, NY, USA). Total ash content was determined by ignition of the sample in a muffle furnace at 600 °C, as is indicated in the official procedure [59]. Total straw and grain mineral concentration (Ca, Fe, Mg, Se and Zn) were determined as follows: straw and grain were finely grounded (<0.45 mm) using an agate ball mill (Retch PM 400 mill); a 1 g was digested with ultra-pure concentrated nitric acid (2 mL) and 30% w/v hydrogen peroxide (2 mL) using a closed-vessel microwave digestion protocol (Mars X, CEM Corp, Matthews, NC, USA) and diluted to 25 mL with ultra-purified water [60]. Sample vessels were thoroughly washed with acid before use. For quality assurance, a blank and a standard reference material (tomato leaf, NIST 1573a) were included in each batch of samples. The digested was determined by ICP-MS. The studied mineral recovery was 95%, compared with certified reference material (CRM) values. To consider the dilution effect in Zn and Se caused by the different straw and grain yield between growing seasons, total nutrient uptake was calculated multiplying grain yield by the total Zn and Se concentration in grain. On the other hand, phytate concentration was estimated by means of phytic acid, whose determination is based on precipitation of ferric phytate and measurement of iron (Fe) remaining in the supernatant [61]. Phytate was extracted from about 0.2 g of ground straw or grain in 10 mL of 0.2 M HCl (pH 0.3) after shaking for 2 h. One ml of supernatant was treated with 2 mL of ferric solution (NH$_4$Fe(SO$_4$)$_2$·12 H$_2$O) in a boiling water bath for 30 min. After cooling, samples were centrifuged, and 1 mL of supernatant was treated with 1.5 mL of 0.064 M bipyridine (2-pyridin-2-ylpyridine, $C_{10}H_8N_2$) to measure Fe. After mixing, the solution was incubated for 10 min at room temperature, and the light absorbance was measured with a spectrophotometer at 419 nm. Finally, the molar ratio between phytate and Ca, Fe, Mg, Se and Zn was calculated to estimate the bioavailability of those nutrients.

4.4. Statistical Analysis

The evolution of the Zn soil treatments on the soil Zn-DTPA was evaluated by a split-plot analysis of variance (ANOVA). The main-plot factor was "sample time" (before starting, in January and harvest of 2018, and in January and harvest of 2019), the subplot factor "Zn application" (0SZn + 0F, 50SZn + 0F, and 50SZn + 8FZn), and its interaction in the model.

Data of mineral concentration (Ca, Fe, Mg, Se and Zn) and phytate/mineral molar ratios in straw and grain, as well as straw and grain yield, thousand grain weight and hectolitre weight, and nutritive value parameters of the straw were subjected to split-split-plot ANOVAs, including the main-plot factor 'study year' (2017/2018 and 2018/2019), the subplot factor 'soil Zn application' (0SZn and 50SZn), the sub-subplot factor 'foliar application' (0F, 8FZn, 10FSe and 8FZn + 10FSe), and their interactions in the model. When significant differences were found in ANOVA, means were compared using Fisher's protected least significant difference (LSD) test at $p \leq 0.05$. All these analyses were performed with the Statistix v. 8.10 package (Analytical Software, Tallahessee, FL, USA). In order to normalise the variable distribution, as well as to stabilise the variance of residues, the transformation $Ln(x + 1)$ was performed for the concentration of Se in grain, total Se content in grain and straw, the molar ratio phytate/Se and the ash content in the straw.

5. Conclusions

The results presented here showed bread-making wheat to be a very suitable crop to be used in biofortification programs with Zn and Se, as it was able to substantially accumulate the Zn and Se applied in combination in the edible parts, to alleviate their deficiency in people when used as staple food, or in livestock when using its straw. In fact, the combined foliar application of Zn and Se increased in straw, 12.3- in Zn and 2.7-fold in Se, and grain, 1.3- and 4.3-fold, respectively. However, the efficiency of the uptake and later accumulation was highly affected by the rainfall. Thus, in Mediterranean climates, characterised by irregularity in precipitations, the years with extensive drought periods could account for lower values, especially in the grain. In addition to the higher Zn and Se concentration in the edible parts, the application of 50 kg Zn ha^{-1} produced on average an increase of around 7% in the grain yield, and 26% in the straw yield, with the remaining productive and nutritive quality parameters almost unaffected.

Author Contributions: Conceptualisation, O.S. and M.J.P.; methodology, D.R.-M., A.R.-M. and M.J.P.; software, O.S.; validation, D.R.-M. and A.R.-M.; formal analysis, D.R.-M. and M.J.P.; investigation, A.R.-M. and M.J.P.; resources, M.J.P.; data curation, D.R.-M. and A.R.-M.; writing—original draft preparation, O.S. and M.J.P.; writing—review and editing, O.S., D.R.-M., A.R.-M. and M.J.P.; visualisation, D.R.-M. and M.J.P.; supervision, O.S. and M.J.P.; project administration, M.J.P.; and funding acquisition, O.S. and M.J.P. All authors have read and agreed to the published version of the manuscript.

Funding: This research was funded by the Regional Government of Extremadura (Spain) and by the European Regional Development Fund (ERDF), grant number IB16093.

Institutional Review Board Statement: Not applicable.

Informed Consent Statement: Not applicable.

Data Availability Statement: Data sharing is not applicable to this article.

Acknowledgments: All authors would like to thank Teodoro Garcia White for his invaluable help with all the laboratory work.

Conflicts of Interest: The authors declare no conflict of interest.

References

1. FAOSTAT. Food and Agriculture Data 2019. Available online: http://www.fao.org/faostat/en/#data/QC (accessed on 10 May 2021).
2. Serna Saldívar, S.O. *Cereal Grains: Properties, Processing and Nutritional Attributes*; CRC Press: Boca Ratón, FL, USA, 2010; p. 747.
3. Shewry, P.; Halford, N. Cereal seed storage proteins: Structures, properties and role in grain utilization. *J. Exp. Bot.* **2002**, *53*, 947–958. [CrossRef]
4. Zúñiga, J.R. Trigo blanco valor nutricional y potencial. *Tierra Adentro* **2007**, *74*. Available online: https://biblioteca.inia.cl/handle/123456789/6331 (accessed on 7 May 2021).
5. Amarakoon, D.; Thavarajah, D.; McPhee, K.; Thavarajah, P. Iron-, zinc-, and magnesium-rich field peas (*Pisum sativum* L.) with naturally low phytic acid: A potential food-based solution to global micronutrient malnutrition. *J. Food Compos. Anal.* **2012**, *27*, 8–13. [CrossRef]
6. Lockyer, S.; White, A.; Buttriss, J.L. Biofortified crops for tackling micronutrient deficiencies—What impact are these having in developing countries and could they be of relevance within Europe? *Nutr. Bull.* **2018**, *43*, 319–357. [CrossRef]
7. Alloway, B.J. Soil factors associated with zinc deficiency in crops and humans. *Environ. Geochem. Health* **2009**, *31*, 537–548. [CrossRef] [PubMed]
8. Dinh, Q.T.; Cui, Z.; Huang, J.; Tran, T.; Wang, D.; Yang, W.; Zhou, F.; Wang, M.; Yu, D.; Liang, D. Selenium distribution in the Chinese environment and its relationship with human health: A review. *Environ. Int.* **2018**, *112*, 294–309. [CrossRef] [PubMed]
9. Poblaciones, M.J.; Santamaria, O.; García-White, T.; Rodrigo, S.M. Selenium biofortification in bread-making wheat under Mediterranean conditions: Influence on grain yield and quality parameters. *Crop. Pasture Sci.* **2014**, *65*, 362–369. [CrossRef]
10. Poblaciones, M.J.; Rodrigo, S.; Santamaria, O. Biofortification of legumes with Selenium in Semiarid Conditions. *Selenium* **2015**, *1*, 324–340.
11. Rodrigo, S.; Santamaria, O.; Perez-Izquierdo, L.; Poblaciones, M.J. Arsenic and selenium levels in rice fields from south-west of Spain: Influence of the years of monoculture. *Plant Soil Environ.* **2017**, *63*, 184–188.
12. Gomez-Coronado, F.; Poblaciones, M.J.; Almeida, A.S.; Cakmak, I. Zinc (Zn) concentration of bread wheat grown under mediterranean conditions as affected by genotype and soil/foliar Zn application. *Plant Soil* **2016**, *401*, 331–346. [CrossRef]
13. Rayman, M.P. Selenium and human health. *Lancet* **2012**, *379*, 1256–1268. [CrossRef]
14. Beckett, G.J.; Arthur, J.R. Selenium and endocrine systems. *J. Endocrinol.* **2005**, *184*, 455–465. [CrossRef]
15. Williams, E.; Harrison, M. Selenium: From health to the biological food chain. *J. Biotech. Res.* **2010**, *2*, 112–120.
16. Reid, M.E.; Duffield-Lillico, A.J.; Slate, E.; Natarajan, N.; Turnbull, B.; Jacobs, E.; Combs, G.F., Jr.; Alberts, D.S.; Clark, L.C.; Marshall, J.R. The nutritional prevention of cancer: 400 mcg per day selenium treatment. *Nutr. Cancer Int. J.* **2008**, *60*, 155–163. [CrossRef]
17. Brown, K.H.; Santizo, M.C.; Peerson, J.M.; Begin, F.; Tonin, B. Nutritional quality of complementary feeding regimens and its relationship to dietary diversity and use of processed foods and animal products in low-income guatemalan communities. *FASEB J.* **2001**, *15*, A732.
18. Roth, D.E.; Richard, S.A.; Black, R.E. Zinc supplementation for the prevention of acute lower respiratory infection in children in developing countries: Meta-analysis and meta-regression of randomized trials. *Int. J. Epidemiol.* **2010**, *39*, 795–808. [CrossRef]
19. Mufarrege, D.J.; Aguilar, D.E. Suplementación con zinc de los bovinos para carne en la provincia de Corrientes (Argentina). *EEA INTA Mercedes Corrientes Not. Coment.* **2001**, *348*, 1–4.
20. Rosa, D.E.; Fazzio, L.E.; Picco, S.J.; Furnus, C.C.; Mattioli, G.A. Metabolismo y deficiencia de zinc en bovinos. Laboratorio de nutrición mineral y fisiología reproductiva. *Analecta Vet.* **2008**, *28*, 34–44.
21. Chasapis, C.T.; Loutsidou, A.C.; Spiliopoulou, C.A.; Steidou, M.E. Zinc and human health: An update. *Arch. Toxicol.* **2012**, *86*, 521–534. [CrossRef] [PubMed]
22. Read, S.; Obeid, S.; Ahlenstiel, C.; Ahlenstiel, G. The role of zinc in antiviral immunity. *Adv. Nutr.* **2019**, *10*, 696–710. [CrossRef] [PubMed]
23. Zhang, J.; Taylor, E.; Bennett, K.; Saad, R.; Rayman, M. Association between regional selenium status and reported outcome of COVID-19 cases in China. *Am. J. Clin. Nutr.* **2020**, *111*, 1297–1299. [CrossRef]
24. Liu, Q.; Zhao, X.; Ma, J.; Mu, Y.; Wang, Y.; Yang, S.; Wu, Y.; Wu, F.; Zhou, Y. Selenium (Se) plays a key role in the biological effects of some viruses: Implications for COVID-19. *Environ. Res.* **2021**, *196*, 110984. [CrossRef]
25. Hoang, B.X.; Hoang, H.Q.; Han, B. Zinc Iodide in combination with Dimethyl Sulfoxide for treatment of SARS-CoV-2 and other viral infections. *Med. Hypotheses* **2020**, *143*, 109866. [CrossRef]
26. Sethuram, R.; Bai, D.; Abu-Soud, H.M. Potential role of zinc in the COVID-19 disease process and its probable impact on reproduction. *Reprod. Sci.* **2021**, *7*, 1–6.
27. White, P.J.; Broadley, M.R. Biofortifying crops with essential mineral elements. *Trends Plant Sci.* **2005**, *10*, 586–593. [CrossRef] [PubMed]
28. Rodrigo, S.; Santamaría, O.; López-Bellido, F.J.; Poblaciones, M.J. Agronomic selenium biofortification of two-rowed barley under mediterranean conditions. *Plant Soil Environ.* **2013**, *59*, 115–120. [CrossRef]

29. Rodrigo, S.; Santamaria, O.; Poblaciones, M.J. Selenium application timing: Influence in wheat grain and flour selenium accumulation under Mediterranean conditions. *J. Agric. Sci.* **2014**, *6*, 23–30. [CrossRef]
30. Cakmak, I.; Kalayci, M.; Kaya, Y.; Torun, A.A.; Aydin, N.; Wang, Y.; Arisoy, Z.; Erdem, H.; Yazici, A.; Gokmen, O.; et al. Biofortification and localization of zinc in wheat grain. *J. Agric. Food Chem.* **2010**, *58*, 9092–9102. [CrossRef] [PubMed]
31. Poblaciones, M.J.; Rengel, Z. Combined foliar selenium and zinc biofortification in field pea (*Pisum sativum* L.): Accumulation and bioavailability in raw and cooked grains. *Crop. Pasture Sci.* **2017**, *68*, 265–271. [CrossRef]
32. Reynolds-Marzal, M.D.; Rivera-Martín, A.M.; Rodrigo, S.M.; Santamaria, O.; Poblaciones, M.J. Biofortification of Forage Peas with Combined Application of Selenium and Zinc under Mediterranean Conditions. *J. Soil Sci. Plant Nutr.* **2021**, *21*, 286–300. [CrossRef]
33. National Research Council (NRC). *Reference Intakes for Vitamin A, vitamin K, As, B, Cr, Cu, I, Fe, Mn, Mo, Ni, Si and Zn*; Institute of Medicine/Food and Nutrition Board, National Academy Press: Washington, DC, USA, 2001.
34. Elmadfa, I.; Meyer, A.; Nowak, V.; Hasenegger, V.; Putz, P.; Verstraeten, R.; Remaut-DeWinter, A.M.; Kolsteren, P.; Dostálová, J.; Dlouhý, P.; et al. The European nutrition and health report 2009. *Forum Nutr.* **2009**, *62*, 1–405.
35. Suttle, N.F. *Mineral Nutrition of Livestock*, 4th ed.; CABI: Cambridge, MA, USA, 2010.
36. Marschner, H. *Mineral Nutrition of Higher Plants*, 2nd ed.; Academic: London, UK, 1995.
37. Longnecker, N.E.; Robson, A.D. Distribution and Transport of Zinc in Plants. In *Zinc in Soils and Plants*; Robson, A.D., Ed.; Kluwer Academic Publishers: Dordrecht, The Netherlands, 1993; pp. 79–91.
38. Ghaderzadeh, S.; Mirzaei Aghjeh-Gheshlagh, F.; Nikbin, S.; Navidshad, B. A Review on properties of selenium in animal nutrition. *Iran. J. Appl. Anim. Sci.* **2016**, *6*, 753–761.
39. Reich, H.J.; Hondal, R.J. Why nature chose selenium. *ACS Chem. Biol.* **2016**, *11*, 821–841. [CrossRef]
40. Terry, N.; Zayed, A.M.; de Souza, M.P.; Tarun, A.S. Selenium in higher plants. *Annu. Rev. Plant Physiol. Plant Mol. Biol.* **2000**, *51*, 401–432. [CrossRef]
41. Ahmad, Z.; Anjum, S.; Skalicky, M.; Waraich, E.A.; Muhammad Sabir Tariq, R.; Ayub, M.A.; Hossain, A.; Hassan, M.M.; Brestic, M.; Sohidul Islam, M.; et al. Selenium Alleviates the Adverse Effect of Drought in Oilseed Crops Camelina (*Camelina sativa* L.) and Canola (*Brassica napus* L.). *Molecules* **2021**, *26*, 1699. [CrossRef] [PubMed]
42. Zhang, M.; Tang, S.; Huang, X.; Zhang, F.; Pang, Y.; Huang, Q.; Yi, Q. Selenium uptake, dynamic changes in selenium content and its influence on photosynthesis and chlorophyll fluorescence in rice (*Oryza sativa* L.). *Environ. Exp. Bot.* **2014**, *107*, 39–45. [CrossRef]
43. Mohtashami, R.; Movahhedi, M.D.; Balouchi, H.; Faraji, H. Improving yield, oil content and water productivity of dryland canola by supplementary irrigation and selenium spraying. *Agric. Water Manag.* **2020**, *232*, 106046. [CrossRef]
44. Manojlović, M.S.; Lončarić, Z.; Cabilovski, R.R.; Popović, B.; Karalić, K.; Ivezić, V.; Ademi, A.; Singh, B.R. Biofortification of wheat cultivars with selenium. *Soil Plant Sci.* **2019**, *69*, 715–724. [CrossRef]
45. Johnson, L. Trends and annual fluctuations in selenium concentrations in wheat grain. *Plant Soil* **1991**, *138*, 67–73. [CrossRef]
46. Moraghan, J.T.; Mascagni, H.J., Jr. Environmental and soil factors affecting micronutrient deficiencies and toxicities. In *Micronutrients in Agriculture*; Mordvedt, J.J., Cox, F.R., Shumann, L.M., Welch, R.M., Eds.; Soil Science Society of America: Madison, WI, USA, 1991; pp. 371–425.
47. Gupta, N.; Ram, H.; Kumar, B. Mechanism of Zinc absorption in plants: Uptake, transport, translocation and accumulation. *Rev. Environ. Sci. Biol. Technol.* **2016**, *15*, 89–109. [CrossRef]
48. Sevanto, S. Phloem transport and drought. *J. Exp. Bot.* **2014**, *65*, 1751–1759. [CrossRef] [PubMed]
49. Fernández, V.; Sotiropoulos, T.; Brown, P. *Foliar Fertilization: Scientific Principles and Field Practices*; International Fertilizer Industry Association (IFA): Paris, France, 2013.
50. Gupta, R.K.; Gangoliya, S.S.; Singh, N.K. Reduction of phytic acid enhancement of bioavailable micronutrients in food grains. *J. Food. Sci. Technol.* **2015**, *52*, 676–684. [CrossRef] [PubMed]
51. Morris, E.R.; Ellis, R. Usefulness of the dietary phytic acid/zinc molar ratio as an index of zinc bioavailability to rats and humans. *Biol. Trace Elem. Res.* **1989**, *19*, 107–117. [CrossRef] [PubMed]
52. Saboor, A.; Ali, M.A.; Ahmed, N.; Skalicky, M.; Danish, S.; Fahad, S.; Hassan, F.; Hassan, M.M.; Brestic, M.; El Sabagh, A.; et al. Biofertilizer-Based Zinc Application Enhances Maize Growth, Gas Exchange Attributes, and Yield in Zinc-Deficient Soil. *Agriculture* **2021**, *11*, 310. [CrossRef]
53. Zinzala, V.N.; Narwade, A.V.; Karmakar, N.; Patel, P.B. Influence of Zinc Applications on Photosynthesis, Transpiration and Stomatal Conductance in Kharif Rice (*Oryza sativa* L.) Genotypes. *Int. J. Curr. Microbiol. Appl. Sci.* **2019**, *8*, 150–168. [CrossRef]
54. Wang, Y.H.; Zou, C.Q.; Mirza, Z.; Li, H.; Zhang, Z.Z.; Li, D.P.; Xu, C.L.; Zhou, X.B.; Shi, X.J.; Xie, D.T.; et al. Cost of agronomic biofortification of wheat with zinc in China. *Agron. Sustain. Dev.* **2016**, *36*, 44–50. [CrossRef]
55. Walkley, A.; Black, I.A. An examination of the Degtjareff method for determining organic carbon in soils: Effect of variations in digestion conditions and of inorganic soil constituents. *Soil Sci.* **1934**, *63*, 251–263. [CrossRef]
56. Bremner, J.M. Nitrogen total. In *Methods of Soil Analysis, Part 3: Chemical Methods*; Sparks, D.L., Ed.; American Society of Agronomy, Inc.: Madison, WI, USA, 1996; pp. 1085–1121.
57. Lindsay, W.L.; Norwell, W.A. Development of a DTPA soil test for zinc, iron, manganese and copper. *Soil Sci. Soc. Am. J.* **1978**, *42*, 421–428. [CrossRef]

58. Zhao, F.J.; McGrath, S.P. Extractable sulphate and organic sulphur in soils and their availability to plants. *Plant Soil* **1994**, *164*, 243–250. [CrossRef]
59. AOCS. *Official Methods of Analysis*; Association of Official Analytical Chemists: Washington, DC, USA, 2006.
60. Adams, M.L.; Lombi, E.; Zhao, F.J.; McGrath, S.P. Evidence of low selenium concentration in UK bread-making wheat grain. *J. Sci. Food Agric.* **2002**, *82*, 1160–1165. [CrossRef]
61. Haug, W.; Lantzsch, H.J. Sensitive method for the rapid determination of phtytate in cereals and cereal products. *J. Sci. Food Agric.* **1983**, *34*, 1423–1426. [CrossRef]

Article

Lamina Cell Shape and Cell Wall Thickness Are Useful Indicators for Metal Tolerance—An Example in Bryophytes

Katharina Petschinger [1], Wolfram Adlassnig [1], Marko S. Sabovljevic [2] and Ingeborg Lang [3,*]

[1] Cell Imaging and Ultrastructure Research, Faculty of Life Sciences, University of Vienna, Althanstrasse 14, A-1090 Vienna, Austria; kpetschinger@gmail.com (K.P.); wolfram.adlassnig@univie.ac.at (W.A.)

[2] Institute of Botany and Botanical Garden, Faculty of Biology, University of Belgrade, Takovska 43, 11000 Belgrade, Serbia; marko@bio.bg.ac.rs

[3] Department of Functional and Evolutionary Ecology, Faculty of Life Sciences, University of Vienna, Althanstrasse 14, A-1090 Vienna, Austria

* Correspondence: Ingeborg.lang@univie.ac.at

Abstract: Bryophytes are widely used to monitor air quality. Due to the lack of a cuticle, their cells can be compared to the roots of crop plants. This study aimed to test a hypothetical relation between metal tolerance and cell shape in biomonitoring mosses (*Hypnum cupressiforme*, *Pleurozium schreberi*, *Pseudoscleropodium purum*) and metal sensitive species (*Physcomitrium patens*, *Plagiomnium affine*). The tolerance experiments were conducted on leafy gametophytes exposed to solutions of ZnSO4, $ZnCl_2$, and $FeSO_4$ in graded concentrations of 1 M to 10^{-8} M. Plasmolysis in D-mannitol (0.8 M) was used as a viability measure. The selected species differed significantly in lamina cell shape, cell wall thickness, and metal tolerance. In those tested mosses, the lamina cell shape correlated significantly with the heavy metal tolerance, and we found differences for ZnSO4 and $ZnCl_2$. Biomonitoring species with long and thin cells proved more tolerant than species with isodiametric cells. For the latter, "death zones" at intermediate metal concentrations were found upon exposure to ZnSO4. Species with a greater tolerance towards $FeSO_4$ and ZnSO4 had thicker cell walls than less tolerant species. Hence, cell shape as a protoplast-to-wall ratio, in combination with cell wall thickness, could be a good marker for metal tolerance.

Keywords: bioindication; bryophytes; moss; zinc; iron; cell shape; particulate matter

Citation: Petschinger, K.; Adlassnig, W.; Sabovljevic, M.S.; Lang, I. Lamina Cell Shape and Cell Wall Thickness Are Useful Indicators for Metal Tolerance—An Example in Bryophytes. *Plants* **2021**, *10*, 274. https://doi.org/10.3390/plants10020274

Academic Editor: Gokhan Hacisalihoglu

Received: 28 November 2020
Accepted: 25 January 2021
Published: 31 January 2021

Publisher's Note: MDPI stays neutral with regard to jurisdictional claims in published maps and institutional affiliations.

1. Introduction

Bryophytes, especially mosses, are widely used for biomonitoring in different environmental studies [1–3]. Similar to primary roots in seed plants, most bryophytes do not possess a cuticle. Their leaflets consist of a monolayer of cells. Thus, bryophytes indiscriminately collect nutrients and other substances from atmospheric, mainly wet deposits. Therefore, they are perfectly suitable organisms to monitor the overall exposure at a given locality over a prolonged time span. Additionally, most bryophyte species are physiologically active over the winter and continue to adsorb deposited elements all year long.

A common method to analyze the air quality is to measure particulate matter (PM)-values [4]. PM may include solid particles and liquid droplets found in the air. PM2.5 are fine inhalable particles with a diameter of up to 2.5 μM, and PM10 includes inhalable particles with a diameter of 10 μM and lower. These particles may contain hundreds of different chemicals [5,6], some of which may seriously affect the human and animal respiratory system [6], resulting in a need for constant PM surveillance.

In Austria, iron and zinc hold the largest proportions of all heavy metals in the PM_{10} and $PM_{2.5}$ range [4,7]. Therefore, the focus is on these two metals as they also play an important political and environmental role regarding air quality monitoring by the Austrian government to ensure policy compliances by the European Union. Furthermore,

159

mosses may exhibit differences in metal uptake behavior and tolerance with respect to the element [8,9]. However, biomonitoring usually considers widely distributed species within the geographic region of interest and neglects possible differences of the species in terms of tolerance levels to elements or compounds.

The over 12,000 moss species are representing a broad morphological diversity. Furthermore, each species manifests itself as protonema, leafy or thallous gametophore, or sporophyte [10]. Here, we focus on the leaflets (lamina, [11]) of the gametophyte since these represent most of the total surface. In spite of a multitude of different lamina shapes, most moss leaflets are composed of a single cell layer. The leaflets may consist of quadrate, rectangular, oblong, fusiform, rhomboidal, hexagonal, linear, flexuous, or vermicular shaped cells of extremely different size, sometimes supplemented by a costa ("midrib"), aberrant cells at the base of the leaf, or by lamellae, papillae and mamillae increasing the leaf surface [11,12]. Usually, lamina cells are classified as parenchymatic (roundish, rectangular, or hexagonal) or prosencymatic (elongated and interleaved; [13]). In this approach, we used a simplified determination of lamina cell form comparing roundish or hexagonal shapes with rectangular or elongated rectangular ones.

Although habitat or life forms have been frequently discussed as predictors of metal tolerance in mosses [8,14,15], we are not aware of studies considering moss morphology or anatomy as related to heavy metal pollutions. The focus is not on molecular differences in cell wall composition across kingdoms, as this has been thoroughly discussed by Sarkar et al. [16] or Fangel et al. [17]. Here, the hypothesis is tested that cell size and/or cell shape is related to tolerance of certain metals in selected moss species. Species commonly used in biomonitoring and species that are not considered as suitable were selected Comparison of the metal tolerance, therefore, contributes to quality assurance in the field of biomonitoring of heavy metals.

2. Results

2.1. Lamina Cell Measurements

The five different moss species (*Physcomitrium patens*, *Plagiomnium affine*, *Hypnum cupressiforme*, *Pleurozium schreberi*, and *Pseudoscleropodium purum*) have distinct leaflets and differ significantly in the size and shape of lamina cells (Figure 1, Table 1). Lamina cells showed a roughly rectangular shape for *P. patens* and a hexagonal shape for *P. affine* *H. cupressiforme*, *P. schreberi*, and *P. purum* had elongated rectangular or linear lamina cells No significant difference in cell shape was found within the same species.

Table 1. Five moss species (*Plagiomnium affine*, *Physcomitrium patens*, *Pseudoscleropodium purum*, *Hypnum cupressiforme*, and *Pleurozium schreberi*) were compared by mean cell length (μm), cell width (μm), length to width ratio, shape, and mean cell area (μM^2) of mid lamina cells ($n = 40$). SD = Standard deviation; μ = mean value; $p = 0.0001$ (Kruskal–Wallis test comparing all five species).

Moss Species	μ Cell Length (SD) (μM)	μ Cell Width (SD) (μM)	μ Cell Length to Width Ratio	Shape of Mid Lamina Cells	μ Cell Area (SD) (μM^2)	μ Cell Wall Thickness (SD) (μM)
Plagiomnium affine	51 (4)	37 (3)	1	hexagonal	1402 (150)	0.65 (0.12)
Physcomitrium patens	63 (12)	31 (6)	2	rectangular	1979 (569)	0.26 (0.05)
Pseudoscleropodium purum	65 (8)	5 (1)	12	rectangular, longish	354 (73)	0.46 (0.07)
Hypnum cupressiforme	72 (11)	3 (1)	25	rectangular, longish	220 (53)	0.88 (0.17)
Pleurozium schreberi	94 (12)	8 (1)	12	rectangular, longish	735 (140)	0.86 (0.18)

P. schreberi had the greatest average cell length (94 μM) followed by *H. cupressiforme* with 72 μM. The latter had the smallest cell width (3 μM), the biggest ratio of cell length to cell width (25), and the smallest cell area with only 220 μM^2 (always respective mean values). The largest average cell width was measured in the moss *P. affine* (37 μM), but the largest cell area was found for *P. patens* (1979 μM^2). With a value of 1, *P. affine* had the smallest ratio of cell length to cell width (Table 1).

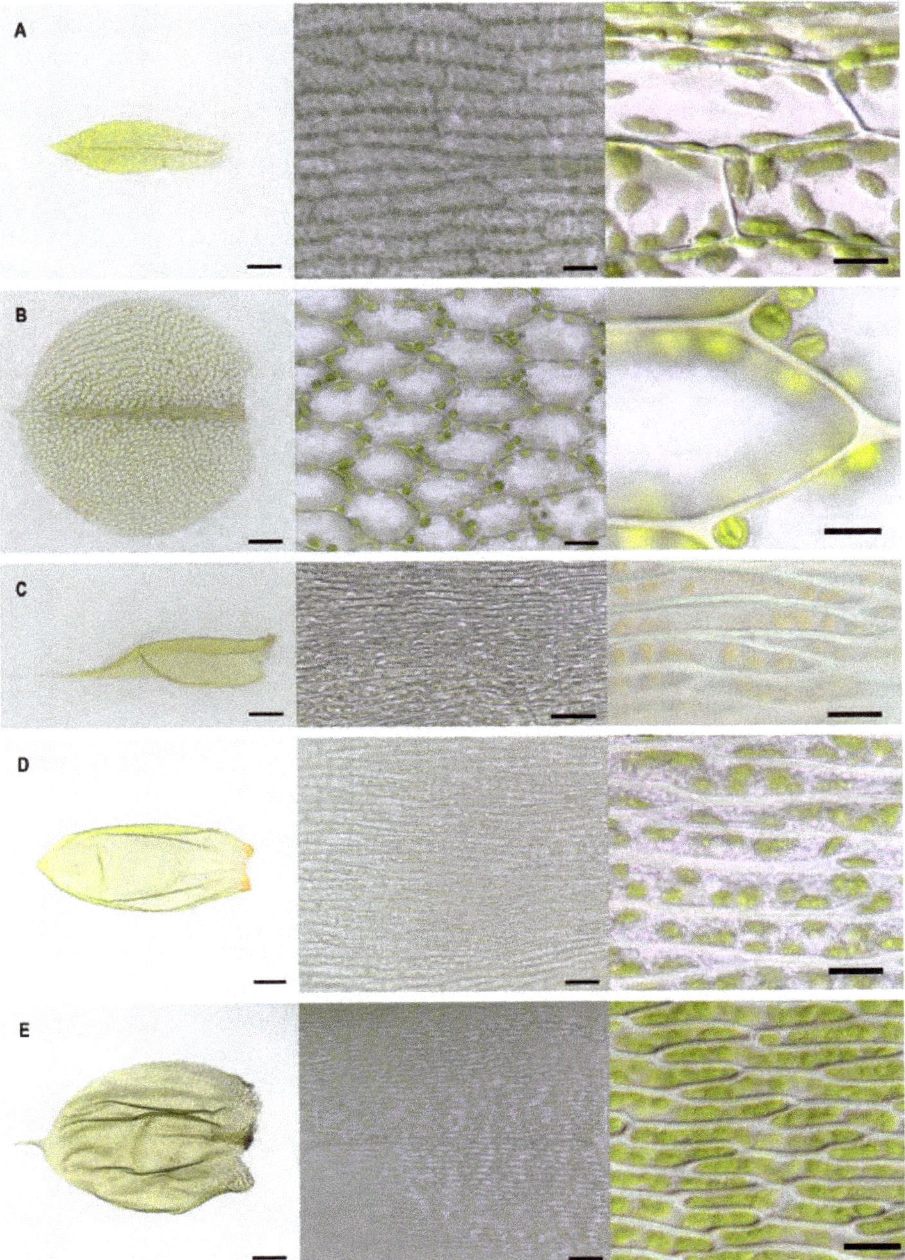

Figure 1. Lamina and cell shapes of five selected bryophyte species (**A**) *Physcomitrium patens*, (**B**) *Plagiomnium affine*, (**C**) *Hypnum cupressiforme*, (**D**) *Pleurozium schreberi*, and (**E**) *Pseudoscleropodium purum*. Bar: 250 µM for leaflet (overview); 25 µM for the respective cell shape (middle panel); 10 µM for close up (right panel).

Cell wall thickness differed significantly between the five species (Dunn's test: $p < 0.05$) except for *H. cupressiforme and P. schreberi* ($p = 0.354$). The thinnest cell walls were found in *P. patens* ($\mu = 0.26$ μM). Wall thickness increased from *P. patens* < *P. purum* < *P. affine* < *H. cupressiforme* to *P. schreberi* with a mean thickness of almost 0.9 μM (Figure 2A). Thus, the biomonitor species *P. schreberi* and *H. cupressiforme* form thick cell walls compared to, e.g., *P. patens*. The tested species showed significant differences in the ratio of cell length to cell wall thickness (Figure 2B; Dunn's text: $p < 0.05$) except for *P. affine* and *H. cupressiforme* ($p = 0.3418$) that both had a similar ratio of lengths to thick walls. This ratio increased from *P. affine* = *H. cupressiforme* < *P. schreberi* < *P. purum* < *P. patens* that had the short cells ($\mu = 63$ μM) and thinnest walls (Figure 2B). Also, the ratio of cell width to cell wall thickness was significantly different in all tested species (Figure 2C; Dunn's test: $p < 0.05$) and increased from *H. cupressiforme* < *P. schreberi* < *P. purum* < *P. affine* < *P. patens*. The thin cells of *H. cupressiforme* ($\mu = 3$ μM) had a width to cell wall thickness ratio of four whereas *P. patens* with its wide cells ($\mu = 31$ μM) had a ratio more than 100 times higher of cell width to cell wall thickness (Figure 2C). The ratio of the cell area to cell wall thickness was similar to the ratio of the width to wall thickness with the same increasing order of species (Figure 2D). The ratio of the cell area to cell wall thickness was 30 times higher in *P. patens* as compared to *H. cupressiforme*.

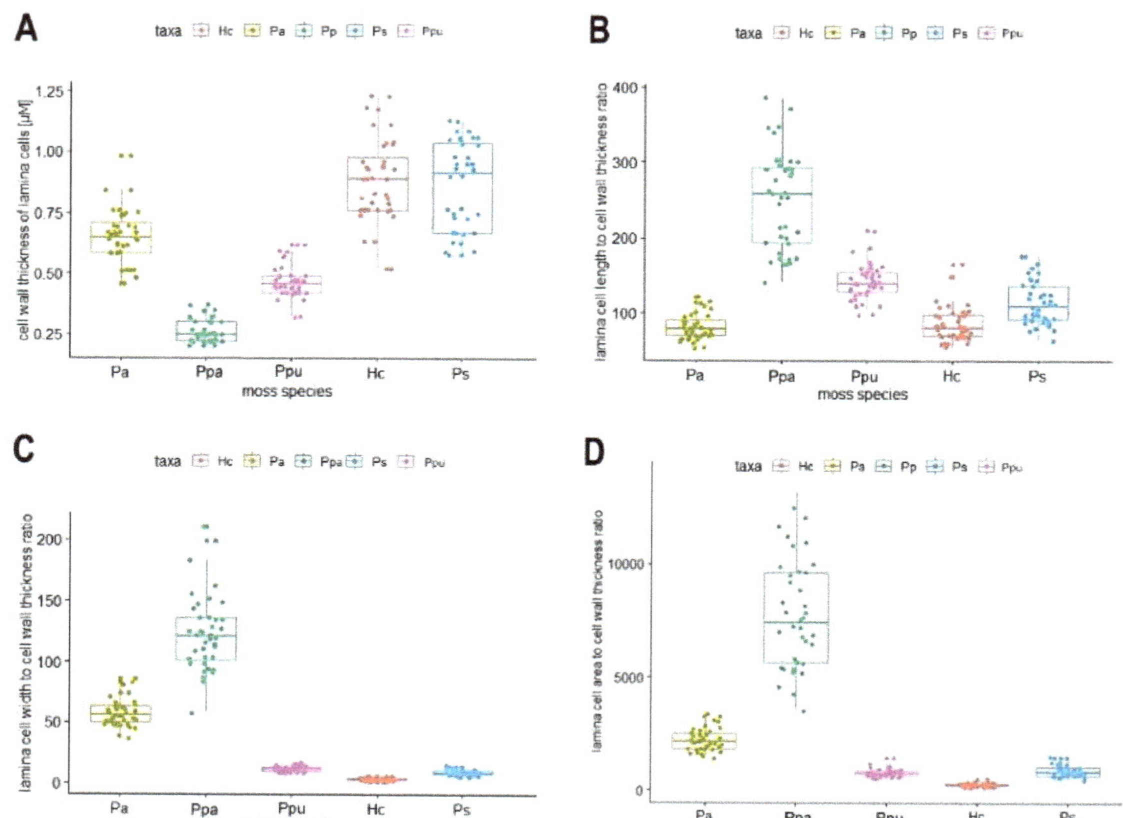

Figure 2. Box plots comparing the five investigated moss species. (**A**) cell wall thickness, (**B**) lamina cell length to cell wall thickness, (**C**) lamina cell width to cell wall thickness, and (**D**) lamina cell area to cell wall thickness. Pa: *Plagiomnium affine*, Ppa: *Physcomitrium patens*, Ppu: *Pseudoscleropodium purum*, Hc: *Hypnum cupressiforme*, and Ps: *Pleurozium schreberi*.

2.2. Metal Tolerance

Metal tolerance was determined by viability tests using plasmolysis in 0.8 M mannitol (Figure 3). Living cells are able to undertake plasmolysis, whereas dead cells have lost semipermeable membrane function and therefore cannot plasmolyze [18]. The "no observed effective concentration" (NOEC) and "lowest observed effective concentration" (LOEC) for the three tested heavy metal solutions ($ZnCl_2$, $ZnSO_4$, and $FeSO_4$) were assessed to achieve a numeric variable of tolerance data (Table 2).

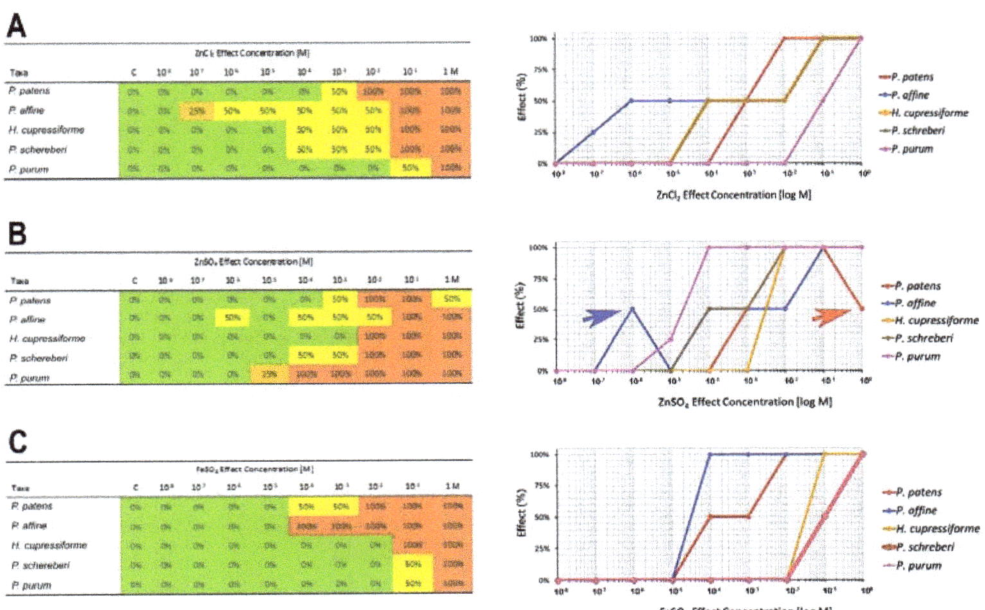

Figure 3. Percentage of dead lamina cells (0%, 25%, 50%, and 100%) and respective dose-response-curves for five moss species (*P. patens, P. affine, H. cupressiforme, P. schreberi, P. purum*) in tenfold dilution series of (**A**) $ZnCl_2$, (**B**) $ZnSO_4$ and (**C**) $FeSO_4$. The arrows in B point to possible "death zones"; blue arrow: *P. affine*; red arrow: *P. patens*.

Table 2. Median of no effect concentration (NOEC) and lowest observed effect concentration (LOEC) of tested substances ($ZnCl_2$, $ZnSO_4$, $FeSO_4$ in Mol) for five moss species. n = 40–80 cells.

	NOEC Median			LOEC Median		
Species	$ZnCl_2$	$ZnSO_4$	$FeSO_4$	$ZnCl_2$	$ZnSO_4$	$FeSO_4$
P. patens	10^{-4}	10^{-4}	10^{-5}	10^{-3}	10^{-3}	10^{-4}
P. affine	5.5×10^{-8}	5.55×10^{-6}	10^{-5}	5.5×10^{-7}	10^{-6}	10^{-4}
H. cupressiforme	10^{-5}	10^{-3}	10^{-2}	10^{-4}	10^{-2}	10^{-1}
P. schreberi	10^{-5}	10^{-5}	10^{-2}	10^{-4}	10^{-4}	10^{-1}
P. purum	5.5×10^{-3}	5.5×10^{-6}	10^{-2}	10^{-1}	5.5×10^{-5}	10^{-1}

For each effect concentration, the median of the tolerance experiments was calculated (n = 40–80 cells per species). Since *P. affine* showed a death zone, the NOEC was formed by an average of three concentrations (10^{-7} M, 10^{-6} M, 10^{-5} M).

For $ZnCl_2$, we observed a decreasing tolerance of moss species *P. purum* > *P. patens* > *P. schreberi* and *H. cupressiforme* > *P. affine*, whereby *P. affine* was the at least tolerant moss of the investigated species. *H. cupressiforme* and *P. schreberi* showed the same LOEC (10^{-4} M $ZnCl_2$). *P. purum* could tolerate the highest observed concentration of 10^{-2} M

ZnCl$_2$ (Figure 3A). Apart from *H. cupressiforme* and *P. schreberi*, the tested species differ significantly with regard to their tolerance to ZnCl$_2$ ($p < 0.05$).

The tolerance of moss species to ZnSO$_4$ dropped from *H. cupressiforme* > *P. patens* > *P. schreberi* > *P. purum* > *P. affine* (concentration range: 10^{-8}–1 M) with *H. cupressiforme* showing the highest NOEC of 10^{-3} M (Figure 3B). There was a significant difference between all tested species to tolerate ZnSO$_4$ ($p < 0.05$). Interestingly, death zones emerged in the case of *P. affine* showing 100 % viable cells at a concentration of 10^{-5} M ZnSO$_4$ and only 50 % of viable cells at a lower concentration of 10^{-6} M ZnSO$_4$. Additionally, a death zone is likely in *P. patens* between a concentration of >10^{-1} M ZnSO$_4$: at 1 M ZnSO$_4$, the tolerance tests showed only 50 % of dead cells, whereas, at a lower concentration (10^{-1} to 10^{-2} M ZnSO$_4$), 100 % of cells died (Figure 3B, arrow).

Thus, visible effects of ZnCl$_2$ (Figure 3A) could be observed at lower concentrations compared to ZnSO$_4$ in *P. affine* and *H. cupressiforme* (Figure 3B) but not in *P. patens*, *P. schreberi*, and *P. purum*. Apparently, some mosses are more sensitive to ZnCl$_2$ than to ZnSO$_4$. *P. patens* and *P. schreberi* had the same LOEC for ZnCl$_2$ and ZnSO$_4$, although the latter species with a slower transition and both tolerated a 10-fold higher concentration of ZnCl$_2$ and ZnSO$_4$ according to the percentage of viability. In contrast, *P. purum* survived a 10^4 higher concentration of ZnCl$_2$ than ZnSO$_4$.

The overall tolerance to iron was greater than to zinc (Figure 3). In the concentration range of 10^{-8}–1 M FeSO$_4$, the tolerance of the tested species decreased as: *P. schreberi* = *P. purum* > *H. cupressiforme* > *P. patens* > *P. affine*. The species used for biomonitoring (*P. schreberi*, *P. purum* and *H. cupressiforme*) had the same LOEC of 10^{-1} FeSO$_4$, whereas *P. patens* and *P. affine* had a LOEC of 10^{-4} M FeSO$_4$. The transition from living to dead *P. affine* and *H. cupressiforme* cells was sudden when compared to the other species (Figure 3C).

2.3. Correlations between Cell Shape and Metal Tolerance

The NOEC and LOEC for the three tested heavy metal solutions (ZnCl$_2$, ZnSO$_4$, FeSO$_4$, respectively) were assessed to achieve a numeric variable of tolerance data (Table 2) and to correlate them to cell shape. The results are shown in Table 3.

Table 3. Spearman correlation between morphometric parameters and maximum no-effect concentrations, shown as ρ (p). Strong ($\rho \geq 0.8$) and highly significant ($p < 0.01$) correlations are highlighted (bold). $n = 200$ cells.

	ZnCl$_2$	ZnSO$_4$	FeSO$_4$
Cell Length	0.16 (0.023)	0.27 (<0.001)	0.60 (<0.001)
Cell Width	−0.34 (<0.001)	−0.24 (<0.001)	**−0.85 (<0.001)**
Ratio Length/Width	0.24 (<0.001)	0.36 (<0.001)	**0.85 (<0.001)**
Cell Wall Thickness	−0.57 (<0.001)	0.24 (0.001)	0.50 (<0.001)
Ratio Length/Cell Wall Thickness	0.68 (<0.001)	−0.06 (0.406)	−0.17 (0.017)
Ratio width/Cell Wall Thickness	0.09 (0.202)	−0.25 (<0.001)	**−0.85 (<0.001)**

A strong, negative correlation occurred between the cell width and the tolerance to FeSO$_4$ ($\rho = -0.85$, $n = 200$, $p \leq 0.001$, Table 3), and also the correlation between the ratio of the cell width to the cell wall thickness was highly negative ($\rho = -0.85$, $n = 200$, $p \leq 0.001$). A strong, positive correlation between the ratio of the cell length to the cell width was also highly significant ($\rho = 0.85$, $n = 200$, $p \leq 0.001$). The same trend, albeit with less strong correlation, was found for both zinc treatments. These data show an increased metal tolerance in species with elongated cells and thick walls.

3. Discussion

Many bryologists are aware of tolerance differences among selected species, their physiological state, life form or even genotypes when comparing or interpreting the results obtained. However, the mechanisms of resistance/tolerance to pollution substances remain obscured. Moreover, with such a huge variation of species, in addition to environmental, physiological, and morphological factors used in biomonitoring studies blur the pattern of pollutant tolerance. Therefore, we aimed to find a "simple" commodity like lamina cell shape that could be linked to metal tolerance.

In general, the monitoring of airborne heavy metal pollutants is a very difficult process [15]. Field receptor measurements are highly expensive, but they supply precise and reliable distribution-estimations about the airborne pollutants. However, they lack information on the effects of these pollutants on biological systems [19]. In this case, other methods were more appropriate, like the biomonitoring of heavy metal pollutants using bryophytes [1,2,15].

Different tolerance levels of the tested moss species to $ZnCl_2$ and $ZnSO_4$ were found but also to $FeSO_4$ (Figure 3). Interspecific differences in the sensitivity were also reported by Tyler [20] as tested by net photosynthesis. We also used alternative tools and found different tolerance levels according to bryophyte species, life forms, or metal applied [9,18,21–23].

The tolerance experiments conducted with the biomonitoring mosses and the cultured mosses represent the tolerance of the species without considering their own background concentration of trace metals. In monitoring surveys, this background concentration of heavy metals in the mosses is usually not determined. However, if a pre-disposition with metals exits in the field samples, the tolerance levels would be lower, which leads to more conservative pollution estimations. The collection sites of the biomonitoring species used here are in the Viennese forest, away from civilization. No particular metal contamination is assumed. Furthermore, it should be noted that mosses have a relatively high, intrinsic concentration of zinc (about 20 µg/g), which is not due to emissions [24]. The same can be assumed for the tested iron samples. In *H. cupressiforme*, the comparison of field samples and tissue culture probes did not result in significant tolerance differences (A. Khan, unpublished data).

Death zones as found in this study for *P. patens* and *P. affine* exposed to $ZnSO_4$ are known from the literature [25–28]. Biebl [25] reported death zones for certain bryophyte species, whereby a low and a high metal concentration resulted in little mortality, but intermediate metal concentrations were lethal. Biebl's observations fit with the results of our study as a high (>1 M) concentration of ZnSO4 caused rather little mortality of lamina cells of *P. patens* (Figure 3). However, further studies are necessary to show if lamina cells of *P. patens* are completely viable at concentrations above 1 M ZnSO4. *P. affine* had a more distinct death zone at a lower concentration of 10^{-6} M $ZnSO_4$. Url [28] also observed death zones in *Nardia scalaris* (a liverwort) for copper and vanadium. The physiological reasons for the death zones are still unclear and would need further investigations.

Metals are positively charged and become first adsorbed to cation exchange sites at the cell wall [29,30]. Sequential elution studies also found most metal retention in the cell wall [31,32]. Hence, metals are deposited to the apoplast, where they have little impact on the living cytoplasm. In mosses with thick cell walls, the apoplast contributes more to the total surface of the leaf compared to mosses with thinner walls. The same is true for species with elongated cells, compared to species with cells of the same volume but more globular or cube-shaped cells. If the cell wall plays a major role in metal retention [21], moss species with such cells or thick walls are therefore more tolerant.

In the present study, cell wall thickness was determined by light microscopy, but even at the highest possible resolution, the edges of cell walls may appear blurry (see Figure 2). To lower this statistical variance and conceive reliable results, a high number of measurements was performed (n = 40–80). Certainly, the molecular composition of the cell wall is also relevant as it has become adapted during evolution and in conquering various ecological niches [16,17]. However, bryophyte species used in biomonitoring appear to have a higher percentage of cell wall within the lamina. This renders them more tolerant, and therefore, they can adsorb more metals. The application of these species in biomonitoring thus results in higher metal amounts measured because other species, mosses or vascular plants, have thinner cell walls with less adsorption capacity. However, for the estimation of toxic pollution levels, we prefer a conservative approach that rather over-estimates the metal levels. This way, the present study confirms that the commonly used species are well suitable for biomonitoring.

4. Material and Methods

4.1. Plant Species

Five species of mosses were selected to study the lamina cell shape in combination with heavy metal tolerance. The chosen species are from five different families and four different orders. *Pleurozium schreberi* (Will. ex Brid.) Mitt. (Hylocomiaceae), *Hypnum cupressiforme* Hedw. (Hypnaceae), and *Pseudoscleropodium purum* (Hedw.) M. Fleisch. (Brachytheciaceae) are commonly used in environmental and biomonitoring studies and were collected in the forest near Vienna, Austria (14 March and 13 April, 2018). Additionally, we selected two species with very different cell shapes and sizes, *Physcomitrium (Physcomitrella) patens* (Hedw.) Bruch and Schimp. (Funariaceae) and *Plagiomnium affine* (Blandow ex Funck) T.J. Kop. (Mniaceae). They were cultured in a growth cabinet (Conviron) at 20 °C with a 14/10 h light/dark cycle. Sterile cultures of *P. patens* were propagated according to [33]. The widely used model species *P. patens* has a shorter life span than the bigger *P. affine*, and they also differ in life forms. None of the species shows a particular preference for metal contaminated sites, although *P. schreberi* has been found at the periphery of mine tailings [34,35]. Table 4 shows further details on the plant material and its cultivation.

Table 4. Taxonomy, origin, and culture of the plant material.

Species	Collection	Culture Conditions
Plagiomnium affine (Blandow ex Funck) T.J. Kop. (Mniaceae)	laboratory	Non-sterile culture 20 °C, 14/10 h day/night
Physcomitrium patens (Hedw.) Bruch and Schimp. (Funariaceae)	laboratory	sterile culture 20 °C, 14/10 h day/night
Pseudoscleropodium purum (Hedw.) M. Fleisch (Brachytheciaceae)	48.183470, 16.067139	Natural habitat
Hypnum cupressiforme Hedw. (Hypnaceae)	48.183470, 16.044465	Natural habitat
Pleurozium schreberi (Will. ex Brid.) Mitt. (Hylocomiaceae)	48.183470, 16.066399	Natural habitat

4.2. Tolerance Tests

Two to three young but fully expanded leaflets of each moss species were placed in 96-well plates filled with serial dilutions (1 M to 10 nM) of $ZnCl_2$ (Carl Roth, Karlsruhe, Germany), $ZnSO_4$ (Merck), and $FeSO_4$ (Merck), respectively. After 48 h, cell viability was tested via plasmolysis [18]; Figure 4B). In brief, after metal exposure, the leaflets were transferred into 0.8 M mannitol (Carl Roth, Germany) solution for 20 to 30 min. Then, the samples were placed in a droplet of the 0.8 M mannitol on a microscope slide, covered with a coverslip, and imaged in the light microscope (see below). The high sugar concentration of the mannitol solution causes osmotic water loss from the cell, mainly the vacuole. The water efflux from the cell eventually leads to a detachment of the protoplast from the cell wall as the vacuole diminishes in size (plasmolysis; Figure 4B). This process works by intact, semipermeable membranes only; damaged or dead cells, e. g. by high metal concentrations, do not plasmolyze. Hence, the plasmolytic viability tests allowed the determination of effect concentrations for the respective metal as well as the generation of dose-response curves. A minimum of 40 lamina cells were assessed per leaf. The leaflet was divided into four quarters; cells at the edges or midrib were not counted. In each quarter, the cells were analyzed individually using higher magnification, and the values summarized into 0%, 25%, 50%, or 100% dead cells, respectively, per quarter. Aiming to evaluate the significance of species differences in tolerance, the "no effect concentration" (NOEC) was analyzed in R Studio using Kruskal–Wallis rank sum test and Dunn's test (see statistical analysis below).

For the interpretation of the tolerance data, "the lowest observed effect concentration" (LOEC) was also used.

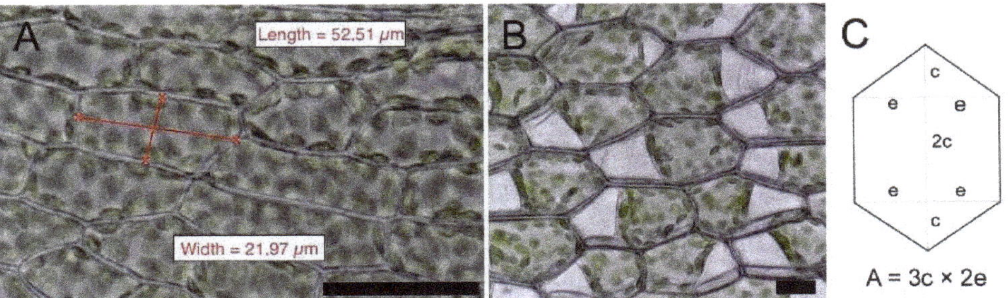

Figure 4. (**A**) measurements of cell length and widths in rectangular cells using the "extended focus" function (Nikon NIS -ElementsBR) of the microscope; bar: 50 µM; and (**B**) hexagonal cell type of *P. affine*, plasmolyzed in 0.8 M mannitol for 20 min; the protoplasts are detached from the cell wall; bar: 25 µM. (**C**) schematic and formula for area calculation of hexagonal cells (changed after Hnilica and Kohout 2018).

4.3. Cell Measurements

Cell lengths were counted parallel to the longitudinal axis of the leaf, and cell width was defined as normally oriented to the longitudinal axis of the leaf (Figure 4). In all species, random measurements were done in fully developed leaves, i.e., leaf four and five from the top of the plantlets. Mid lamina cells are defined as those cells situated in the middle of the leaflets but not next to the margins and not next to the costa ("midrib"). This was done to reflect the cells that were covering the biggest leaf surface in each species and to avoid artifacts of different cell types present in some leaves. For cell wall measurements, at least two typical midleaf cells per lamina were randomly chosen, and at least 40 measurements per cell were performed towards all cell neighbors to reduce possible variabilities of wall thickness.

4.4. Microscopy

An upright light microscope (Nikon Eclipse Ni-U) was used in bright field and interference contrast mode. The instrument was equipped with the objectives Plan Fluor 4× (NA 0.13), Plan Apo 10× (NA 0.45), Plan Apo 20× (NA 0.75), Plan Fluor 40× (NA 0.75), Plan Fluor 60× oil immersion (NA 0.50–1.25), Plan Fluor 100x oil immersion (NA 1.30) and an attached camera (Nikon DS-Ri2). For picture processing, the software NIS-Element BR (Nikon), including an "extended focus" tool, was used. The calibrated measurements were directly exported to Excel (Microsoft Office 365).

The cell areas of *P. patens*, *H. cupressiforme*, *P. schreberi*, and *P. purum* were calculated using the formula of a rectangle. For *P. affine*, we used the area of a hexagon [36] as it fitted best to the cell shape of this species (Figure 4C).

4.5. Statistical Analysis

Statistical analyses were conducted in STATA, version 14.2 (StataCorp LLC, College Station, TX, USA), and documented by archival of files containing all commands. Cell size was characterized by arithmetic mean (μ), standard deviation (SD), and sample size (n). Since samples differed significantly from a normal distribution, differences between the species were tested for significance by using the Kruskal–Wallis test with the post hoc Dunn's test for pairwise comparison. Possible correlations between cell size and metal resistance were characterized by Spearman's Rho (ρ). $\rho > 0.8$ was regarded as a strong, $\rho > 0.5$ as a moderate, and $\rho > 0.2$ as a weak correlation. $p < 0.05$ was regarded as significant throughout the study. Insignificant correlations were considered as meaningless, regardless of ρ. Figures were generated in R Studio, version 1.1.456 (RStudio, Boston, MA, USA).

5. Conclusions

In bryophytes, metal tolerance is species-specific, but the reasons for the different tolerance levels are not clear. Our data confirm a hypothetic relation of lamina cell shape and metal tolerance in the tested mosses. Those species with long and thin lamina cells cope better with high levels of metal than species with isodiametric cells. In the tested species, this correlation is particularly strong for iron, but a similar trend is shown for zinc. Apart from the cell shape, the thickness of the cell wall plays an important role in metal tolerance, most likely due to its adsorption capacity for positively charged ions. Although more bryophyte species should be tested in the future, plant cell anatomy, as in the case of lamina cells described here, is a helpful tool to indicate the metal tolerance of a moss.

Author Contributions: I.L. designed and supervised the work; K.P. conducted the experiments during her master thesis; W.A. was responsible for statistical data and analyses; M.S.S. supervised the work and contributed to the collection of field samples; all authors discussed the data and wrote the manuscript. All authors have read and agreed to the published version of the manuscript.

Funding: The work was supported by a bilateral grant for Scientific and Technological Cooperation between Austria and Serbia (OeAD; SRB 15/2018) to M.S.S. and I.L. This article was funded by Open Access Publishing Funds of the University of Vienna.

Institutional Review Board Statement: Not applicable.

Informed Consent Statement: Not applicable.

Data Availability Statement: Data are available upon request.

Acknowledgments: We thank Helmuth Goldammer for his tireless ambition to always aim for the best picture. This article was funded by Open Access Publishing Funds of the University of Vienna.

References

1. Harmens, H.; Norris, D.A.; Sharps, K.; Mills, G.; Alber, R.; Aleksiayenak, Y.; Blum, O.; Cucu-Man, S.M.; Dam, M.; De Temmerman, L.; et al. Heavy metal and nitrogen concentrations in mosses are declining across Europe whilst some "hotspots" remain in 2010. *Environ. Pollut.* **2015**, *200*, 93–104. [CrossRef] [PubMed]
2. Zechmeister, H.; Hohenwallner, D.; Riss, A.; Hanus-Illar, I. Estimation of element deposition derived from road traffic sources by using mosses. *Environ. Pollut.* **2005**, *138*, 238–249. [CrossRef] [PubMed]
3. Zechmeister, H.; Hohenwallner, D.; Riss, A.; Hanus-Illnar, A. Variations in heavy metal concentrations in the moss species *Abietinella abietina* (Hedw.) Fleisch according to sampling time, within site variability and increase in biomass. *Sci. Total Environ.* **2003**, *301*, 55–65. [CrossRef]
4. Danninger, E. Inspektionsbericht: Luftgüteüberwachung. DVR: 0069264. 2017. Available online: https://www.land-oberoesterreich.gv.at/.
5. Particulate Matter (PM) Basics. Available online: https://www.epa.gov/pm-pollution/particulate-matter-pm-basics (accessed on 23 January 2020).
6. Xing, Y.-F.; Xu, Y.-H.; Shi, M.-H.; Lian, Y.-X. The impact of PM2.5 on the human respiratory system. *J. Thorac. Dis.* **2016**, *8*, E69–E74.
7. Pongratz, T. Luftgütemessungen in der Steiermark, Jahresbericht. 2017. Bericht Nr. Lu-07-2018. Available online: https://www.umwelt.steiermark.at.
8. Sassmann, S.; Adlassnig, W.; Puschenreiter, M.; Cadenas, E.J.P.; Leyvas, M.; Lichtscheidl, I.K.; Lang, I. Free metal ion availability is a major factor for tolerance and growth in *Physcomitrella patens*. *Environ. Exp. Bot.* **2015**, *110*, 1–10. [CrossRef]
9. Sassmann, S.; Weidinger, M.; Adlassnig, W.; Hofhansl, F.; Bock, B.; Lang, I. Zinc and copper uptake in *Physcomitrella patens*: Limitations and effects on growth and morphology. *Environ. Exp. Bot.* **2015**, *118*, 12–20. [CrossRef]
10. Vanderpoorten, A.; Goffinet, B. *Introduction to Bryophytes*; Cambridge University Press: Cambridge, UK, 2009.
11. Shevock, J.R. The amazing design of a moss leaf. *BryoString* **2015**, *3*, 9–19.
12. Flowers, S. *Mosses: Utah and the West*; University Press: Provo, UT, USA, 1973.
13. Probst, W. *Biologie der Moos- und Farnpflanzen*, 2nd ed.; Quelle & Meyer: Heidelberg, Germany, 1987.
14. Boquete, T.; Lang, I.; Weidinger, M.; Richards, C.L.; Alonso, C. Patterns and mechanisms of heavy metal accumulation and tolerance in two terrestrial moss species with contrasting habitat specialization. *Environ. Exp. Bot.* **2021**, *182*, 104336. [CrossRef]

15. Stanković, J.D.; Sabovljević, A.D.; Sabovljević, M.S. Bryophytes and heavy metals: A review. *Acta Bot. Croat.* **2018**, *77*, 109–118. [CrossRef]

16. Sarkar, P.; Bosneaga, E.; Auer, M. Plant cell walls throughout evolution: Towards a molecular understanding of their design principles. *J. Exp. Bot.* **2009**, *60*, 3615–3635. [CrossRef]

17. Fangel, J.U.; Ulvskov, P.; Knox, J.P.; Mikkelsen, M.D.; Harholt, J.; Popper, Z.A.; Willats, W.G.T. Cell wall evolution and diversity. *Front. Plant Sci.* **2012**, *3*, 152. [CrossRef] [PubMed]

18. Sassmann, S.; Wernitznig, S.; Lichtscheidl, I.K.; Lang, I. Comparing copper resistance in two bryophytes: *Mielichhoferia elongata* Hornsch. versus *Physcomitrella patens* Hedw. *Protoplasma* **2010**, *246*, 119–123. [CrossRef] [PubMed]

19. Wolterbeek, H.T.; Verburg, T.G. Judging Survey Quality: Local Variances. *Environ. Monit. Assess* **2002**, *73*, 7–16. [CrossRef] [PubMed]

20. Tyler, G. Bryophytes and heavy metals: A literature review. *Bot. J. Linn. Soc.* **1990**, *104*, 231–253. [CrossRef]

21. Lang, I.; Wernitznig, S. Sequestration at the cell wall and plasma membrane facilitates zinc tolerance in the moss *Pohlia drummondii*. *Environ. Exp. Bot.* **2011**, *74*, 186–193. [CrossRef]

22. Sabovljević, M.S.; Weidinger, M.; Sabovljević, A.D.; Adlassnig, W.; Lang, I. Is the Binding Pattern of Zinc(II) Equal in Different Bryophyte Species? *Microsc. Microanal.* **2018**, *24*, 69–74. [CrossRef]

23. Sabovljevic, M.S.; Weidinger, M.; Sabovljevic, A.D.; Stankovic, J.; Adlassnig, W.; Lang, I. Metal accumulation in the acrocarp moss Atrichum undulatum undercontrolled conditions. *Environ. Pollut.* **2020**, *256*, 113397. [CrossRef]

24. Zechmeister, H.; Kropic, M.; Moser, D.; Denner, M.; Hohenwallner, D.; Hanus-Illnar, A.; Scharf, S.; Riss, A.; Mirtl, M. *Moos-Monitoring in Österreich. Aufsammlung 2015. Umweltbundesamt Report REP-0595:1-180*; Umweltbundesamt GmbH: Vienna, Austria, 2016.

25. Biebl, R. Über die gegensätzliche Wirkung der Spurenelemente Zink und Bor auf die Blattzellen von *Mnium rostratum. Oesterreichische Bot. Z.* **1947**, *4*, 61–73. [CrossRef]

26. Shaw, A.J. Metal tolerance in bryophytes. In *Heavy Metal Tolerance in Plants: Evolutionary Aspects*; Shaw, A.J., Ed.; CRC Press Inc.: Boca Raton, FL, USA, 1990; pp. 133–152.

27. Url, W. Zur Kenntnis der Todeszonen im konzentrationsgestuften Resistenzversuch. *Physiol. Plant* **1957**, *10*, 318–327. [CrossRef]

28. Url, W.G. Über Schwermetalle-, zumal Kupferresistenz einiger Moose. *Protoplasma* **1956**, *46*, 768–793. [CrossRef]

29. Büscher, P.; Koedam, N.; Van Speybroeck, D. Cation-exchange properties and adaption to soil acidity in bryophytes. *New Phytol.* **1990**, *115*, 177–186. [CrossRef]

30. Soudzilovskaia, N.A.; Cornelissen, J.H.C.; During, H.J.; van Logtestijn, R.S.P.; Lang, S.I.; Aerts, R. Similar cation exchange capacities among bryophyte species refute a presumed mechanism of peatland acidification. *Ecology* **2010**, *91*, 2716–2726. [CrossRef] [PubMed]

31. Fernandez, J.A.; Vazquez, M.D.; Lopez, J.; Carballeira, A. Modelling the extra and intracellular uptake and discharge of heavy metals in Fontinalis antipyretica transplanted along a heavy metal and pH contamination gradient. *Environ. Pollut.* **2006**, *139*, 21–31. [CrossRef] [PubMed]

32. Vazquez, M.D.; Lopez, J.; Carballeira, A. Modification of the sequential elution technique for the extraction of heavy metals from bryophytes. *Sci. Total Environ.* **1999**, *241*, 53–62. [CrossRef]

33. Vidali, L.; Bezanilla, M. Physcomitrella patens: A model for tip cell growth and differentiation. *Curr. Opin. Plant Biol.* **2012**, *15*, 625–631. [CrossRef]

34. Adlassnig, W.; Weiss, Y.S.; Sassmann, S.; Steinhauser, G.; Hofhansl, F.; Baumann, N.; Lichtscheidl, I.K.; Lang, I. The copper spoil heap Knappenberg, Austria, as a model for metal habitats—Vegetation, substrate and contamination. *Sci. Total Environ.* **2016**, *563–564*, 1037–1049. [CrossRef]

35. Sirka, P.; Kubesova, S.; Miskova, K. Bryophytes of spoil heaps rich in toxic metals in Central Slovakia. *Thaiszia-J. Bot.* **2018**, *28*, 059–077.

36. Hnilica, E.; Kohout, R. Mathe-Lexicon.at. 2018. Available online: https://:www.mathe-lexikon.at (accessed on 10 June 2019).

 *plants*

Article

Flax and Sorghum: Multi-Element Contents and Nutritional Values within 210 Varieties and Potential Selection for Future Climates to Sustain Food Security

Gokhan Hacisalihoglu [1,*] and Paul R. Armstrong [2]

1 Department of Biological Sciences, Florida A&M University, Tallahassee, FL 32307, USA
2 USDA-ARS Center for Grain and Animal Health Research, Manhattan, KS 66502, USA; paul.armstrong@usda.gov
* Correspondence: gokhan.h@famu.edu

Abstract: The Dietary Guidelines for Americans recommends giving priority to nutrient-dense foods while decreasing energy-dense foods. Although both flax (*Linum usitatissimum*) and sorghum (*Sorghum bicolor*) are rich in various essential minerals, their ionomes have yet to be investigated. Furthermore, previous studies have shown that elevated CO_2 levels could reduce key nutrients in crops. In this study, we analyzed 102 flax and 108 sorghum varieties to investigate their ionomic variations (N, P, K, Ca, Mg, S, B, Zn, Mn, Fe, Cu, and Mo), elemental level interactions, and nutritional value. The results showed substantial genetic variations and elemental correlations in flax and sorghum. While a serving size of 28 g of flax delivers 37% daily value (DV) of Cu, 31% of Mn, 28% of Mg, and 19% of Zn, sorghum delivers 24% of Mn, 16% of Cu, 11% of Mg, and 10% of Zn of the recommended daily value (DV). We identified a set of promising flax and sorghum varieties with superior seed mineral composition that could complement breeding programs for improving the nutritional quality of flax and sorghum. Overall, we demonstrate additional minerals data and their corresponding health and food security benefits within flax and sorghum that could be considered by consumers and breeding programs to facilitate improving seed nutritional content and to help mitigate human malnutrition as well as the effects of rising CO_2 stress.

Keywords: food security; nutrient dense; superfood; multi minerals; health benefits; zinc; iron; gluten free; percent daily value; elevated CO_2

Citation: Hacisalihoglu, G.; Armstrong, P.R. Flax and Sorghum: Multi-Element Contents and Nutritional Values within 210 Varieties and Potential Selection for Future Climates to Sustain Food Security. *Plants* **2022**, *11*, 451. https://doi.org/10.3390/plants11030451

Academic Editor: Dimitris L. Bouranis

Received: 4 January 2022
Accepted: 2 February 2022
Published: 6 February 2022

Publisher's Note: MDPI stays neutral with regard to jurisdictional claims in published maps and institutional affiliations.

1. Introduction

There are several diet-related chronic diseases (e.g., diabetics, heart disease, obesity, and cancer), and therefore, specifically plant-based nutrition is expected to be increasingly important worldwide for prevention and control of these diseases [1]. Therefore, one of the utmost research areas of plant biology has been plant nutritional values for the human diet.

Due to the growing popularity and demand for plant nutrition, there is an increasing need for research on the improvement of yield and quality of crop plants. Flax (*Linum usitatissimum*) is an annual crop and an important source of alpha linolenic acid (ALA) omega-3 fats as well as protein with all nine essential amino acids, except lysine (Figure 1) [2]. Flax is grown in cool climates including Canada, China, Russia, and the United States (North Dakota and Minnesota) [1]. Sorghum (*Sorghum bicolor*) is one of the Poaceae family cereal crops with important antidiabetic, anticholesterol, and low glycemic index (GI) features and is grown in the United States, India, Mexico, and China (Figure 1) [3]. Flax and sorghum seeds both contain a diverse set of mineral nutrients together with protein, oil, and carbohydrate (Figure 1). Ionomic profiling of the accumulated elements in living organisms has been successfully applied to study leaves, roots, whole plants, and seeds [4]. Furthermore, mineral and trace elements have been successfully determined in many other crop species including common beans, peas, soybeans, and maize [5–13].

Atmospheric carbon dioxide (CO_2) levels have increased from 278 µmol/mol to 417 µmol/mol (present, 2021) and are expected to reach 550 µmol/mol and 800+ µmol/mol by 2050 and 2100, respectively, with the current average increase rate of 2.5 µmol/mol [14]. A continuous rise in the levels of atmospheric CO_2 is expected to potentially affect plant life negatively. A study with soybean found that elevated CO_2 levels influenced seed nutritional levels by decreasing most mineral content including the concentrations of K, Mg, Fe, and B [15]. Furthermore, a reduction in seed nitrogen (N) has been reported, and therefore, protein levels under elevated CO_2 levels [16].

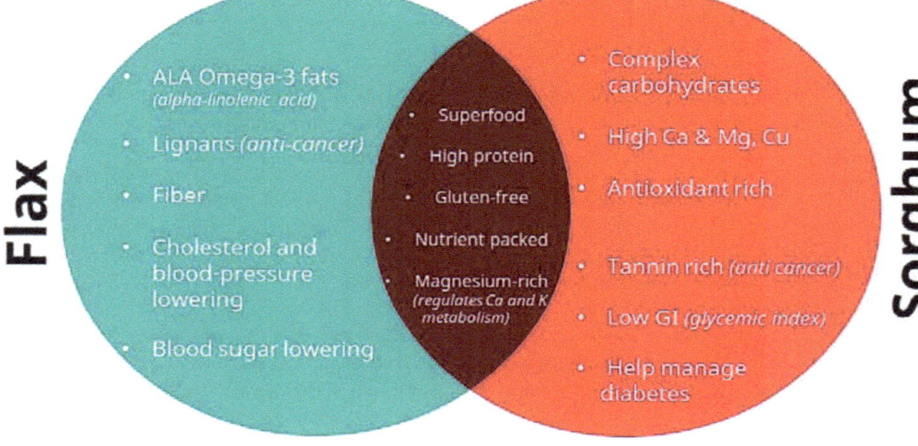

Figure 1. Potential health and food security benefits of flax and sorghum seeds.

One way to mitigate future climate effects and maintain food sustainability is to screen and identify top varieties that can naturally offer superior mineral concentrations. The importance of seeds in maintaining human health and diet could be determined by their nutritional content through a recommended percent daily value (% DV, how much it contributes to a daily 2000 calorie diet) [6]. Therefore, measurement of mineral element contents can provide valuable information for consumers and crop breeders.

Among several studies that have been conducted on seed ionome, most of them have been carried out in major staple crop plants. The specific aims of this study were: (1) to determine the variability of macronutrient and trace-element concentrations among 102 flax and 108 sorghum varieties; (2) to analyze the elemental interrelationships and % DV of nutrients; (3) to identify superior varieties that could be used to improve the nutritional value potential in flax and sorghum.

2. Materials and Methods

2.1. Flax and Sorghum Seeds

A total collection of diverse global varieties including 102 flax (*Linum usitatissimum*) and 108 sorghum (*Sorghum bicolor*) varieties were selected based on their maximum geographic diversity obtained from the USDA National Germplasm Center. All seeds used in this study were field grown with standard agronomical practices (Tables 1 and 2).

2.2. Multi-Elemental Seed Analysis

Elemental concentrations of macro- and micronutrients were quantified by Waters Agricultural Labs Inc. (Camilla, GA, USA). Minerals were analyzed by open vessel wet digestion using an inductively coupled argon plasma spectrometer (ICAP, DigiBlock 3000

ICP-MS). Briefly, seeds were dried at 80 °C in an oven overnight, and then ground in a Wiley mill. A 0.5 g dried sample was mixed with 5 mL concentrated nitric acid and incubated at 95 °C for 90 min. Then, 4 mL of 30% H2O2 was added to each tube and incubated at 95 °C for 20 min. Samples were cooled for 2 min, brought to 50 mL with distilled H2O, and mixed. Samples were transferred to ICP tubes for analysis, in accordance with manufacturer's specifications. The ICP-MS was calibrated using distilled H2O as a blank and two plant standards.

Total nitrogen (N) determination was performed by U.S. Department of Agriculture (Manhattan, KS, USA) using combustion gas analysis (LECO FP-628, St. Joseph, MI, USA), following the manufacturer's instructions as previously described in Hacisalihoglu et al. [13].

2.3. Estimating Nutritional Value (% DV)

The nutritional values of flax and sorghum seeds were estimated using a 28 g dry weight serving portion. The U.S. recommended daily value indices (2000 calorie diet for adults) were as follows: Mn (2 mg), P (1000 mg), Cu (2 mg), Fe (18 mg), Mg (400 mg), Zn (15 mg), and Ca (1000 mg) [17,18]. Percent daily values (% DV) were estimated from a 28 g of seeds serving (dry weight basis) by using the following formula:

$$\% \text{ DV} = (\text{amount of nutrient mg} / \text{recommended DV mg}) * 100 \tag{1}$$

%DV; Percent daily values; mg; milligrams.

2.4. Data Analysis

All lab analyses were completed with three replications. Elemental statistical correlation analysis was performed using SigmaPlot (SPSS Inc., Chicago, IL, USA), as described previously [9]. Descriptive statistics for each macro- and micronutrients and varieties were determined using the average of the ICP-MS results from the three biological replications. Graphs were made with SigmaPlot software (SPSS Inc., Chicago, IL, USA).

3. Results

3.1. Variations in Flax and Sorghum Multi-Element Contents

The 102 flax varieties showed a wide variation in seed multi-element contents (Tables 1 and 2 and Figure 2). There was a 5.7-fold range of copper (Cu) content, 4.5-fold range of iron (Fe) content, 4.2-fold range of boron (B) content, 3.3-fold range of zinc (Zn) content, 2.6-fold range of manganese (Mn) content, 2.1-fold range of calcium (Ca) content, and 2-fold range of potassium (K) and molybdenum (Mo) contents (Table 2).

The 108 flax varieties showed a wide variation in seed multi-element contents (Table 2 and Figure 2). There was a 46-fold range of Fe content, 12-fold range of Cu content, 6.6-fold range of B content, 6.3-fold range of Mn content, 9.7-fold range of Zn content, 4.5-fold range of Mo content, 5-fold range of Ca, 2.7-fold range of P, and 2.6-fold range of potassium (K) (Table 2).

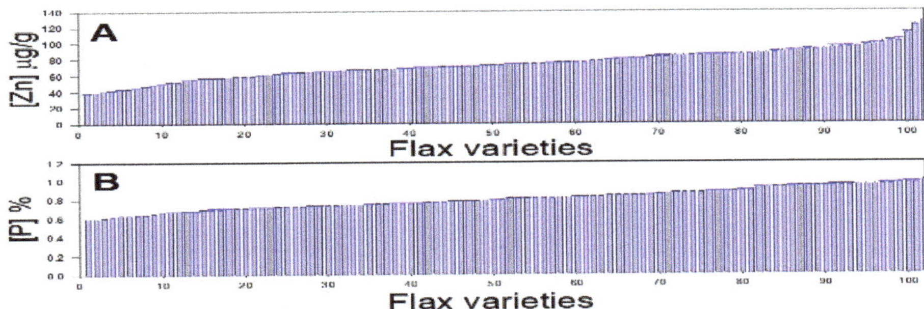

Figure 2. *Cont.*

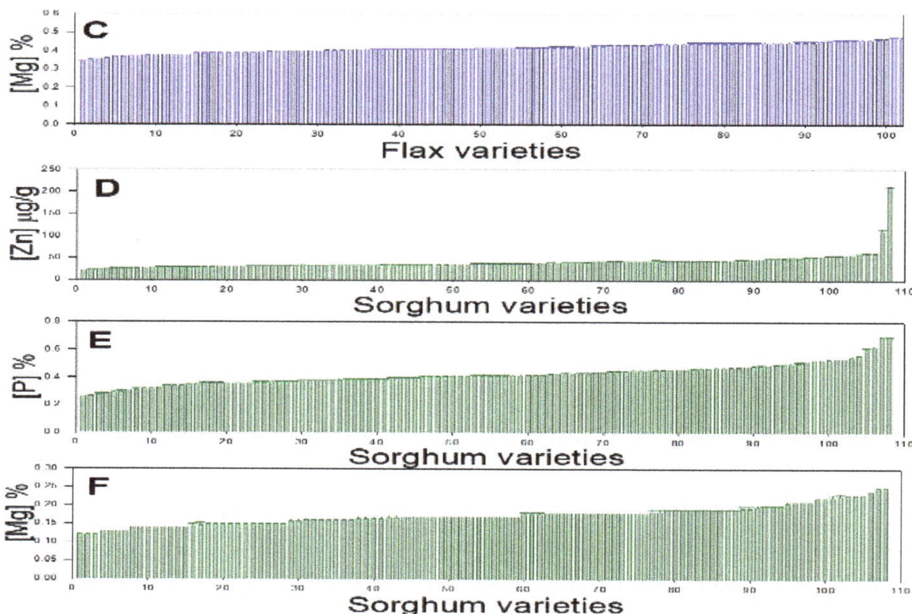

Figure 2. Average concentration of three replicates: (**A**) Flax Zn; (**B**) flax P; (**C**) flax Mg; (**D**) sorghum Zn; (**E**) sorghum P; and (**F**) sorghum Mg. (see Table 1 for all others).

Table 1. Mean elemental concentrations of 102 flax varieties as % (N, P, K, Mg, Ca, and S) and µg/g (Zn, Mn, Fe, Cu, and Mo) obtained from ICP-MS.

Flax Variety	N	P	K	Mg	Ca	S	B	Zn	Mn	Fe	Cu	Mo
Ames8040	3.36	0.82	0.75	0.46	0.23	0.27	17.7	93.5	23.8	87.4	10.4	3.06
Ariane	4.37	0.93	0.86	0.44	0.20	0.30	18.7	93.4	21.7	71.5	15.5	2.64
Beladiy6903	3.82	0.81	0.78	0.44	0.21	0.29	18.5	70.4	21.1	67.1	9.53	2.91
Benvenotolabrador	3.91	0.76	0.73	0.45	0.22	0.27	18.0	72.1	28.9	75.0	8.38	2.85
Charurraolajlen19	3.66	0.77	0.72	0.43	0.22	0.26	21.0	78.8	23.7	69.4	11.9	2.79
Charurraolajlen29	3.87	0.71	0.77	0.41	0.20	0.25	15.7	70.7	20.9	67.5	7.54	3.11
CIli1319	3.41	0.81	0.81	0.45	0.19	0.25	15.8	67.7	19.9	62.6	8.97	2.85
CIli1339	4.00	0.78	0.81	0.42	0.20	0.28	14.9	49.3	21.8	65.6	12.0	2.89
CIli1340	3.52	0.83	0.74	0.44	0.24	0.29	20.7	71.2	26.4	71.5	14.9	2.72
CIli1341	4.14	0.71	0.68	0.41	0.22	0.29	20.6	64.3	20.6	65.8	9.05	2.79
CIli1350	3.73	0.73	0.71	0.41	0.22	0.27	15.5	38.0	20.1	67.3	9.19	2.97
CIli1351	4.16	0.75	0.72	0.43	0.24	0.29	18.2	64.1	21.7	71.1	8.13	2.92
CIli1354	4.11	0.73	0.71	0.41	0.21	0.27	15.7	41.1	21.0	65.2	9.30	3.05
CIli1369	4.08	0.74	0.68	0.42	0.22	0.29	20.5	73.5	19.9	62.5	14.2	2.88
CIli1370	3.47	0.86	1.08	0.38	0.30	0.27	12.3	75.6	24.3	81.4	12.2	2.60
CIli1373	3.37	0.94	1.18	0.42	0.32	0.26	10.2	56.9	23.6	83.1	11.4	3.37
CIli1374	3.58	0.87	0.87	0.47	0.24	0.27	17.4	60.4	38.8	64.3	12.8	3.81
CIli1395	3.61	0.82	0.87	0.43	0.22	0.28	16.3	43.8	19.9	64.4	8.65	3.24
CIli1397	3.59	0.95	1.07	0.40	0.34	0.26	9.61	52.1	28.2	74.2	10.5	3.25
CIli1404	3.73	0.84	0.82	0.44	0.24	0.28	20.8	63.2	23.9	54.5	15.1	2.82
CIli1418	4.34	0.97	0.88	0.39	0.38	0.29	11.7	95.2	30.5	95.0	16.2	3.05
CIli1426	4.20	0.78	0.81	0.41	0.21	0.28	16.8	43.6	21.7	59.9	8.85	3.23
CIli1427	3.91	0.83	0.89	0.42	0.27	0.29	11.2	63.1	18.8	71.7	8.38	3.84
CIli1429	3.88	0.93	1.10	0.40	0.32	0.27	11.0	83.1	26.3	84.1	14.38	2.27
CIli1431	4.20	0.81	0.82	0.42	0.23	0.29	15.6	44.8	21.8	65.5	9.49	2.75
CIli1436	4.36	0.74	0.76	0.39	0.24	0.30	25.3	81.5	28.5	74.0	15.5	2.96

Table 1. *Cont.*

Flax Variety	N	P	K	Mg	Ca	S	B	Zn	Mn	Fe	Cu	Mo
CIIi1449	3.69	0.87	0.86	0.44	0.25	0.26	15.5	54.9	21.8	43.7	12.3	2.69
CIIi1452	4.55	0.70	0.73	0.41	0.22	0.29	20.2	63.7	22.7	73.4	9.96	3.22
CIIi1458	3.94	0.81	0.78	0.43	0.22	0.26	17.6	52.2	22.0	49.5	12.5	3.04
CIIi1476	4.11	0.89	0.79	0.46	0.18	0.29	16.5	86.4	23.9	81.8	9.75	2.35
CIIi1492	3.86	0.64	0.72	0.40	0.23	0.27	19.2	39.0	17.0	48.3	10.2	2.92
CIIi1669	4.26	0.67	0.81	0.40	0.22	0.30	23.0	57.1	28.4	70.6	6.79	2.86
CIIi1751	3.63	0.71	0.80	0.38	0.22	0.28	22.7	84.9	28.9	64.3	15.6	3.47
CIIi1763	4.32	0.94	0.93	0.42	0.25	0.30	16.5	87.1	27.9	68.7	9.51	2.87
CIIi1821	3.26	0.62	0.60	0.36	0.22	0.26	40.5	78.0	30.3	54.1	24.2	3.17
CIIi1836	3.75	0.74	0.83	0.41	0.22	0.28	18.9	47.2	23.8	51.7	10.7	2.98
CIIi1931	3.40	0.67	0.68	0.39	0.19	0.26	34.0	79.7	25.8	67.8	17.7	3.21
CIIi1938	3.50	0.60	0.67	0.38	0.22	0.25	32.0	74.3	30.2	63.0	14.7	3.11
CIIi1943	3.69	0.70	0.70	0.40	0.23	0.26	31.2	83.2	30.1	62.7	15.5	3.29
CIIi1955	3.30	0.71	0.73	0.39	0.24	0.26	31.0	86.7	31.1	62.2	16.4	3.34
CIIi1980	3.27	0.76	0.85	0.43	0.26	0.25	27.5	68.4	24.8	61.8	12.3	2.87
CIIi1983	3.87	0.89	1.12	0.41	0.28	0.26	21.4	91.7	21.0	76.3	5.17	3.47
CIIi1989	3.65	0.72	0.81	0.43	0.22	0.26	22.1	69.6	24.5	63.9	10.0	3.41
CIIi1990	3.32	0.78	0.75	0.46	0.20	0.26	23.7	76.9	23.7	71.9	9.41	2.95
CIIi2010	3.79	0.93	0.83	0.47	0.24	0.26	14.4	73.7	27.6	65.8	12.7	3.18
CIIi2033	3.88	0.82	0.76	0.44	0.24	0.27	20.4	67.1	34.5	196.1	14.1	3.01
CIIi2070	3.98	0.81	0.75	0.40	0.22	0.27	14.6	71.0	19.1	70.7	12.4	3.11
CIIi2424	4.21	0.82	0.76	0.45	0.24	0.28	18.4	75.5	22.3	72.1	8.70	2.78
CIIi2443	4.27	0.74	0.79	0.41	0.23	0.28	18.6	58.6	21.9	61.8	8.52	3.00
CIIi2444	3.94	0.81	0.83	0.42	0.29	0.30	22.1	82.3	26.3	84.4	15.4	2.72
CIIi2446	4.18	0.87	0.77	0.46	0.25	0.26	14.1	55.9	24.9	69.0	8.60	2.76
CIIi2534	4.41	0.74	0.71	0.40	0.23	0.28	15.5	67.5	19.6	77.8	11.0	2.50
CIIi3242	4.22	0.93	0.84	0.42	0.24	0.30	17.9	93.8	31.4	89.1	11.6	2.70
CIIi3246	4.48	0.86	0.90	0.41	0.28	0.27	15.5	73.2	31.1	57.7	7.74	2.83
CIIi3303	3.96	0.77	0.78	0.42	0.25	0.28	21.2	79.2	32.3	76.0	8.50	2.74
CIIi3310	4.33	0.73	0.70	0.42	0.20	0.28	17.4	49.7	21.9	56.7	12.5	3.01
CIIi3312	3.96	0.68	0.69	0.39	0.23	0.27	21.8	70.5	27.4	71.2	8.90	2.39
CIIi3314	4.49	0.75	0.80	0.42	0.23	0.28	20.2	69.8	29.3	70.8	10.5	2.57
CIIi3317	4.11	0.61	0.70	0.38	0.27	0.27	19.4	73.9	19.7	61.3	11.2	3.43
CIIi3318	4.36	0.69	0.80	0.37	0.22	0.27	27.7	79.5	21.8	75.3	11.0	3.39
CIIi452	4.41	0.88	0.87	0.44	0.24	0.29	13.7	100.4	34.0	92.9	7.53	2.41
CIIi641	3.87	0.97	0.88	0.46	0.22	0.28	13.7	69.6	27.3	63.1	12.9	3.30
CIIi642	4.11	0.79	0.77	0.41	0.25	0.27	13.0	61.2	23.0	53.3	12.2	3.09
CIIi643	4.01	0.94	1.01	0.42	0.25	0.28	13.6	127.1	15.5	80.6	6.53	3.70
Coeruleum	4.02	0.92	0.95	0.40	0.30	0.28	14.1	91.2	28.1	95.3	15.4	2.54
CrownCanada	3.69	0.97	0.96	0.40	0.33	0.27	13.2	110.4	25.0	94.8	16.8	3.03
Danese129a	4.08	0.71	0.73	0.39	0.21	0.28	22.9	70.1	25.6	69.6	9.65	2.75
Dufferin	4.11	0.74	0.70	0.42	0.23	0.29	19.1	75.0	26.9	79.7	13.8	2.67
Flanders	3.72	0.72	0.67	0.41	0.23	0.27	19.0	41.8	19.4	47.4	15.1	2.91
FP966	3.61	0.63	0.72	0.39	0.25	0.25	29.3	66.2	24.5	69.7	13.6	3.52
Gercello	4.77	0.95	0.91	0.44	0.26	0.28	16.8	62.1	28.1	69.2	14.8	3.10
Giza139a	4.11	0.64	0.65	0.38	0.26	0.28	19.4	58.8	22.7	66.6	6.17	2.72
Gujrat2	2.90	0.63	0.66	0.37	0.24	0.24	29.7	77.3	37.3	57.9	15.8	3.16
H39seln	3.97	0.70	0.68	0.39	0.25	0.28	18.9	58.7	24.8	61.9	6.13	2.44
Hazeldean	4.44	0.95	0.81	0.44	0.21	0.28	14.8	94.7	23.2	82.7	13.1	2.93
Jalaun	3.90	0.76	1.00	0.37	0.29	0.26	15.9	67.6	21.5	59.6	5.72	2.39
Jalomita	4.27	0.87	0.75	0.44	0.23	0.28	16.3	84.5	20.9	73.9	14.2	2.86
Katan92	4.22	0.77	0.67	0.43	0.21	0.27	17.8	68.0	20.1	57.4	11.0	2.96
Katan93	3.63	0.81	0.71	0.41	0.21	0.27	17.5	83.6	24.7	67.1	12.4	3.08
Kenyaci709	3.90	0.73	0.62	0.34	0.22	0.27	21.1	90.3	41.0	66.5	14.0	2.92
Mcduffp900	3.58	0.65	0.66	0.38	0.28	0.27	17.4	67.5	23.2	57.7	12.0	3.39

Table 1. *Cont.*

Flax Variety	N	P	K	Mg	Ca	S	B	Zn	Mn	Fe	Cu	Mo
Moose	3.97	0.93	0.87	0.46	0.24	0.28	15.1	71.6	26.3	60.5	11.9	3.26
NP80	3.74	0.60	0.59	0.35	0.20	0.25	25.0	85.6	30.8	53.5	16.9	2.97
NO1040	3.73	0.81	0.74	0.47	0.23	0.28	20.1	75.0	32.4	65.7	14.3	3.05
NO11	3.58	0.95	0.88	0.39	0.37	0.27	11.0	89.3	33.8	70.0	16.2	2.86
NO129	4.98	0.78	0.74	0.41	0.24	0.30	19.7	57.1	21.4	65.9	7.79	3.12
Norland	4.05	0.95	0.87	0.44	0.24	0.27	16.2	101.3	20.4	81.5	13.2	2.60
NP121	4.32	0.91	1.11	0.40	0.25	0.27	23.9	121.1	18.9	74.8	4.21	3.58
NP124	3.99	0.79	0.82	0.44	0.23	0.26	20.0	64.7	23.0	56.5	10.2	2.53
Omega	3.86	0.99	0.95	0.43	0.37	0.27	13.0	99.1	28.6	80.3	19.4	4.58
Pasrur2	2.91	0.67	0.63	0.37	0.24	0.24	26.0	84.5	34.5	65.9	16.0	3.05
Rembrandt	3.63	0.83	0.74	0.44	0.22	0.28	19.1	90.9	23.1	78.8	12.7	2.38
Saidabad	4.19	0.75	0.70	0.41	0.22	0.29	19.8	96.8	32.3	88.4	18.9	2.99
Somme	3.84	0.86	0.81	0.44	0.23	0.26	17.8	84.5	23.6	68.9	17.4	3.08
SzeepiOlajlen	4.11	0.92	0.87	0.36	0.37	0.27	11.4	73.3	25.2	70.3	13.9	2.85
Tomagaon	3.53	0.76	0.80	0.42	0.24	0.25	11.0	66.8	20.7	59.3	7.90	2.39
Uruguay	3.80	0.84	0.72	0.43	0.21	0.27	17.3	84.2	25.5	64.2	13.6	2.80
Verin	3.69	0.84	0.82	0.44	0.23	0.27	18.0	87.9	24.2	81.0	12.1	2.83
Viking	4.17	0.85	0.69	0.44	0.27	0.27	15.8	57.7	36.6	49.7	8.78	3.12
Vimy	4.00	0.68	0.97	0.41	0.27	0.28	23.0	64.9	35.8	76.3	9.59	2.93
W62611FKA14	3.74	0.92	0.79	0.45	0.20	0.27	15.8	83.6	22.0	74.3	11.4	2.98
WickingHeggenen	4.15	0.87	0.71	0.45	0.25	0.27	19.0	82.5	34.4	66.3	11.5	2.96

Table 2. Mean elemental concentrations of 108 sorghum varieties as % (N, P, K, Mg, Ca, and S) and μg/g (Zn, Mn, Fe, Cu, and Mo) obtained from ICP-MS.

Sorghum Variety	N	P	K	Mg	Ca	S	B	Zn	Mn	Fe	Cu	Mo
52	2.99	0.70	0.58	0.25	0.05	0.43	10.3	43.3	50.9	91.5	11.3	2.65
282	2.19	0.48	0.46	0.19	0.02	0.17	5.07	47.3	25.5	267	8.41	2.32
434	2.28	0.70	0.54	0.25	0.05	0.39	9.97	60.9	50.2	235	16.6	3.52
1398	2.78	0.61	0.74	0.23	0.05	0.38	13.0	59.8	61.1	643	13.4	2.62
1491	1.94	0.47	0.47	0.20	0.01	0.15	2.80	45.7	26.1	269	3.37	2.02
1728	1.80	0.46	0.60	0.16	0.02	0.14	7.27	211	34.6	49.4		2.09
3967	1.51	0.41	0.47	0.18	0.01	0.14	3.05	43.3	18.1	35.6	3.83	2.10
4058	1.66	0.40	0.44	0.16	0.02	0.15	2.53	44.5	19.2	33.6	3.40	1.15
4080	2.02	0.49	0.53	0.19	0.02	0.15	3.16	36.7	20.8	32.6	3.15	1.46
4116	1.80	0.36	0.37	0.14	0.02	0.15	2.94	37.5	16.6	30.6	3.23	1.79
93447	2.02	0.41	0.53	0.17	0.04	0.12	3.54	55.7	21.9	44.2	27.3	2.06
54K94	2.21	0.63	0.72	0.21	0.02	0.16	4.69	43.4	35.1	55.3	6.37	1.81
88-07095	1.51	0.34	0.47	0.12	0.01	0.11	9.37	21.8	17.4	21.8	3.41	1.47
88-07105	1.60	0.26	0.37	0.12	0.02	0.14	2.40	29.7	15.6	35.8	4.92	1.34
88-07108	1.78	0.32	0.40	0.16	0.02	0.14	2.91	112	22.6	45.3	6.37	1.22
88-07197	2.01	0.41	0.41	0.19	0.02	0.16	2.89	31.4	17.5	44.7	6.55	1.76
88-07207	2.06	0.53	0.54	0.18	0.02	0.14	2.73	37.2	26.7	34.0	5.13	1.66
88-07225	1.99	0.54	0.63	0.17	0.02	0.16	2.64	48.1	27.8	45.2	6.82	1.99
A7774	2.78	0.56	0.57	0.21	0.02	0.18	3.31	63.0	18.7	53.5	4.58	2.43
A84	1.45	0.42	0.53	0.18	0.02	0.13	4.12	37.6	15.9	37.5	3.18	2.92
A96	2.12	0.46	0.48	0.19	0.03	0.14	3.64	35.4	18.8	42.1	4.11	2.60
ABTx631	1.48	0.38	0.41	0.17	0.02	0.12	3.16	30.0	15.2	29.6	4.22	2.96
AcchoKaruha	2.22	0.47	0.42	0.18	0.02	0.12	3.04	44.8	25.2	55.3	4.61	3.19
AS4055	1.49	0.37	0.49	0.15	0.01	0.12	3.11	27.3	11.7	24.9	3.31	2.17
AS4136	1.96	0.42	0.49	0.16	0.01	0.15	2.61	44.6	20.0	34.6	4.42	3.72
AS5826	2.06	0.40	0.43	0.18	0.02	0.14	3.22	42.7	15.9	20.6	5.89	3.35
Barking119	2.01	0.44	0.53	0.18	0.01	0.13	3.23	44.6	19.0	36.6	5.22	2.38

Table 2. *Cont.*

Sorghum Variety	N	P	K	Mg	Ca	S	B	Zn	Mn	Fe	Cu	Mo
BE25	2.19	0.48	0.48	0.18	0.02	0.14	3.55	44.2	21.8	45.0	7.27	3.09
Bok11	1.68	0.43	0.44	0.19	0.02	0.15	3.60	36.0	17.2	38.0	4.00	3.91
BrownKaoliang	2.17	0.42	0.47	0.17	0.02	0.15	5.06	43.6	16.3	51.7	5.47	2.40
BTx623	1.81	0.37	0.49	0.15	0.02	0.10	3.78	23.3	13.0	39.0	5.15	1.48
ChananSingoo	1.99	0.43	0.43	0.18	0.03	0.14	3.76	51.0	20.5	33.8	6.16	3.14
ChineseAmber	2.63	0.53	0.45	0.23	0.02	0.17	5.76	57.1	28.6	46.1	7.17	2.94
Collier	1.73	0.41	0.39	0.17	0.01	0.17	4.04	24.2	21.1	32.3	3.97	2.23
Cowley	1.83	0.31	0.31	0.15	0.03	0.15	5.42	31.3	14.5	29.6	4.61	4.10
DaShanDong	1.69	0.47	0.53	0.18	0.02	0.13	3.80	39.9	17.9	35.0	4.24	2.52
Dokhnah	2.17	0.40	0.36	0.17	0.01	0.15	6.74	44.6	22.2	38.0	4.66	1.82
Elmota	1.92	0.46	0.43	0.20	0.02	0.14	5.13	43.1	18.7	38.8	6.29	2.46
ERJieZi	1.58	0.34	0.43	0.13	0.01	0.12	3.13	33.5	18.5	29.3	3.93	1.72
FAO54919	2.11	0.46	0.40	0.19	0.01	0.16	2.96	43.4	18.8	33.8	4.18	2.00
Grif534	1.63	0.26	0.34	0.13	0.02	0.13	2.08	35.7	16.6	40.0	5.67	1.48
Grif539	1.87	0.30	0.32	0.14	0.03	0.14	2.52	37.4	21.5	42.0	5.75	1.55
Grif553	1.89	0.32	0.33	0.14	0.01	0.15	2.08	28.4	19.6	35.7	5.83	1.48
Grif574	1.95	0.48	0.41	0.20	0.03	0.16	3.95	55.2	26.0	48.4	6.99	1.53
Grif604	1.74	0.29	0.28	0.14	0.02	0.14	2.28	31.2	9.7	49.9	5.02	1.61
Grif610	1.63	0.28	0.34	0.14	0.03	0.14	2.59	48.9	15.2	57.0	7.19	1.49
Grif7260	1.57	0.42	0.58	0.15	0.02	0.13	4.94	30.2	15.4	33.0	3.80	2.01
Grif7263	1.82	0.39	0.44	0.17	0.01	0.13	2.66	33.0	13.7	38.6	5.07	2.27
IS1019	2.25	0.31	0.41	0.14	0.02	0.14	5.73	34.0	17.7	28.7	4.58	1.85
IS10931	2.31	0.46	0.44	0.18	0.02	0.17	3.02	33.2	19.1	43.2	4.32	2.91
IS1213C	1.98	0.36	0.36	0.17	0.02	0.13	4.66	29.1	20.0	35.7	4.13	2.59
IS12684C	1.88	0.37	0.54	0.16	0.03	0.14	4.98	37.5	18.7	40.6	6.01	2.35
IS12845	2.06	0.44	0.44	0.19	0.04	0.13	5.76	39.2	29.2	38.6	6.52	1.43
IS13232	1.76	0.49	0.56	0.17	0.02	0.14	3.68	30.9	24.9	42.5	4.07	2.06
IS13236	2.20	0.50	0.62	0.17	0.03	0.15	3.63	32.6	32.0	41.9	6.32	1.92
IS14098	1.67	0.36	0.43	0.15	0.03	0.13	3.48	35.0	13.3	32.8	3.80	2.02
IS24424	2.24	0.52	0.49	0.23	0.03	0.14	4.94	47.5	25.4	47.7	4.39	1.78
IS24449	1.77	0.40	0.42	0.17	0.02	0.12	3.40	33.0	13.0	32.8	3.15	2.30
IS24451	1.97	0.41	0.34	0.19	0.01	0.14	2.68	37.4	16.3	32.8	4.20	2.27
IS27569	1.69	0.42	0.47	0.16	0.01	0.12	2.95	36.7	15.5	48.8	4.14	2.36
IS27601	1.43	0.37	0.52	0.15	0.02	0.14	3.67	38.7	19.2	14.0	4.24	1.88
IS28214	1.76	0.43	0.54	0.16	0.02	0.13	2.92	49.6	18.8	35.4	4.02	1.98
IS2871C	2.23	0.53	0.50	0.22	0.02	0.14	5.50	52.7	17.1	42.0	5.36	2.30
IS2874	1.87	0.45	0.46	0.19	0.03	0.14	3.04	48.0	13.9	45.0	4.16	2.26
IS3098	1.43	0.38	0.47	0.15	0.02	0.12	3.37	29.6	11.5	28.5	2.32	1.83
IS5168C	2.16	0.45	0.41	0.21	0.03	0.14	4.19	51.4	22.9	38.3	5.54	2.67
IS6541	2.13	0.41	0.32	0.18	0.02	0.14	3.25	35.6	14.6	37.1	3.67	1.59
IS6733C	1.79	0.37	0.47	0.15	0.01	0.13	3.50	40.4	16.2	34.5	4.03	2.58
IS8120C	1.91	0.34	0.38	0.14	0.01	0.12	2.72	22.2	12.7	19.3	3.43	2.42
JolaNandyal	1.72	0.36	0.41	0.13	0.02	0.12	3.63	29.1	13.4	25.2	2.71	1.78
JowarRedJan	2.06	0.38	0.39	0.16	0.02	0.12	6.00	38.1	22.3	27.2	3.35	2.24
KA12Janjari	1.84	0.41	0.69	0.18	0.02	0.14	5.03	32.6	15.5	28.0	5.38	2.73
Kabutuwa	1.96	0.51	0.56	0.21	0.03	0.14	5.24	57.6	25.2	24.4	7.93	5.12
Kaoliang	2.05	0.41	0.46	0.16	0.02	0.13	3.70	34.2	12.8	27.4	3.90	3.10
Kaoliangwx	1.48	0.38	0.43	0.15	0.02	0.13	3.16	35.1	13.9	34.7	4.11	3.26
KharuthWara	2.18	0.36	0.33	0.18	0.03	0.13	4.31	41.3	15.9	42.9	3.10	2.99
Kulum	1.61	0.38	0.45	0.14	0.02	0.12	3.27	27.1	12.9	35.1	3.61	1.75
Kuyuma	1.94	0.35	0.43	0.13	0.01	0.11	3.41	28.2	12.4	23.2	2.92	2.35
Leoti	1.70	0.38	0.44	0.18	0.01	0.13	4.52	32.7	25.2	22.1	3.33	2.06
LianTouSan	1.97	0.48	0.55	0.18	0.01	0.14	2.47	42.8	17.6	35.0	5.31	2.15
Lula	1.61	0.38	0.39	0.17	0.02	0.13	3.60	38.8	18.5	30.2	4.69	3.03

Table 2. *Cont.*

Sorghum Variety	N	P	K	Mg	Ca	S	B	Zn	Mn	Fe	Cu	Mo
M35-1	1.68	0.32	0.36	0.16	0.01	0.12	3.42	26.7	10.5	24.9	2.19	2.24
ManfrediMinu	1.99	0.43	0.45	0.18	0.02	0.15	4.62	30.0	42.3	36.7	5.43	1.38
Marupantse	1.68	0.41	0.46	0.16	0.02	0.14	3.09	33.6	20.5	35.4	4.98	2.58
Mashica	1.97	0.43	0.52	0.17	0.02	0.13	3.75	34.1	14.9	26.5	2.51	2.79
MN1592	1.67	0.36	0.41	0.17	0.02	0.14	3.46	33.8	18.5	33.5	3.54	3.91
MN4315	2.41	0.54	0.44	0.23	0.02	0.16	7.76	44.8	22.5	42.3	6.86	2.08
MN586	1.73	0.41	0.46	0.17	0.01	0.17	5.47	38.2	16.6	34.5	6.12	3.03
MN707	1.93	0.41	0.48	0.15	0.02	0.14	3.28	26.5	20.3	21.1	4.27	3.82
MsumbjiSB117	2.07	0.47	0.41	0.18	0.02	0.14	2.62	43.5	32.7	39.0	4.56	1.62
N290b	1.71	0.36	0.47	0.16	0.02	0.12	2.63	28.5	14.0	29.7	4.06	2.86
OrangeNo1	2.16	0.28	0.31	0.12	0.02	0.14	2.53	28.4	20.7	22.6	4.27	3.92
P9517	1.83	0.39	0.42	0.17	0.02	0.14	2.03	35.9	19.5	37.1	6.55	2.33
R3	1.95	0.45	0.43	0.18	0.02	0.15	2.83	40.8	19.7	32.2	5.42	3.51
S1049	2.66	0.46	0.43	0.19	0.03	0.16	1.96	49.9	19.3	36.2	6.07	2.20
S24	2.85	0.50	0.44	0.22	0.02	0.15	4.66	43.2	21.3	37.7	9.02	1.93
SAP155	1.82	0.35	0.36	0.15	0.01	0.11	2.51	35.8	15.9	33.3	4.06	1.60
SAP157	1.55	0.36	0.38	0.17	0.02	0.12	3.74	26.3	16.1	24.6	4.32	2.37
SAP158	1.72	0.36	0.54	0.15	0.03	0.12	2.69	27.5	17.3	27.8	3.34	2.30
SAP172	1.80	0.43	0.42	0.18	0.02	0.13	2.25	37.2	14.7	36.7	4.46	2.82
SDSL87046T	1.52	0.38	0.41	0.16	0.02	0.11	2.71	31.0	15.1	29.2	4.59	1.75
Shangani9356	1.65	0.39	0.47	0.14	0.02	0.13	2.37	30.5	16.8	31.4	2.91	1.84
SO85	1.96	0.49	0.47	0.19	0.02	0.15	2.75	52.5	20.1	18.5	7.41	3.78
StFederita	2.22	0.45	0.43	0.17	0.03	0.13	3.64	29.3	22.3	27.2	6.49	1.72
Takanda	2.37	0.55	0.47	0.24	0.02	0.14	3.33	39.1	24.3	50.1	5.50	1.90
Texas660	1.93	0.44	0.35	0.20	0.02	0.14	2.05	44.0	15.9	42.1	5.54	2.52
UI4822	1.99	0.44	0.49	0.19	0.01	0.16	2.47	56.1	15.0	47.4	4.62	2.37
Wray	1.78	0.41	0.35	0.22	0.01	0.13	5.62	36.1	15.0	29.8	4.03	3.17

3.2. Elemental Correlations among Nutrients in Seeds

There was a positive relationship between flax seed P and Mg (Figure 3A) and a good positive correlation between P and K (Figure 3B). There was a strong positive relationship between sorghum seed P and Mg (Figure 3C) and S and Mn (Figure 3D,E). A weak relationship was observed between sorghum seed Zn and Cu (Figure 3F,G) when excluding the very high singular point. Sorghum seed P and K had a reasonable positive correlation (Figure 3H).

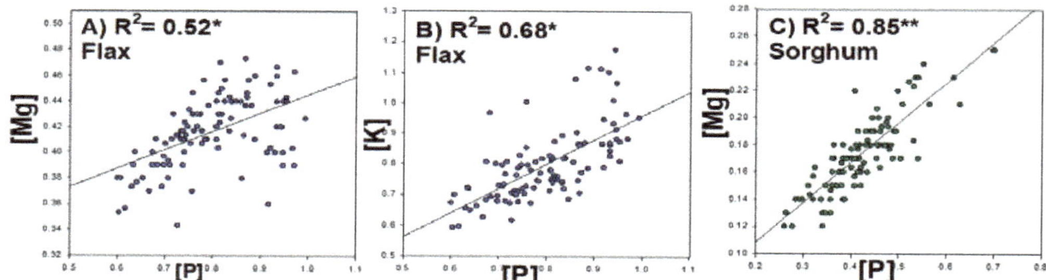

Figure 3. *Cont.*

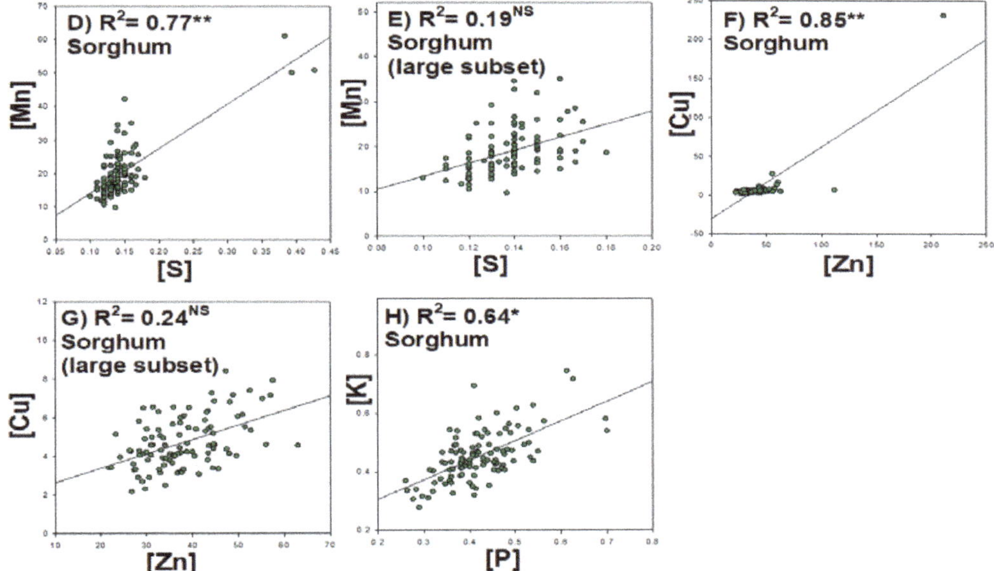

Figure 3. Selected correlations among seed elemental concentrations: (**A**) Flax P and Mg; (**B**) flax P and K; (**C**) sorghum P and Mg; (**D**) sorghum S and Mn; (**E**) sorghum large subset S and Mn; (**F**) sorghum Zn and Cu; (**G**) sorghum large subset Zn and Cu; (**H**) sorghum P and K. * and **, significant at the $p < 0.05$ and $p < 0.01$, respectively. NS, not significant, as determined using linear regression; R^2, linear regression coefficient squared.

3.3. Nutritional Value of Flax and Sorghum

Eight minerals were analyzed based on the United States Department of Agriculture (USDA) [16,17] recommended percentage daily value (% DV), which is a calculation of nutritional content in a 28 g (1 oz) serving of food and contribution to a 2000 calorie daily diet (USDA, 2020) [15]. As shown in Figure 4, daily consumption of 28 g of flax seeds could provide 37% DV of Cu, 31% DV of Mn, 28% DV of Mg, 19% DV of Zn, 19% DV of Zn, 18% DV of P, 11% DV of Fe, and 5% DV of Ca and K.

Our analysis of sorghum showed that daily consumption of 28 g seeds could provide 24% DV of Mn, 16% DV of Cu, 11% DV of Mg, 10% DV of Zn, 9% DV of P, 7% DV of Fe, 4% DV of Ca, and 3% DV of K (Figure 4).

Figure 4. Daily value (% DV) provided by the 28 g serving size of flax and sorghum.

3.4. Identification of Promising Top Flax and Sorghum Varieties

Based on the results of screening the 102 flax varieties, six varieties were chosen for their superior mineral content performance: Omega, Clli1374, Clli1418, Clli1821, Clli643 and Clli2033 (Table 3). Similarly for screening of 108 sorghum varieties, six varieties were chosen for their superior mineral content performance: PI529799, PI365024, PI185574, PI266958, PI534144, and PI550850 (Table 3).

Table 3. A set of top six superior varieties of flax and sorghum with superior seed mineral concentration. Highest, the highest mean for that specific element; High, considerably higher mean than other varieties.

Flax Variety	Performance
Omega	Highest: P; Mo. High: Ca, Zn, Cu, Ni
Clli1374	Highest: Mg. High: Mn, Mo
Clli1418	Highest: Ca. High: Ni, Fe, P
Clli1821	Highest: B, Cu
Clli643	Highest: Zn. High: Mo, K
Clli2033	Highest: Fe, High: Ni, Mn
Sorghum Variety	**Performance**
PI520799	Highest: P. High: Mg, Ca, S, B, Zn, Mn, Fe, Zn, Cu, Mo
PI365024	Highest: K, B, Mn, Fe. High: Mg, Ca, S, Zn, Cu
PI185574	Highest: Mg, Ca, S, Ni. High: P, K, B, Mn, Fe, Cu
PI266958	Highest: Zn. High: K, B, Mn, Ni
PI534144	Highest: Mo. High: Zn, Cu, P
PI550850	Highest: Cu, High: Ca

4. Discussion

The expected rising carbon dioxide (CO_2) levels (from 416 ppm in 2021 to 550 ppm in 2050) could cause stress to food crop plants. Furthermore, this could cause a reduction in nutritional quality or fewer nutritious crops, and therefore, trigger malnutrition [15,16]. One approach to minimize this issue is to identify food crop varieties with higher natural nutrient composition potential.

In this study, we analyzed ionomic data from 210 flax and sorghum varieties collected from around the world. Our results demonstrated substantial genetic variation in both crop species (Figure 2 and Table 4). This is consistent and follows a number of recent findings in peas, soybean and common beans, pearl millet, and sweet potato [7,9,13,19,20].

A better understanding of the relationships among various mineral nutrients is also critically important. A small number of correlated element pairs were detected in the current study. The selected correlations in Figure 3 were the best examples of elemental pair correlations in both flax and sorghum. The results of our analysis of macronutrients (N, P, K, Ca, Mg, and S) and micronutrients (Zn, Fe, Cu, B, Mn, and Mo) are summarized in Figure 3. In particular, the correlation analysis showed that seed P had a positive correlation with K and Mg in both crop species, therefore, their accumulation may be related. Furthermore, sorghum seed Zn had a positive correlation with Cu, which suggests that accumulation of these trace elements is related. Furthermore, the positive correlations between element pairs suggest potential shared transport systems in flax and sorghum systems. Similar positive associations between Zn and Cu have been reported in previous studies in soybean and common beans [7,9]. These findings are also consistent with previous studies in sweet potatoes that showed medium to high correlations among minerals such as Fe, Zn, Ca, and Mg [20]. This may suggest that elemental correlations may simplify selection for future breeding efforts. Furthermore, this is consistent with studies in pearl millet that have reported good elemental correlation and the possibility of simultaneous improvement of those nutrients [19].

Overall, among the 210 total varieties, six unique flax varieties and six unique sorghum varieties were identified with superior seed nutrient composition (Table 3). In flax, the highest P and Mo were observed in variety Omega, while the highest Cu and B were observed in variety Clli1821. In sorghum, the highest K, Mn, Fe, and B were observed in variety PI365024, while the highest Mg, Ca, S, and Ni were observed in variety PI185574. In addition, four more flax and sorghum varieties were identified as superior varieties (Table 3). These sets of a few selected superior varieties show significantly higher mineral concentration, which suggests that there is a potential of further improving mineral nutrient content in both flax and sorghum. Similarly, Gorindaraj et al. [19] explored the genetic variability of pearl millet for seed nutritional traits and reported the top 10 pearl millet accessions that could be used to develop nutritionally superior cultivars.

Table 4. Descriptive statistics in seed ionomic concentrations of 102 diverse flax genotypes and 108 sorghum genotypes. SD, standard deviation. Each value is the mean of three replicates.

		Macronutrients					
		N	P	K	Ca	Mg	S
		%	%	%	%	%	%
Flax	Avg.	3.92	0.80	0.80	0.24	0.42	0.27
	SD	0.37	0.10	0.12	0.04	0.03	0.01
	Min	2.90	0.60	0.59	0.18	0.34	0.24
	Max	4.98	0.99	1.18	0.38	0.47	0.30
Sorghum	Avg.	1.93	0.42	0.45	0.02	0.17	0.15
	SD	0.32	0.08	0.09	0.01	0.03	0.05
	Min	1.35	0.26	0.28	0.01	0.12	0.10
	Max	2.99	0.70	0.74	0.05	0.25	0.43
		Micronutrients (Trace elements)					
		Zn	Fe	Cu	B	Mn	Mo
		µg/g	µg/g	µg/g	µg/g	µg/g	µg/g
Flax	Avg.	73.40	70.10	11.80	18.80	25.50	2.98
	SD	16.90	16.50	3.45	5.36	5.11	0.35
	Min	38.00	43.70	4.21	9.61	15.50	2.27
	Max	127.00	196.00	24.20	40.50	41.00	4.58
Sorghum	Avg.	40.80	48.10	5.25	3.89	20.10	2.35
	SD	20.1	69.10	2.96	1.79	7.98	0.73
	Min	21.80	14.00	2.19	1.96	9.70	1.15
	Max	211	643	27.0	13.0	61.1	5.12

5. Conclusions

In this study, the multi-element contents and nutritional values of 102 flax and 108 sorghum varieties were evaluated. Our results revealed that there is substantial genetic variation of seed mineral nutrient traits both in flax and in sorghum. We elaborated on the six superior flax varieties and the corresponding six superior sorghum varieties that seem to hold promise for mitigating rising CO_2 stress as well as malnutrition. This study also provides an opportunity for future genetic studies to further efforts in biofortification efforts of flax and sorghum.

Author Contributions: Conceptualization, G.H.; methodology, G.H. and P.R.A.; formal analysis, G.H. and P.R.A.; writing—original draft preparation, G.H.; writing—review and editing, G.H. and P.R.A.; funding acquisition, G.H. and P.R.A. All authors have read and agreed to the published version of the manuscript.

Funding: This research was funded by the United States Department of Agriculture (USDA-ARS) 1890 Research Sabbatical grant to G.H. (FAMU) in collaboration with P.A. (USDA).

Data Availability Statement: Not applicable.

Acknowledgments: The authors thank the USDA National Plant Germplasm Center for providing.

Conflicts of Interest: The authors declare no conflict of interest.

References

1. FAOSTAT. FAO Statistical Databases. Food and Agriculture Organization of the United Nations, Rome. Available online: http://www.fao.org/faostat (accessed on 13 December 2021).
2. Muir, A.D.; Westcott, N.D. *Flax: The Genius Linum*; CRC Press: Boca Raton, FL, USA, 2003.
3. Awika, J.M. *Sorghum: Its Unique Nutritional and Health-Promoting Attributes*; Woodhead Publishing: Cambridge, MA, USA, 2017.
4. Salt, D.E.; Baxter, I.; Lahner, B. Ionomics and the study of the plant ionome. *Annu. Rev. Plant Biol.* **2008**, *59*, 709–733. [CrossRef] [PubMed]
5. Baxter, I.R.; Gustin, J.L.; Settles, A.M.; Hoekenga, O.A. Ionomic characterization of maize kernels in the intermated B73 x Mo17 population. *Crop Sci.* **2013**, *53*, 208–210. [CrossRef]
6. Hacisalihoglu, G.; Kochian, L.V.; Vallejos, E. Distribution of Seed Mineral Nutrients and Their Correlation in *P. vulgaris*. *Proc. Fla. State Hort. Soc.* **2005**, *118*, 102–105.
7. Hacisalihoglu, G.; Settles, A.M. Natural Variation of Seed Composition of 91 Common Bean Genotypes and Their Possible Assoc with Seed Coat Color. *J. Plant Nutr.* **2013**, *36*, 772–780. [CrossRef]
8. Hacisalihoglu, G.; Gustin, J.; Louisma, J.; Armstrong, P.; Peter, G.; Settles, A.M. Enhanced Single Seed Trait Predictions in Soybean and Robust Calibration Model Transfer. *J. Agric. Food Chem.* **2016**, *64*, 1079–1086. [CrossRef] [PubMed]
9. Hacisalihoglu, G.; Settles, A.M. Quantification of Seed Ionome Variation in 90 Diverse Soybean (*Glycine max*) Lines. *J. Plant Nutr.* **2017**, *40*, 2808–2817. [CrossRef]
10. Hacisalihoglu, G.; Kantanka, S.; Miller, N.; Gustin, J.L.; Settles, A.M. Modulation of early maize seedling performance via priming under sub-optimal temperatures. *PLoS ONE* **2018**, *13*, e0206861. [CrossRef] [PubMed]
11. Hacisalihoglu, G.; Burton, A.L.; Gustin, J.; Eker, S.; Asikli, S.; Heybet, E.; Ozturk, L.; Cakmak, I.; Yazici, A.; Burkey, K.O.; et al. Quantitative trait loci associated with soybean seed weight and composition under different phosphorus levels. *J. Integ. Plant Bio.* **2018**, *60*, 232–241. [CrossRef] [PubMed]
12. Hacisalihoglu, G. Zinc (Zn): The Last Nutrient in the Alphabet and Shedding Light on Zn Efficiency for Future of Crop Production under Suboptimal Zn. *Plants* **2020**, *9*, 1471. [CrossRef] [PubMed]
13. Hacisalihoglu, G.; Beisel, N.; Settles, A.M. Characterization of pea seed nutritional value within a diverse population of *Pisum sativum*. *PLoS ONE* **2021**, *16*, e0259565. [CrossRef] [PubMed]
14. IPCC. Intergovernmental Panel on Climate Change. Climate Change 2021: The Physical Science Basis. Available online: https://www.ipcc.ch/report/ar6/wg1/#FullReport (accessed on 13 December 2021).
15. Zheng, G.; Chen, J.; Li, W. Impacts of CO_2 elevation on physiology and seed quality of soybean. *Plant Div.* **2020**, *42*, 44–51 [CrossRef] [PubMed]
16. Jablonski, L.M.; Wang, X.; Curtis, P.S. Plant reproduction under elevated CO_2 conditions: A meta-analysis of reports 79 crop and wild species. *New Phytol.* **2002**, *156*, 9–26. [CrossRef]
17. USDA. Dietary Guidelines for Americans. Available online: www.dietaryguidelines.gov (accessed on 13 December 2021).
18. FDA. U.S. Food and Drug Administration: Nutrient Content Claims for "Good Source," "High," "More," and "High Potency" Code of Federal Regulations. 2014, Subpart D, Title 21, 101.54. Available online: https://www.accessdata.fda.gov/scripts/cdrh/cfdocs/cfcfr/cfrsearch.cfm?fr=101.54 (accessed on 13 December 2021).
19. Govindaraj, M.; Rai, K.N.; Kanatti, A.; Upadhyaya, H.D.; Shivade, H.; Rao, A.S. Exploring the genetic variability and diversity of pearl millet core collection germplasm for grain nutritional traits improvement. *Sci. Rep.* **2020**, *10*, 21177. [CrossRef] [PubMed]
20. Tumwegamire, S.; Kapinga, R.; Rubaihayo, P.R.; Labonte, D.R.; Grüneberg, W.J.; Burgos, G.; Zum Felde, T.; Caprio, R.; Pawelzik, E.; Mwanga, R.O.M. Evaluation of dry matter, protein, starch, sucrose, β-carotene, iron, zinc, calcium, and magnesium in East African sweetpotato [*Ipomoea batatas*] germplasm. *HortScience* **2011**, *46*, 348–357. [CrossRef]

MDPI
St. Alban-Anlage 66
4052 Basel
Switzerland
Tel. +41 61 683 77 34
Fax +41 61 302 89 18
www.mdpi.com

Plants Editorial Office
E-mail: plants@mdpi.com
www.mdpi.com/journal/plants